ILLUSTRATED DICTIONARY OF

ENTOMOLOGY

NIPA® GENX ELECTRONIC RESOURCES & SOLUTIONS P. LTD.
New Delhi-110 034

ILLUSTRATED DICTIONARY OF ENTOMOLOGY

Dr. Paras Nath

***Reader*, M. Sc. (Ag.) Ent. & Agril. Zool., Ph.D., P.D.F. (C.S.I.R.),**
FESI, FERA, SPPSF, FSPE, FSES, FBRS.
Department of Entomology, Institute of Agricultural Sciences
Banaras Hindu University, Varanasi – 221 005 (India)

NIPA® GENX ELECTRONIC RESOURCES & SOLUTIONS P. LTD.

New Delhi-110 034

NIPA® GENX ELECTRONIC RESOURCES & SOLUTIONS P. LTD.

101,103, Vikas Surya Plaza, CU Block
L.S.C. Market, Pitam Pura, New Delhi-110 034
Ph : +91 11 27341616, 27341717, 27341718
E-mail: newindiapublishingagency@gmail.com
web: www.nipabooks.com

For customer assistance, please contact
Phone: + 91-11-27 34 17 17
Fax: + 91-11- 27 34 16 16
E-Mail: feedbacks@nipabooks.com

ISBN: 978-81-19215-29-4

Composed and Designed by NIPA®.

This dictionary is affectionately dedicated to my
Maa (Late) Smt. Anmola Devi who taught me the value of hard work and
wisdom and Guru Maa Smt. Shital Mishra who patiently cared and groomed my future for which I shall always be indebted.

प्रोफेसर पंजाब सिंह
Prof. Panjab Singh
FNAAS, FIAS, FNIE, FBRSI, FSEE
(EX-SECRETARY, DARE, GOI & DG, ICAR)
कुलपति
VICE-CHANCELLOR

काशी हिन्दू विश्वविद्यालय
BANARAS HINDU UNIVERSITY
VARANASI - 221 005 (INDIA)
Phones : (0542) 2368938/230-7220 (O)
: (0542) 2368339/230-7209 (R)
Fax : (0542) 2369951
e-mail : vc_bhu@sify.com

VC/
July 16, 2006

Foreword

The science of entomology, the amazing world of insects, with its quick development necessitates a dictionary of technical terms used by the teachers, scientists, technologists, extension workers and students of entomology and allied fields. This dictionary is certainly of interest for the competence, experience and tenacity of its author and will certainly be of a great service at various levels of knowledge of entomology in teaching, and also research, whether basic or applied.

I am very happy that Dr. Paras Nath has made sincere efforts to compile the technical terms generally encountered while going through the entomological literature. This publication *"Illustrated Dictionary of Entomology"* would, I am sure, go a long way in helping the students, teachers, researchers and all those interested in the field of entomology as a ready reckoner of technical terms.

I congratulate Dr. Paras Nath of the Department of Entomology and Agricultural Zoology, Institute of Agricultural Sciences, Banaras Hindu University, for his *"Illustrated Dictionary of Entomology"* which I believe will serve the purpose of teachers, students and all those who are directly or indirectly concerned with the science of entomology and I hope they will find this publication highly useful.

(Panjab Singh)

Preface

I feel great pleasure to bring out this Illustrated Dictionary of Entomology for the benefit of primarily amateur entomologists, naturalists and agricultural scientists with an interest in insects and for all those who have special interest to understand the world of insects. The technical words related to the various disciplines of entomology have been included in this Dictionary and where ever possible the technical meaning of these words have been tried to make it more clear with the help of labeled illustrations. Efforts have been made by the author to define the entomological terms in a simple manner in order to make them understandable by the students of entomology and all those who are not the experts of entomology rather they encounter such technical words while dealing with the related literature and fails to understand its real meaning as the meaning provided in general English dictionaries may not give the clear picture of these technical terms. The university students of biology and agriculture sciences in the beginning face lot of problems in understanding the subject because of poor knowledge of such technical words which are also not included in the general English dictionaries. About 4016 technical terms related to the morphology, anatomy, physiology, systematics, ecology, pest management and general entomology have been defined. An individual technical word having different applications have been incorporated in a convincible manner with the help of illustrated diagram. Therefore this dictionary will serve as a ready reckoner to the students of such discipline. This dictionary will also be useful for all those who are to appear in some competitive examinations to understand and solve the objective type questions.

It is believed that this dictionary will find equal favour with the teachers, students and all those who deal insects in one way or the other. Any suggestion received from any quarter would be given due consideration and may be incorporated while reprinting / revising the dictionary.

The author is grateful to Dr. S. R. Singh, Director, Institute of Agricultural Sciences, Banaras Hindu University for moral support and to Professor D. S. Mishra, Head, Department of Entomology & Agril. Zoology, B. H. U. for constant encouragement, support and blessing throughout the preparation of this manuscript.

In my academic pursuits I have faced tough times and it was strenuous to pass over such periods. Dr. Prabhakar Jha, Professor of French in Banaras Hindu

University and Formerly Ambassador of India to Madagascar because of whose blessings I could not bother for such arduous situations. Really I am very lucky to have his blessings.

Preparation of this dictionary has been a time consuming task that demanded long hours. I thank Mrs Usha Nath, my life partner for her everlasting support of my career, and especially during the years when I was surrounded by ill wills. I cannot forget her life giving moral support which gave me the strength to think and do positive. Certainly I consumed the time of my children Divya and Sanskar which they deserved to understand this complex world. I thank both of them as they bothered me not and shaped their destiny as they wished.

I thank Mr. S. G. Sunderarajan, my colleague who always helped me to come out of the web of ill will designes and encouraged me to remain bussy. I cannot forget his contributions and express my heartful gratitude.

I thank Mr. Virendra Nath Vishwakarma for typing the manuscript and Mr. Anand Kumar Pandey, my Ph. D. student, for help in the preparation of manuscript, his help have added greatly to the value and reliability of this work.

June, 28th, 2006

PARAS NATH
Banaras Hindu University
Varanasi - 221 005

Contents

A

"Age and Area": The hypothesis (by Willis) that the older a species is, the wider will be its geographical distribution.

A- Without; lacking.

A.I.: See active ingredient.

A.O.U. Code: A code of nomenclature published in 1885 (revised 1908) by the American Ornithologists' Union for the standardization of bird nomenclature.

Ab- Away from; off.

Abdomen: The posterior most of the three main body divisions of insects (Fig. 1). This is the third major division (**tagma**) of an insect body usually composed of 10-11 segments. It contains reproductive organs in adults and sometimes locomotor appendages on immatures.

Abdominal legs: See prolegs (Fig. 44).

Abiotic disease: A disease caused by nonliving factors, such as nutrient deficiency. Sometimes referred to as an abiotic condition.

Abiotic: Nonliving, both in characteristics and origin. This is the nonliving components of a system, such as temperature, mineral nutrients, water, soil type, sunlight, and air pollutants.

Abscission: The natural process by which leaves, fruits, or other organs become detached from plants.

Absconding: Honey bee colony abandons hive because of enemies (ants, waxmoth etc.)

Absolute population estimate: An estimate of population size expressed as absolute number per ground-surface area (e.g., number per hectare) or per unit of volume (e.g., number per litre of soil).

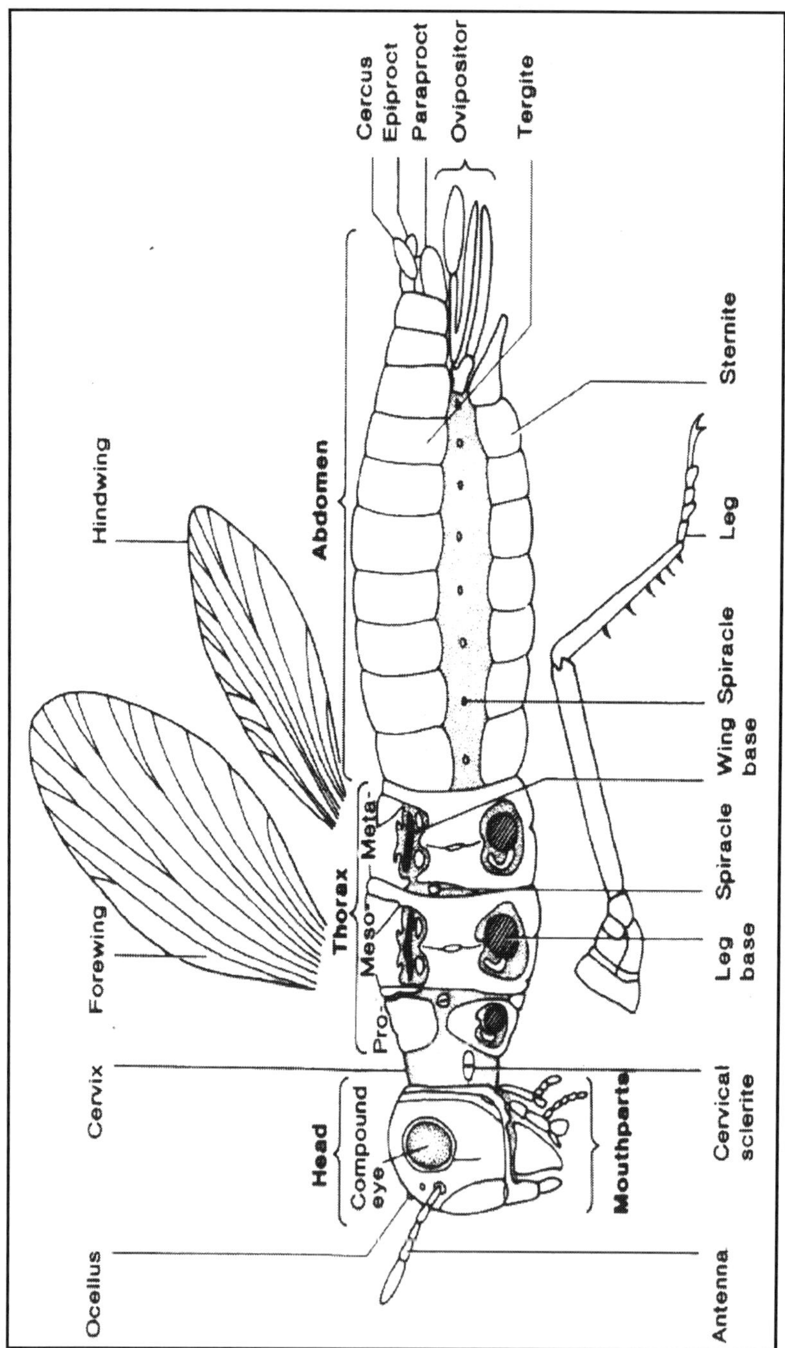

Fig. 1. Main body parts of a generalized pterygote insect.

Absorption: The assimilation of molecules into cells, e.g., the process by which plants take up nutrients through their roots; the selective taking up of fluids or substances in solution by cells or absorbed vessels (L. *ab* = from + *sorbeo* = suck in).

Acalypterae: Cyclorrhaphous flies having wings without posterior lobes covering halteres.

Acanthae: Fine, unicellular, cuticular extensions (Fig. 2).

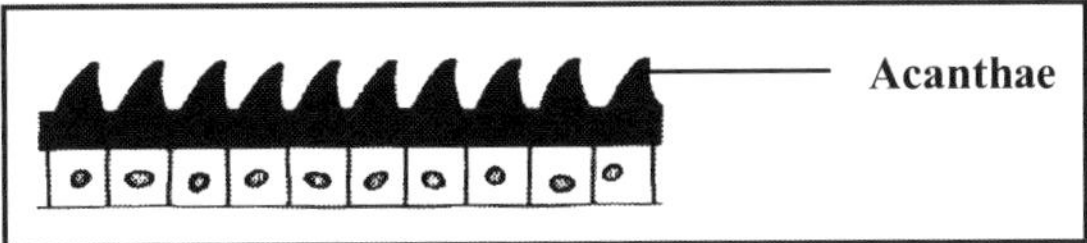

Fig. 2. Acanthae.

Acari: The order of arthropods within the class Arachnida that comprises the mites and ticks.

Acaricide (miticide): A chemical pesticide toxic to mites and ticks (pest arthropods that belong to the class Acarina). It kills acarines, i.e., ticks and mites at recommended doses.

Acarology: Branch of zoology dealing with ticks and mites.

Acceptable daily intake (ADI): The level of a chemical residue to which daily exposure over a lifetime is not thought to cause appreciable risk.

Accessory cell: A closed cell in the front wing of Lepidoptera formed by the fusion of two branches of the radius, usually the R_3 cell.

Accessory cells: Cells involved with or secreting the cuticular structures that make up a sensory receptor; for example, trichogen and tormogen cells. (Fig. 3)

Accessory gland: Any secondary or subsidiary gland of a major glandular system; more specially, a gland opening into the genital chamber. Secretary cells/tissues associated with the internal reproductive organs of either males or females and producing substances accompanying the sperms or eggs; gland associated with the insect reproductive system but fulfilling some function secondary to the reproductive process. Present in both sexes of most insects (Fig. 4 a & 5 a).

Accessory pulsatile organs: Valved pumps aiding the circulation of haemolymph, often lying close to the dorsal vessel.

Accessory sexual characters: The structures and organs (except the gonads) of which the genital tract is composed, including accessory glands and external genitalia (cf. Primary, Secondary sexual characters).

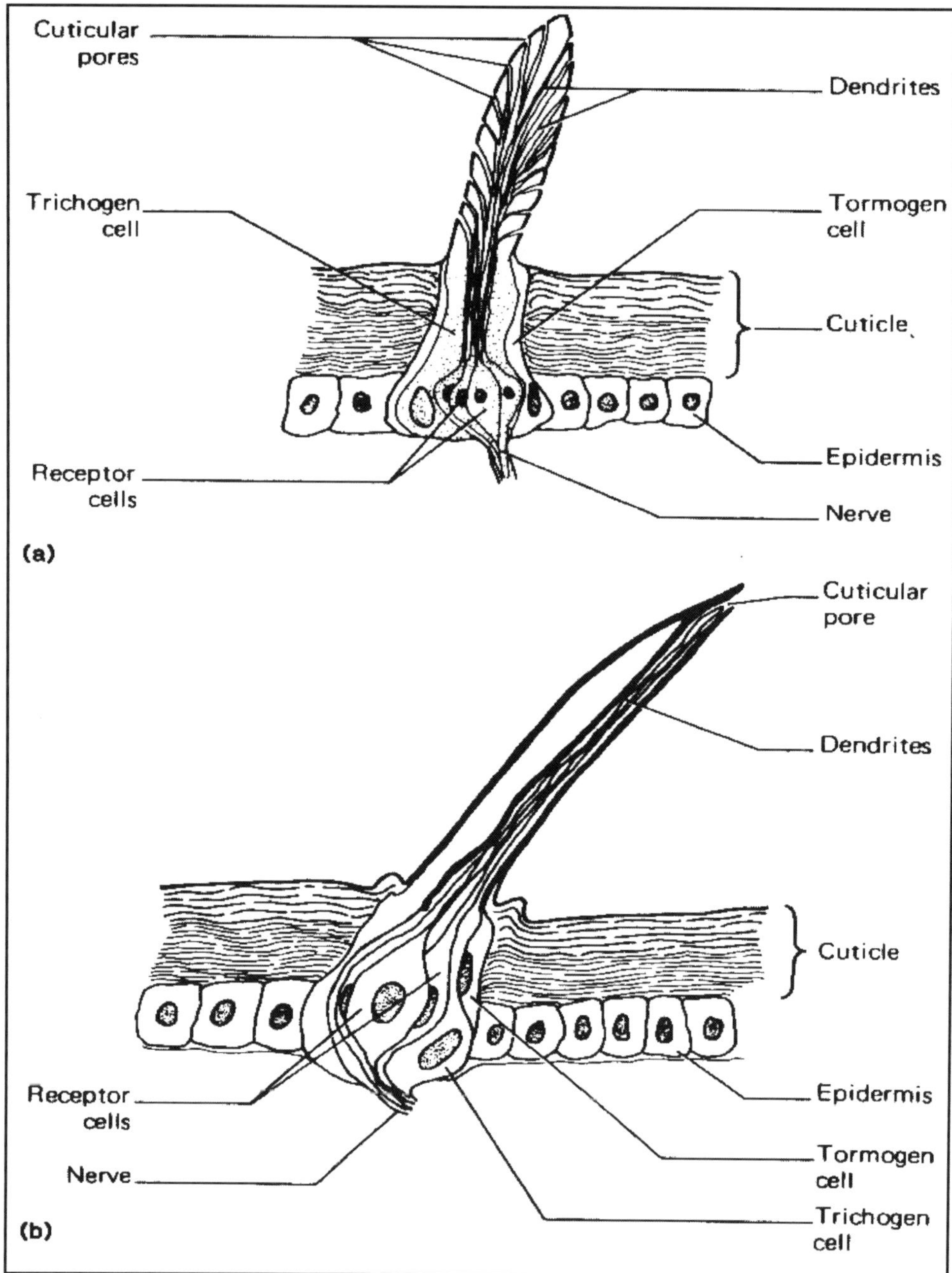

Fig. 3. Chemosensilla. (a) Olfactory sensillum. (b) Contact chemosensillum.

Accessory vein: An extra branch of a longitudinal vein (indicated by a subscript a; for example, an accessory of M_1 is designated M_{1a}); the most posterior vein in the anal area of the front wing of certain Hymenoptera.

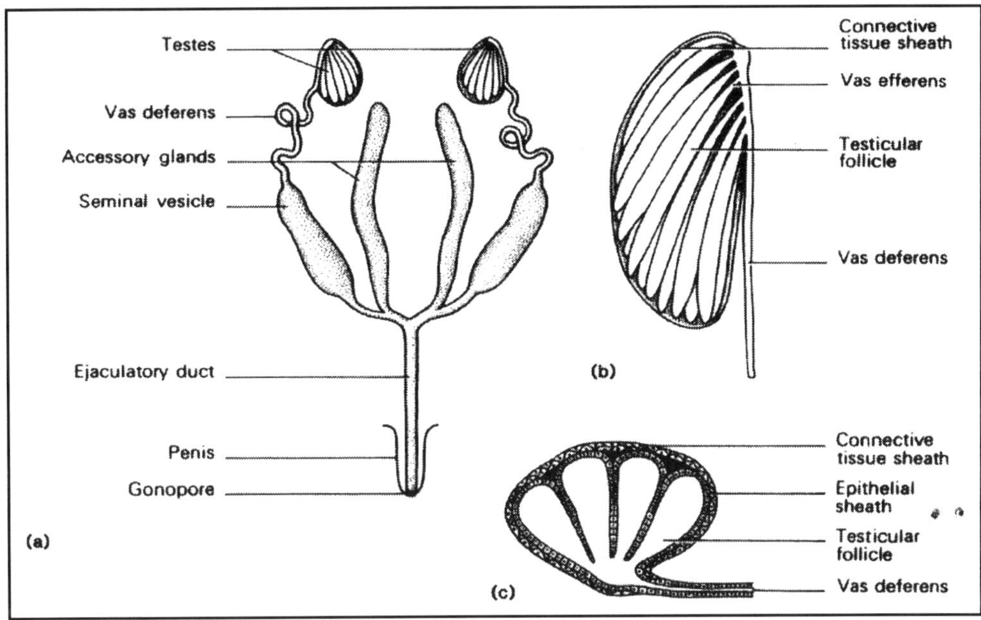

Fig. 4. Generalized male reproductive system. (a) Principal male organ. (b) Detailed structure of a testis. (c) Section of a testis. After *Snodgrass,* 1935.

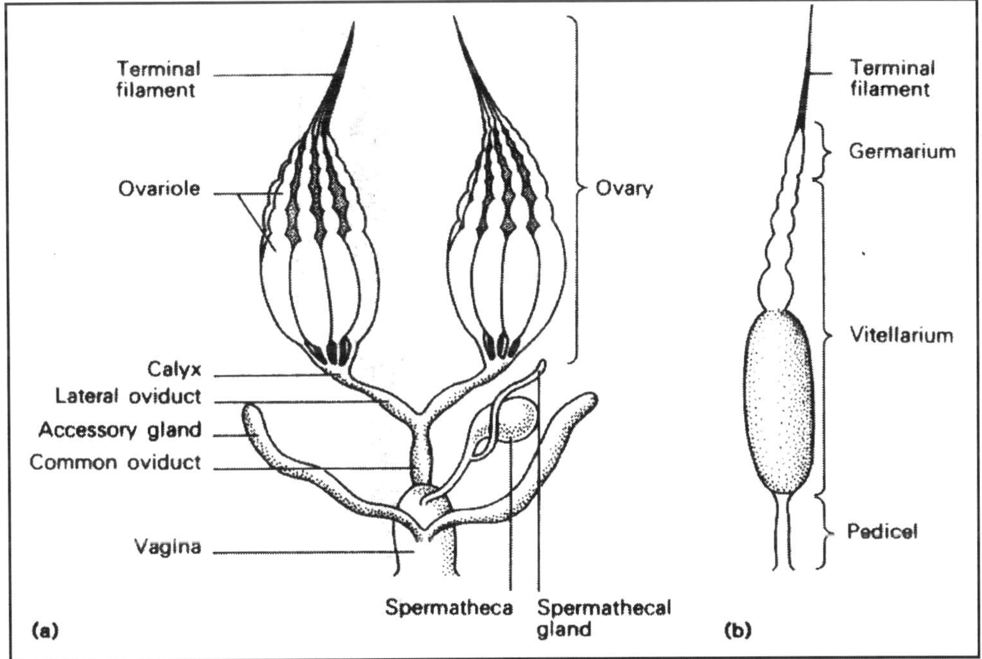

Fig. 5. Generalized female reproductive system. (a) Principal female organ. (b) Single ovariole. After *Snodgrass,* 1935.

Acclimation/acclimatization: Response of organisms to seasonal and climatic changes such as response of an insect to temperature according to the previous experience.

Accumulative pesticide: See *biomagnification.*

Acephalous. Headless; larva without a distinct head capsule; applies to certain Diptera.

Acephate: Synthtic contact/systemic organophosphorous insecticide.

Acetylcholine (Ach): A chemical substance produced at nerve connections (synapses) each time a nerve impulse is transmitted; it is a chemical $CH_3COOCH_2CH_2N\ (CH_3)_3^+$ that functions as a synaptic neurotransmitter in the nervous system of animals.

Acetylcholinesterase: An enzyme that catalyzes the breakdown of acetylcholine.

Acid equivalent: The theoretical yield of parent acid from the active ingredient of a week acid pesticide formulation.

Acid gland: Principal poison gland of bees whose secretion contains apitoxin.

Acidic: Acid in reaction resulting from excess of hydrogen ions over hydroxyl (OH) ions in solution: pH less than 7.

Acinar: Resembling a cluster.

Acinus: Resembling a single grape in a cluster.

Acoustical communication: Communication by sound and hearing.

Acquired active immunity: Immunity induced by injection of toxoids or killed or attenuated microorganisms.

Acquired character: A character which arises during the life of an individual, either in response to the environment or form a functional cause (cf. Lamarckism).

Acre: A unit of measurement for area that is equal to 43,560 square feet or 0.414 hectares, 640 A is equal to 1 square mile or a section of land.

Acron: Anterior, preoral body cap of a segmented animal, also called prostomium. (Fig. 6).

Acrosternite: Small sclerite between the antecostal suture and intersegmental membrane on the abdominal sternum; compare with acrotergite.

Acrostichal bristles: One or more longitudinal rows of small bristles along the center of the mesonotum (Diptera).

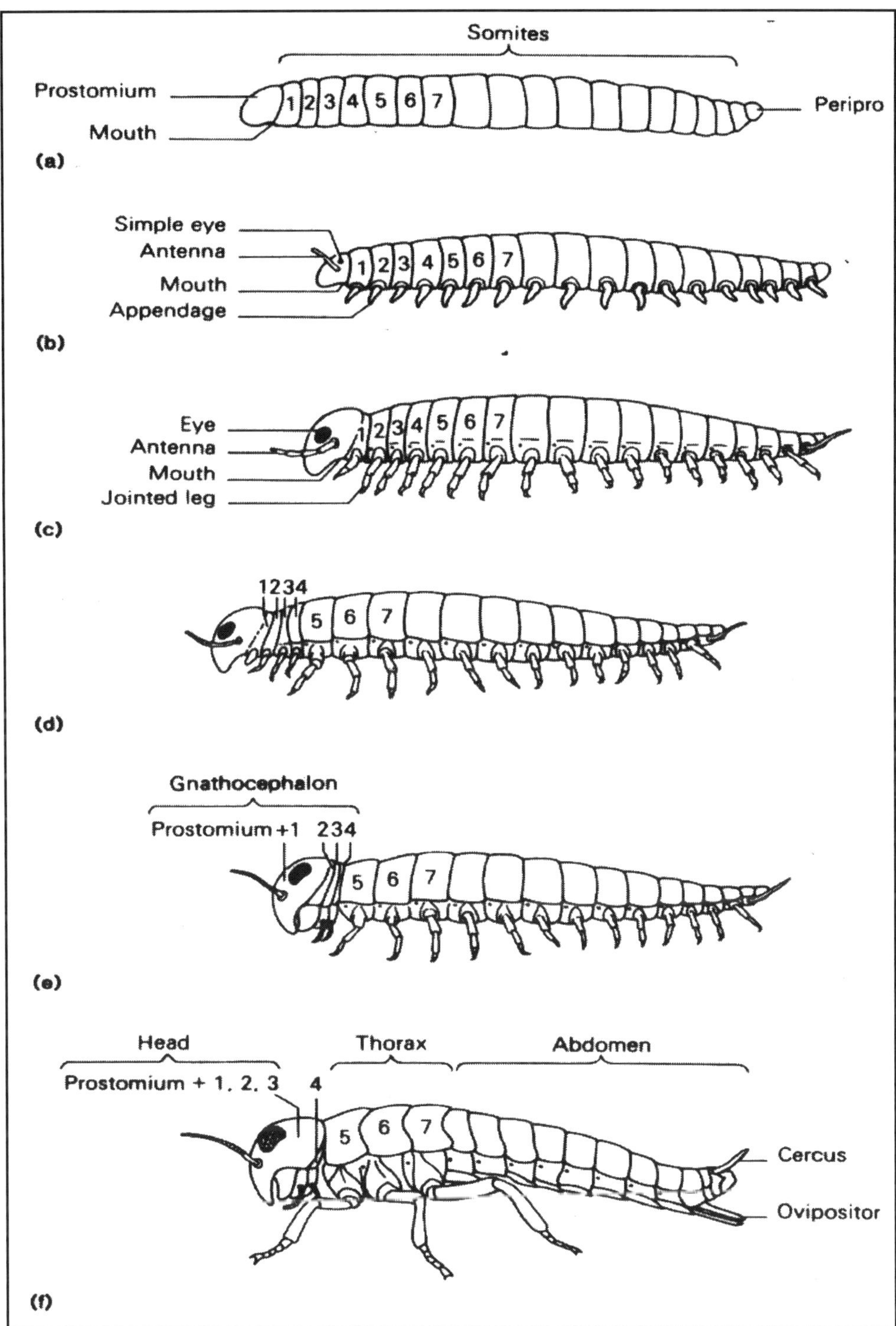

Fig. 6. Hypothetical stages in the evolution of the insect form. After *Ross,* 1965.

Acrotergite: The anterior part of a secondary segment, sometimes large (then called **postnotum**), often reduced. This is the small sclerite between the antecostal suture and intersegmental membrane on the abdominal tergum (Fig. 20).

Acrotrophic ovariole: *See* Telotrophic ovariole.

Action potential: A brief electrical depolarization that propagates along an axon or muscle fiber. Depolarization of a nerve cell, shown as a spike on an oscilloscope.

Action threshold: The population density at which pest control measures are applied to avoid economic loss.

Activation reduce space center: A region in a fertilized egg where cleavage cells are induced to undergo further growth and development.

Activation: In embryology, the commencement of embryonic development within the egg.

Activator: A Chemical compound or material added to a pesticide to increase its toxicity either directly or indirectly by increasing the potency or activity of the pesticide. See also *synergist.*

Active ingredient (ai): The component of a pesticide formulation responsible for the toxic effect is known as active ingredient (a.i.). These chemicals are biologically active as toxins. A given pesticide formulation may have more than one active ingredient.

Active space: The zone of pheromone concentration within which a response is elicited.

Active ventilation: Ventilation aided by movements such as muscular pumping.

Aculea (pl. aculeae): With a sting (Hymenoptera).

Aculea (pl., aculeae): Minute spines on the wing membrane (Lepidoptera).

Aculeate: With aculeae (Lepidoptera); with a sting belonging to the aculeate Hymenoptera - wasps (including all bees) in which the ovipositor is modified as a sting.

Acuminate: Tapering to a long point.

Acute poisoning: Illness or death from a single does of a toxicant.

Acute toxicity: An expression of the degree to which a toxicant is poisonous to an organism after a single exposure, contact, inhalation, or ingestion (cf., chronic toxicity).

Acute: Pointed; forming an angle of less than 90 degrees.

Ad- Next to; at.

Adaptation: The fitness of a structure, or an organism for a particular environment. (L. *ad* = to + *apto* = fit). An animal species for its own survival makes full use of the available nutrient pool and other environmental conditions prevailing for its growth and development and thus, takes care of its own protection against ravages of environmental conditions in any of its habitat. These all require some characteristics to be possessed. Any feature of the organism or its parts, which is of definite significance in allowing that organism to exist under the condition of its habitat, is called adaptation. Thus, adaptation is a quality of the organism, population or species that increases its chance of leaving viable progeny.

The condition of showing fitness for a particular environment, as applied to characteristics of a structure, function, or entire organism; also the process by which such fitness is acquired (cf. Preadaptation, Environment). Thus, adaptation is (i) Any genetically determined characteristic of an organism that enhances the fitness (survival and reproduction) of that organism. (ii) The result of natural selection and evolution that leads to organisms becoming better suited to their environments.

Adaptive: Fitted for a particular environment (cf. Environment).

Adaptive color change: A fairly rapid change of colour in response to a change in the color characteristics of the environment.

Adaptive convergence: The evolution, or presence within a series of comparable ecological niches, of only distantly related forms which superficially resemble one another in morphological and other characters, correlated with very similar or identical environmental conditions (cf. Adaptation, Adaptive radiation).

Adaptive radiation: Evolution and spread of a single phyletic line of organisms into several distinctive ecological niches resulting in a series of sometimes strikingly different forms, each adapted to a particular niche.

It is evolutionary divergence of members of a single phyletic line into a series of rather different ecological niches. It is believed by many experts to be the major pattern of evolution. It involves the sudden (in geologic time) diversification of a group of organisms that share a common ancestor, often itself own movements. Thus, adaptive radiation is the diversification through evolution of organisms of common origin in response to a variety of new environmental opportunities.

Additive effect: The efficacy of a pesticide mixture that is equal to the sum of the toxicities of the individual pesticides (cf., antagonism and synergism).

Adecticous pupa: A pupa with immovable mandibles (Fig. 7).

Adecticous: Without functional mandibles (movable mouthparts) used for escape from a cocoon or pupal cell; *compare with* decticous (Fig. 7).

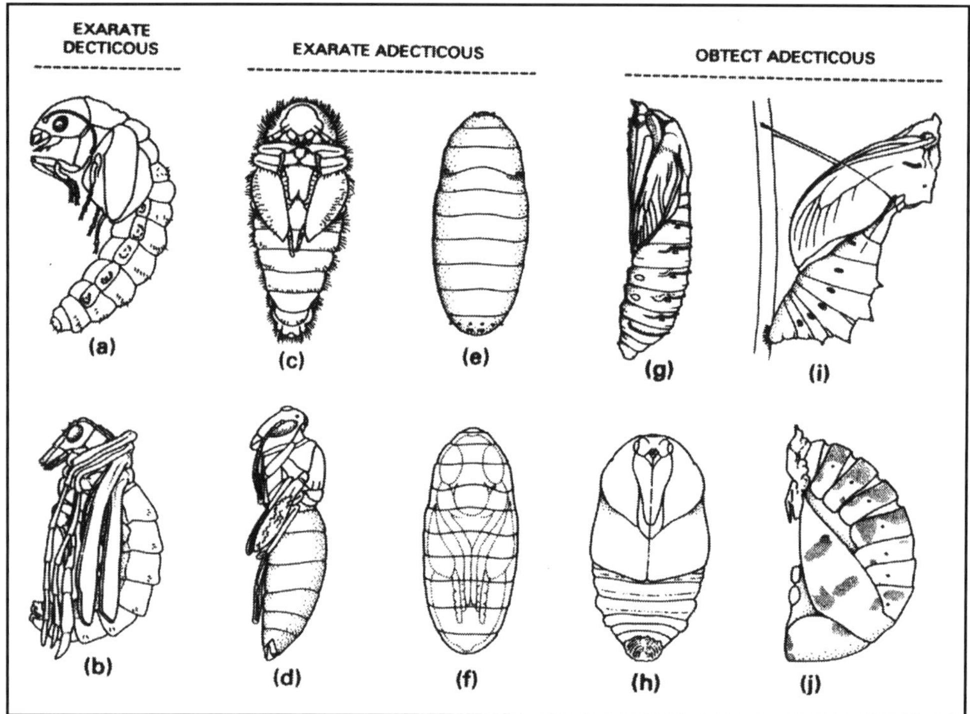

Fig. 7. Examples of pupal types. Exarate decticous pupae: (a) Megaloptera: Sialidae; (b) Mecoptera: Bittacidae. Exarate adecticous pupae: (c) Coleoptera: Dermestidae; (d) Hymenoptera: Vespidae; (e), (f) Diptera: Calliphoridae, puparium and pupa within. Obtect adecticous pupae: (g) Lepidoptera: Cossidae; (h) Lepidoptera: Saturniidae; (i) Lepidoptera: Papilionidae, chrysalis; (j) Coleoptera: Coccinellidae. After *Common* and *Waterhouse,* 1972.

Adenotrophic viviparity: Viviparity (producing living offspring) in which there is no free-living larval stage; eggs develop within the female uterus, nourished by special **milk glands** until the larvae mature, at which stage they are laid and immediately pupate; occurring only in some Diptera (Hippoboscidae and *Glossina*).

Adfrontal areas: A pair of narrow oblique sclerites on the head of a lepidopterous larva.

Adherence: The ability of a material to stick to a particular surface.

Adhesive: (=Sticker) Material added to increase pesticide retention; different commercial preparations of methyl cellulose are available for this purpose.

Adipocytes: Cell composing fat body.

Adjuvant: Any ingredient that improves the properties of a pesticide formulation. These are the substances that are added to an insecticide to increase their effectiveness such as dusts, solvents, emulsifiers, and adhesives. Thus, adjuvant is any nonpesticidal substance in a pesticides formulation that improves the physical, chemical, or biological properties of the active ingredient, or improves application efficacy e.g., supplement, dusts, solvents, sticker, adhesive, spreader, wetter, emulsifiers, and adhesives that are combined with insecticides to increase their effectiveness.

Adergenic blocking agent: Agents that block the nervefibres which liberate sympathin-an afrenaline-like substance.

Adsorption: The adhesion of molecules to the surface of a solid. Pesticide binding to soil particles is an example.

Adsorption: Adhesion of an extremely thin layer of gas molecules, dissolved substance, or liquid to a solid surface (. *Ad* = to + *sorbeo* = suck in).

Adult: The reproductive, sexually mature, final stage in the life cycle of an organism.

Adulticide: A pesticide that targets adult insect pests.

Adventitious vein: A secondary vein, neither accessory nor intercalary, usually the result of a number of cross veins lined up to form a continous vein.

Adventitious: Tissues or organs that arise in other than the usual location on a plant. Usually with reference to the production of lateral roots or buds.

Aeciospore: A type of spore produced by a rust fungus.

Aedeagus: The male copulatory organ of an insect variably constructed usually taken to include accessory structures besides the penis but some times refers to the penis alone; the sclerotized median intromittent organ of a male insect that is the distal part of the phallus which is used for copulation and is also called penis (Fig. 8).

Aerobic: Respiration dependent on oxygen (Gr. *aeros* = air + *bios* = life) *e.g.,* in, man; a process or an organism that requires oxygen.

Aeropile (s): The opening in the chorion (egg shell) through which air enters, often covered by a plastron; channels in insect eggs by which the gas-filled layer communicates with the outside of the egg.

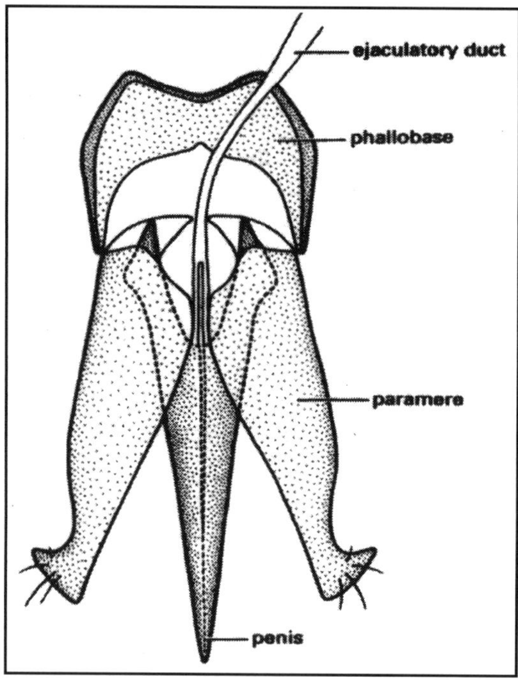

Fig. 8. Male external genitalia: dorsal view of Aedeagus (penis) of a click beetle (Coleoptera: Elateridae). After *Snodgrass,* 1957.

Aerosol: Pesticide formulation in which propulsion from a pressurized container or by aerosol generators such as fogging machines or ultra-low-volume (ULV) equipment suspends extremely fine pesticide particles or droplets (0.1 to 5 um in diameter) in the air; very finely divided particles (of insecticide solution) in the air, often applied to the pressurized canisters which generate an aerosol.

Aesthetic pest: A pest whose mere presence is objectionable for psychological reasons.

Aestivation: Dormancy during a warm or hot or dry season usually the summer. Also called summer dormancy; the act of passing in the summer in a quiet, torpid condition. (L. *aestas*=summer) *e.g.,* in case of frog.

Afferent neuron: Usually bipolar neuron with cell body located peripherally in the insect and dendrites associated with sensory structures; also called sensory neuron.

Aflatoxin: Potent mycotoxins that are carcinogenic and are naturally produced by the fungus *Aspergillus flavus* when it is growing on peanuts, corn, cereals, and other crops.

African sleeping sickness: A disease caused by infection with *Trypanosoma* species protozoans; transmitted by the bite of the tsetse fly.

Agamic: Reproducing parthenogenetically, that is, without mating.

Age distribution: The proportions of individuals in different age groups of a population at any given time.

Age-grading: Determination of the physiological age of an insect.

Aggregation pheromone: An intraspecific chemical messenger that cause insect congregation at food sites, reproductive habitats, or hibernation sites.

Aggregation: It is tendency among organisms to concentrate in numbers larger than found in normal distribution and which congregate under various circumstances mostly by their own movements.

Aging: All changes in structure and function that occur from the beginning to the termination of life.

Agitate: Stirring or shaking a pesticide mixture so that the components will not separate or settle in the application tank.

Agitator: A mechanical device in the spray tank to ensure uniform distribution of toxicant and to prevent sedimentation.

Agroecology: The study of ecology in relation to agricultural systems.

Agroecosystem: A relatively simple ecosystem created by humans for agricultural purpose, in which the major producer-the crop-is usually a single plant species. This is a crop ecosystem managed by man to maximize the production of an agricultural commodity largely created and maintained to satisfy a human want or need. This is constituted by a complex of organisms and the environmental factors found in an agricultural field. It differs from a natural ecosystem in that it is relatively artificial. This type of ecosystem is manipulated by a human that typically has few common or major species (crops) and numerous rare or minor species (mainly pests).

Air assisted sprayer: A sprayer in which an air stream is employed to help carry spray droplets to the object being sprayed.

Air-blast sprayer: Spray equipment that uses a fan to create a high-speed air stream to transport and deposit pesticide to target. See *air-blast sprayer.*

Air-sac: Thin-walled, dilated sections of the **tracheae** (Fig. 9).

Akinesis: A state of immobility, due to lack of any stimuli.

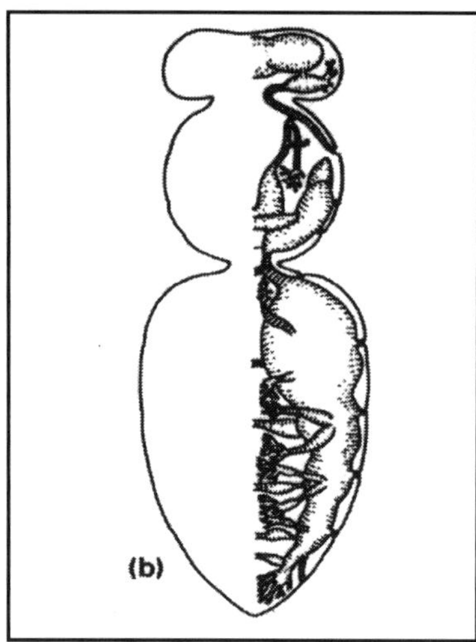

Fig. 9. Tracheae with mechanically ventilated air-sacs, as in honey bees. After *Snodgrass*, 1956.

Akipokinetic hormone: A peptide hormone released from the corpora cardiaca and influencing the fat body and the flight musculature.

Alarm pheromones: Chemicals released by one insect that elicit dispersal or aggressive behavior by conspecifics. Thus, it is an intraspecific chemical messenger produced to elicit aggressive or defensive behavior.

Alary muscles: A wing-shaped paired muscles that support the heart by connecting the heart to lateral portions of the terga. It is segmented fibro muscular structures that is associated with the heart and having a role in its functioning (Fig. 10). These are the muscles in the dorsal diaphragm, the contractions of which induce the flow of blood into the hemocoel surrounding the heart.

Alatae. Winged insects; usually refers to winged aphids. cf. apterous.

Albinism: In zoology, the absence of pigmentation, and particularly of melanins, in an animal (cf. melanism).

Aldicarb: Synthetic carbamate insecticide.

Aldrin: A synthetic chlorinated hydrocarbon insecticides.

Aleukia: The absence or extremely decreased number of leukocytes (white blood cells) in circulating-blood.

Aleuritic acid: An organic acid present in lac.

Algicide: Pesticide used to kill or control algae.

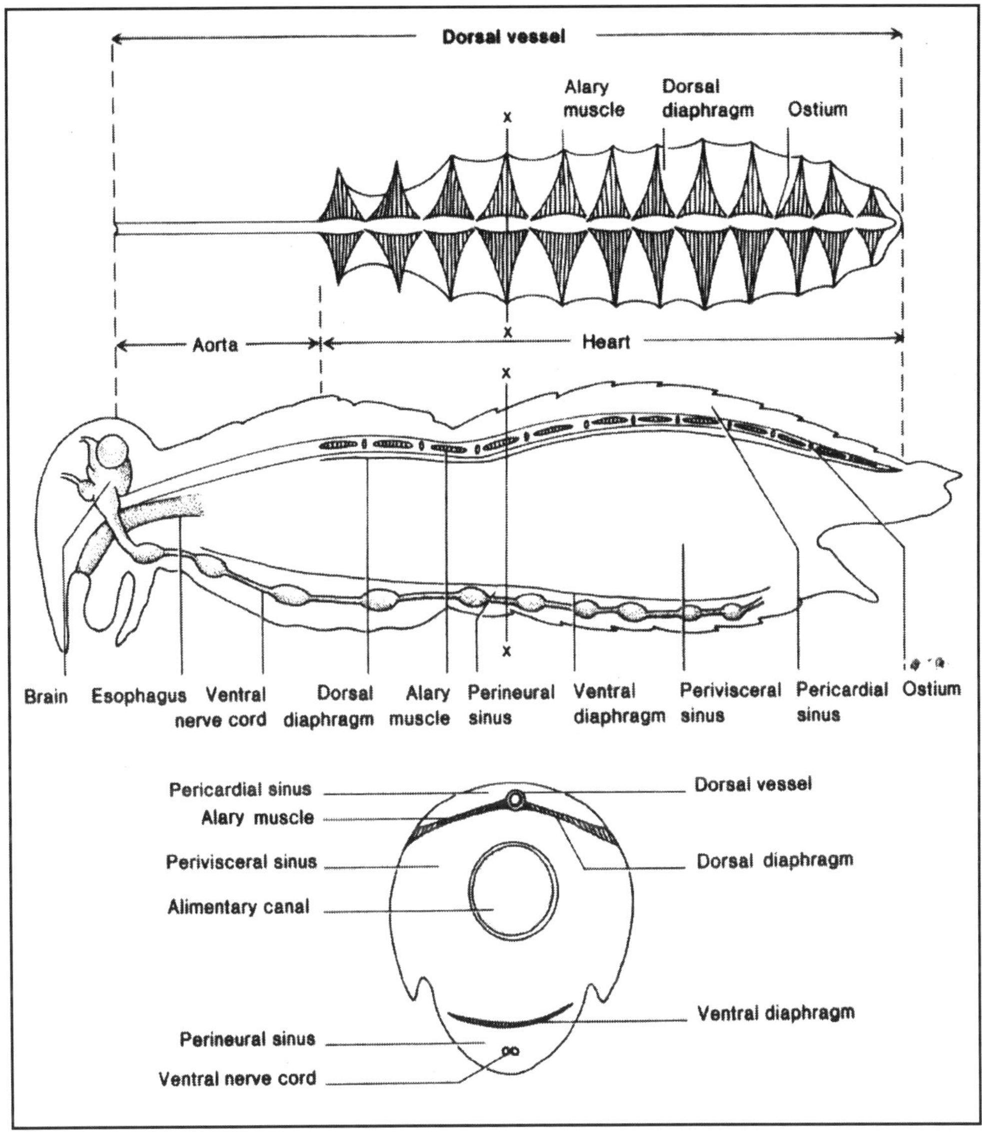

Fig. 10. Generalized insect circulatory system. x – x's top two diagrams indicate cross-sectional plane of bottom diagram.

Algorithm: A precisely defined sequence of rules stating how to produce specific outputs from specific inputs through a given number of steps.

Alienicole. Refers to Aphididae; parthenogenetic viviparous females, usually on the secondary host plant, and often differing from the other forms.

Alimentary canal: The tube of the digestive system, extending from the mouth to the anus and including the foregut, midgut, and hindgut (Fig. 41).

Alinotum: The wing-bearing plate on the dorsum (notal plate) of the meso- or metathorax of a pterygote insect (Fig. 11).

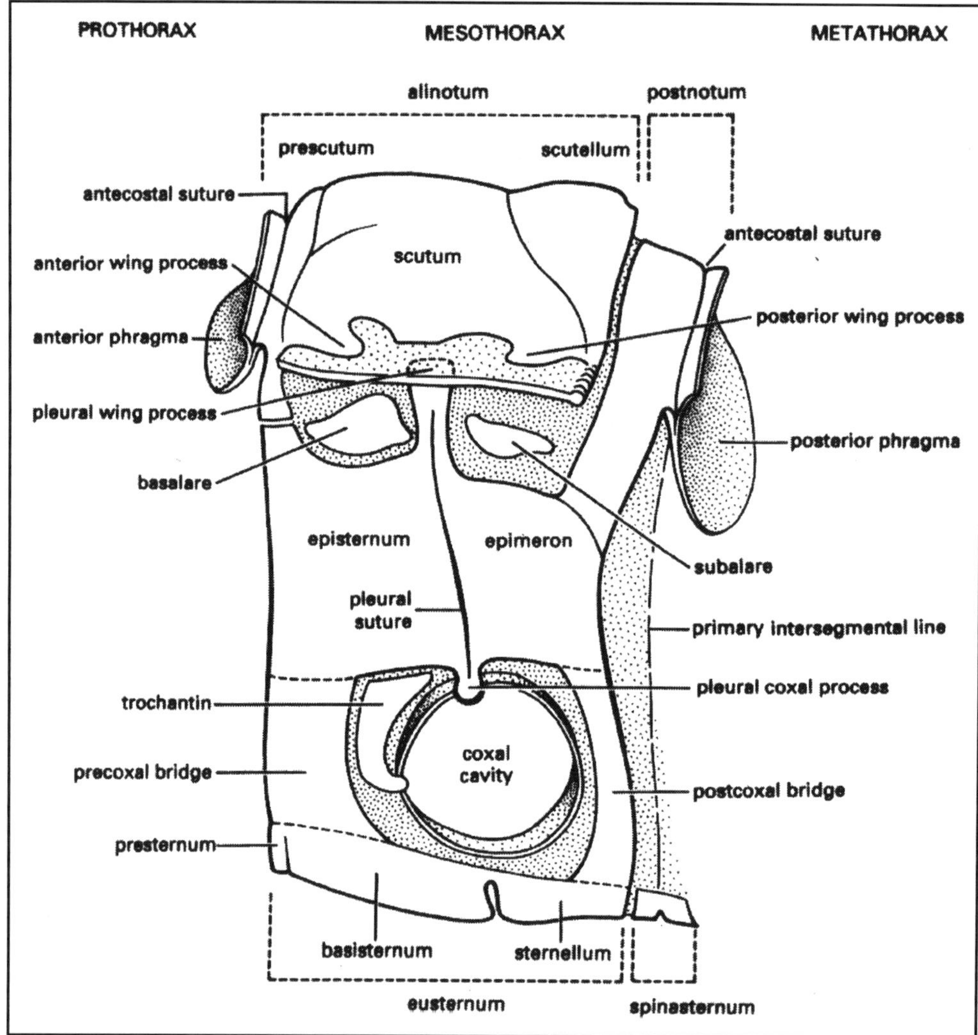

Fig. 11. Diagrammatic lateral view of a wing-bearing thoracic segment, showing the typical sclerites and their subdivisions. After *Snodgrass*, 1935.

Aliphatic: Organic chemical compound in which the carbon atoms are linked in open chains rather than rings.

Alitrunk: The fused thorax and first abdominal segment (**propodeum**) of adult ants *See also* **Mesosoma.**

Alkaline: Having the reaction of an alkali resulting from an excess of hydroxyl ions (OH) over hydrogen ions in solution: pH more than 7.

Alkaloids: Nitrogen-containing chemical compounds produced by some plants that are toxic or inhibitory to herbivores. These are the chemicals found in plants, many with important pharmacological actions.

All-, Allo-: Other; different.

Allantois: An embryonic membrane common to reptiles, birds and mammals; the membrane is associated with the diffusion of CO2 and O2 between the embryo and the environment of the embryo.

Allatectomy: An operation resulting in removal of the corpora allata.

Allele: Any of several alternative gene forms at a given chromosomal locus. It is an alternative expression of a gene having the some locus in homologous chromosomes (cf. Gene). Or, one of two or more alternative forms of a gene at corresponding sites on homologous chromosomes. Defensive chemicals. *See also* secondary plant substances.

Allelochemic (s): A substance involved in communication between members of different species.

Allelochemical: A chemical functioning in interspecific communication. *See also* **Allomone; Kairomone; Pheromone; Synomone.**

Allelopathy: Direct inhibition of one organism by another organism, mediated through the release of toxic or noxious chemical compounds. Originally applied to the inhibition of one plant by another plant, as mediated by toxic compounds released by plant roots.

Allergens: Insect secretions, usually proteins, whose effect depends on the physiological response of the victim.

Allethrin: The first pyrethroids insecticide synthesized.

Allochronic species: Species which do not occur at the same time level (cf. Synchronic species).

Allochthonous: Not aboriginal; exotic; introduced; acquired from elsewhere (opp. autochthonous).

Allometric development: Differential growth rate of one part of an individual in relation to another part or to the individual as a whole.

Allometric growth: Genetically determined tendency for a certain body part to grow at a more rapid rate than other parts; also called heterogonic growth. This is the abnormal development of one part of the body.

Allomone: A chemical messenger produced by one species that has an effect on another species, with a result that benefits the donor but not the recipient (e.g., repellent). This is a communication chemical that benefits the producer by the effect it invokes in the receiver. Or, an interspecific chemical messenger that benefits the releaser but not the receiver (e.g., repellents, toxicants). Or, chemical emitted by an individual of one species and received by one of another species that induces a response favourable to the emitter. Or, an external chemical signal that acts between different species to benefit the producer- for example, to repel a predator.

Allopatric: Having ranges that do not overlap. It is a term applied to two or more populations which occupy mutually exclusive (but usually adjacent) geographical areas (cf. Sympatric). Or, non-overlapping geographic distributions of organisms or taxa. Or, having separate and mutually exclusive areas of geographical distribution.

Allopatric hybridization: Hybridization between two allopatric populations (species or subspecies) along a well-defined contact zone (cf. Sympatric hybridization).

Allopatric speciation: Formation of species during geographical isolation (cf. Sympatric speciation).

Allopatric species: Species which occupy different geographical ranges are termed allopatric species.

Allotype: A paratype of the opposite sex to the holotype (cf. Paratype).

Alpha diversity: The variety of organisms occurring in a particular location.

Alpha taxonomy: The level of taxonomy concerned with the characterization and naming of species (cf. Beta taxonomy, Gamma taxonomy).

Alternate host: Either of the two unlike host-plants of a pest or pathogen that requires both to complete its full life-cycle. A plant host that is different from the primary host, on which a pathogen must develop to complete its life cycle (e.g., a heteroecious rust such as wheat rust). Often used in the entomological literature as synonymous with alternative host.

Alternation of generations: The alternation of a bisexual with a unisexual generation.

Alternative hosts: Host-plants other than the main host on which an organism (bacterium, fungus, insect, mite etc.) can develop. Weeds and wild plants are often alternative hosts of certain pests and diseases. Alternative host is used by a pest or pathogen when the primary or preferred host (crop) is not present. Alternative hosts are not required for the completion of the pest or parasite life cycle.

Altruism: Behaviour costly to the individual but beneficial to others.

Altruistic behavior: Self-sacrifice.

Alula (pl., alulae): A lobe at the base of the wing (Diptera; see calypter).

Ambulatory legs: Legs adapted for walking.

Ametabola-(alternative term, Apterygota)**:** Subclass of primitive wingless insects that do not undergo metamorphosis. Comprises the orders Thysanura, Diplura, Protura and Collembolan. cf. Metabola.

Ametabolous: Development without change of form. Lacking metamorphosis, i.e. with no change in body form during development to adult, with immature stages lacking only genitalia. Insect that go through three life stages: egg, young and adult. The young stages look the same as adults but do not have reproductive organs. Examples include silverfish and firebrats. Ametabolous insect develops without metamorphosis; that is, changing little in form during the course of growth and molting.

Amine: Any organic compound containing nitrogen cerived from ammonia.

Amino acid: Any organic compound containing an amino group (NH_2) and a carboxylic acid group (COOH).

Amnion: In embryology, the layer covering the germ band. It is inner membrane around the developing embryo (Fig. 12 and 13).

Amnion: The highest group of vertebrates; includes reptiles birds and mammals, so called because of the presence of the embryonic membrane amnion in these three groups of vertebrates.

Amoebiasis: A disease caused by the infection of the protozoan **Entomoeba histolytica**; general symptom is the dysentery commonly known as amoebic dysentery *e.g.*, in man.

Amoebocyte: A cell having the shape or properties of ***Amoeba*** *e.g.*, in Coelenterata (Gr. *Amoibe*=change, *kytes*= hollow).

Amoeboid: Like amoeba putting forth pseudopodia.

Amphibian: A class of vertebrates capable of living in land and water both *e.g.*, frog (Gr. *amphi*=on both sides+*bios*= life).

Amphiploid: A polyploid produced by the chromosome doubling of a species hybrid, that is, of an individual with two rather different chromosome sets.

Amphipneustic. Respiratory system in which only the first (thoracic) and the last one or two pairs of spiracles (abdominal) are functional; usually refers to insect larvae.

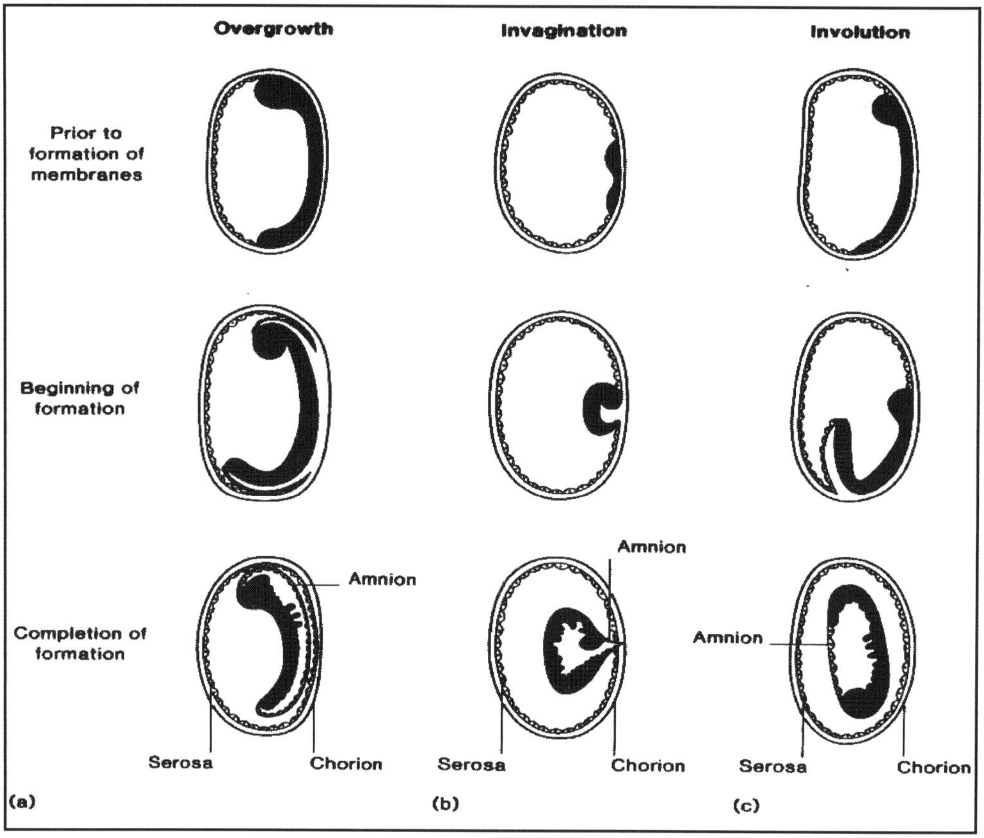

Overgrowth
Invagination
Involution
Prior to formation of membranes
Beginning of formation
Completion of formation
Amnion
Amnion
Amnion
Serosa
Chorion
Serosa
Chorion
Serosa
Chorion
(a)
(b)
(c)

Fig. 12. Formation of extraembryonic membranes.

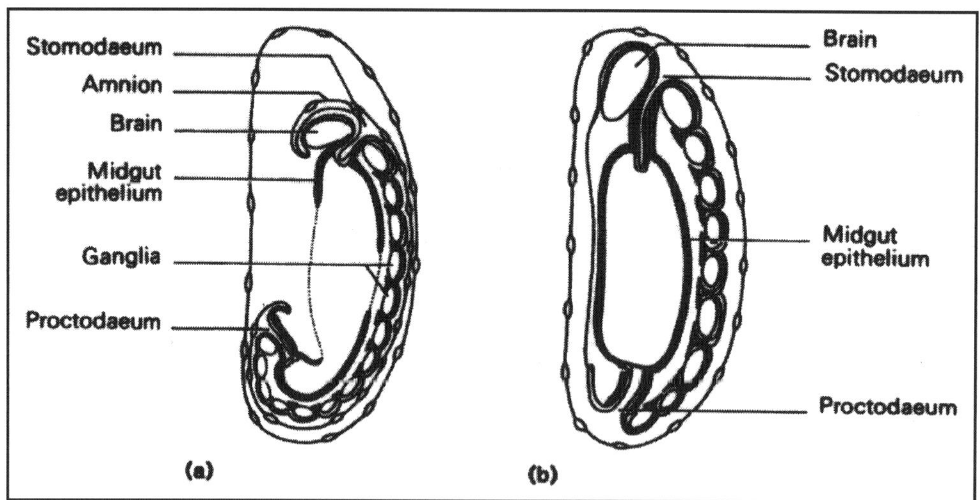

Fig. 13. Sagittal sections of an insect embryo at successive stages in the development of the alimentary canal.

Amphitoky (also **amphitokous parthenogenesis; deuterotoky):** Parthenogenesis in which the female produces offspring of both sexes.

Amplexiform: A form of wing-coupling in which there is extensive overlap between the fore and hind wing, but without any specific coupling mechanism.

Amylase: An enzyme that catalyzes the splitting of starch into maltose units.

An-: Without.

Anaemia: Any condition in which the number of red blood cells and the amount of haemoglobin are less than normal. It is frequently manifested by pallor of the skin and mucous membranes.

Anaerobic: Respiration not dependent on oxygen (Gr. a*n* = without + *ceros* = air + *bios* = life); refers to chemical processes or organisms that occur or live, respectively, without oxygen.

Anal area: The posterior part of the wing, supported by the anal vein (s).

Anal cell: A cell in the anal area of the wing; cell 1 A (Diptera).

Anal crossing: Where vein branches posteriorly from Cu + A (Odonata).

Anal lobe: A lobe in the posterior basal part of the wing.

Anal loop: A group of cells in the hind wing of drangonflies, between Cu2, 1 A, 2A, which may be rounded elongate or foot-shaped.

Anal margin: The posterior edge of an insect wing (Fig. 14).

Anal or vannal fold: A distinctive fold in the anal area of wing (Fig. 14).

Anal ring: An elevated ring like structure surrounding the anus in coccids.

Anal veins: The longitudinal veins posterior to the cubitus, usually three in number.

Anal: In the direction or position of the anus or on the last abdominal segment (which bears the anus); the posterior basal part (for example, of the wing).

Analogous: Similar in external features or function but not in essential structural pattern or origin (cf. homologous).

Analogy: Similarity of external features of function, but having a different evolutionary origin. (Gr. *ana* = according to + *logos* = ratio).

Anamnia: Vertebrates in which the embryo is not enclosed by an amnion during development in aquatic vertebrates *e.g.,* the cyclostomes, fishes and amphibians. (Gr. *an* = without + *amnion*).

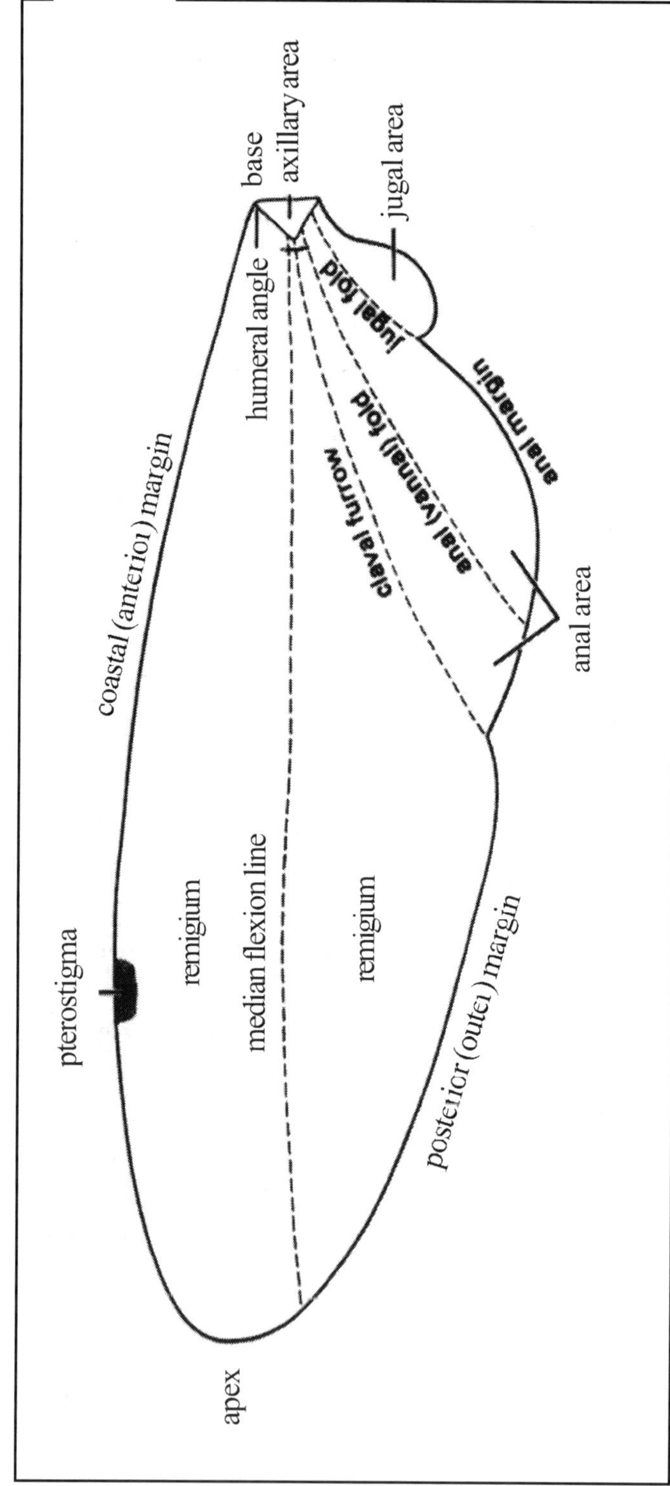

Fig. 14. Nomenclature for the main areas, folds and margins of a generalized insect wing.

Anamorphic development: Development in which the immature stages have fewer abdominal segments than the adult. *See also* **Epimorphic development.**

Anamorphosis: Development of an organism in which one or more body segments are added posteriorly at the time of molting. The additional abdominal segments are added to the body as development progresses.

Anapleurite: One of three theoretical, primitive subcoxal sclerites.

Anatomy: The science of internal morphology, as revealed by dissection.

Anautogenous: Requiring an exogenous source of protein for vitellogenesis.

Androconia: Specialised scales and associated sex-pheromone-producing glands among the wing scales of some male moths and butterflies (male Lepidoptera).

Anemophilous: Plants which are pollinated by the wind.

Anemophily: Wind pollination.

Anemotaxis: Directed response relative to an air current. This is orientation with respect to currents in air.

Angiosperm: Flowering plants that produce seed in a closed ovary.

Anholocycly: Refers to Aphididae; species without winged sexual forms, reproduction parthenogenetic and viviparously, hosts herbaceous and no alternation of generations.

Anhydrobiosis: Life without water. Some nematodes and other organisms can survive desiccation by eliminating water from their bodies and becoming quiescent with no detectable metabolism; when water again becomes available, the organisms rehydrate and resume normal activities.

Aniline: Synthetic dyes produced from oily liquid (aniline) extracted from coal tar.

Anionic surfactant: Salt of an organic acid, the structure of which determines its surface activity.

Anisomorphal: A defensive allomone of the walkingstick *Anisomorpha.*

Anlage, anlagen: Precursor; the first recognizable sign of a developing organ in an embryo.

Annelida: The most highly evolved of all the worms, with ring-like segments (annuli). Distinct metameric segmentation; includes all types of earthworms and leeches.

Annual: A plant that completes its life cycle in less than one year and then dies. Sometimes classified as summer and winter annuals, depending on the time of year that the plant is growing and reproducing.

Annulate: Ringed; usually refers to narrow ring like segments in the funicle of the antenna (Fig. 15).

Annulated: With ringlike segments or divisions.

Annulus: Narrow, ringlike segments in the antenna, follow after the pedicel (Fig. 15).

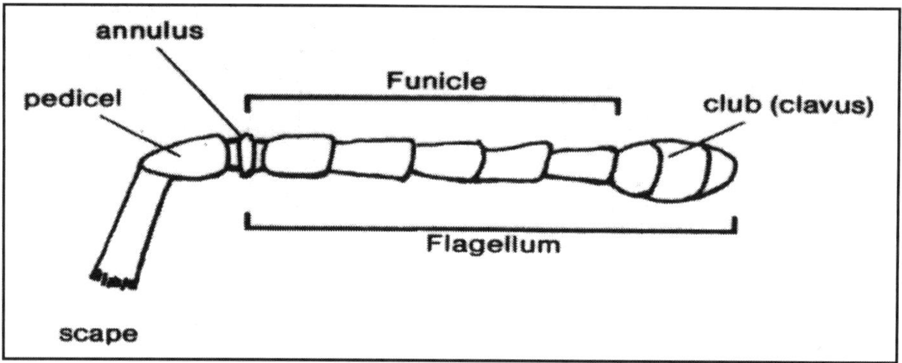

Fig. 15. Insect antenna.

Anoplura: The order of insects comprising the sucking lice (*anopl,* means unarmed; *ura,* means tail) (Fig. 16); an alternative name for the insect order Siphunculata.

Anoxic: Lacking oxygen.

Anoxybiosis: Life without oxygen. Some organisms can survive oxygen-depleted conditions by lowering their metabolism and becoming quiescent until oxygen again becomes available.

Anphylaxis: Hypersensitivity to foreign protein or poison (such as venom) marked by an intense systemic reaction.

Antagonism: When two or more chemicals have opposing actions, such that the action of one is impaired or the total effect of both is less than that of one chemical when used separately.

Ante-: Before; in front of.

Anteapical cell: A cell in the distal part of the wing (leafhoppers).

Anteapical: Just proximal of the apex.

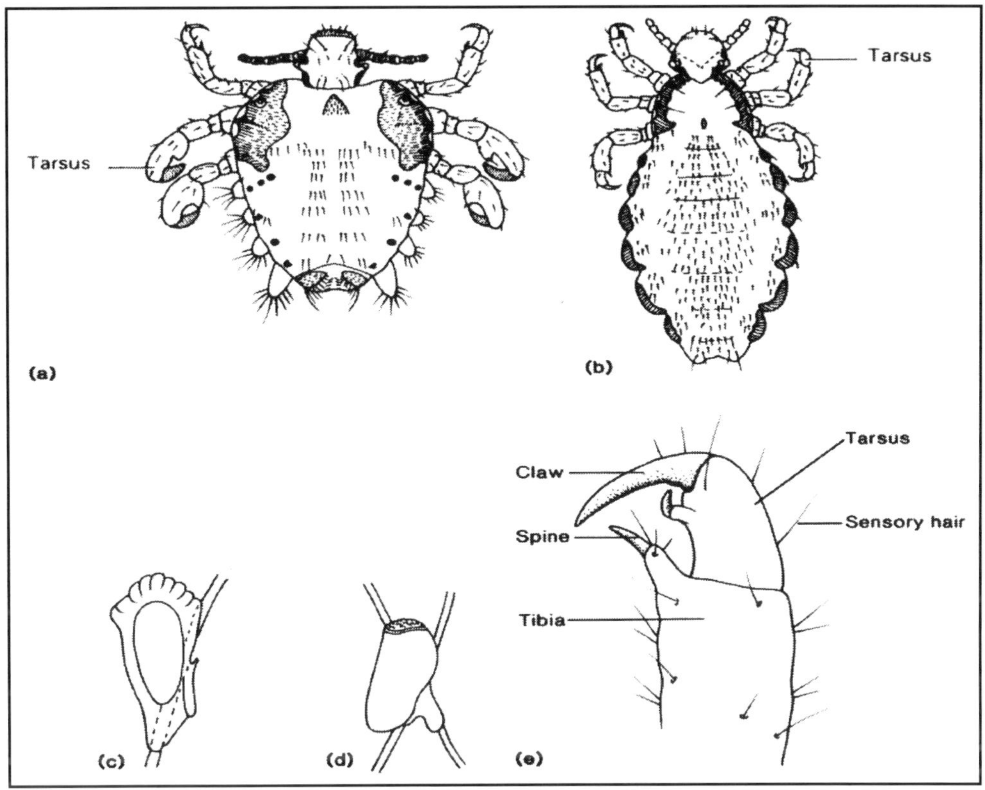

Fig.16. Anoplura, sucking lice. (a) Crab louse, *Pthirus pubis.* (b) Human louse, *Pediculus humanus.* (c) Egg or nit of crab louse attached to host hair. (d) Egg of human head louse attached to host hair. (e) Foreleg of human louse showing tibiotarsal claw.

Anteclypeus: An anterior division of the clypeus.

Antecosta: An internal ridge, associated with the tergum and sternum of an abdominal segment (Fig. 17).

Antecostal suture (*also* intersegmental groove): A groove marking the position of the intersegmental fold between the primary segments. (Fig. 11 and 18).

Antecoxal sclerite: A sclerite of the metasternum, just anterior to the hind coxae.

Antenna (pl. antennae): One of a pair of elongate segmented appendages located on the head of an insect or mollusk, and two pair on crustacean heads lying usually anterodorsally above the mouth parts, and usually bearing many sensillae and functioning as olfactory and tactile receptors sensory in function derived from the second head segment. They are often long, thread-like and conspicuous (as in cockroaches, grasshoppers and aphids) but may be small and relatively inconspicuous in many groups of insects. Also called feelers and smeller (Fig. 15 and 19).

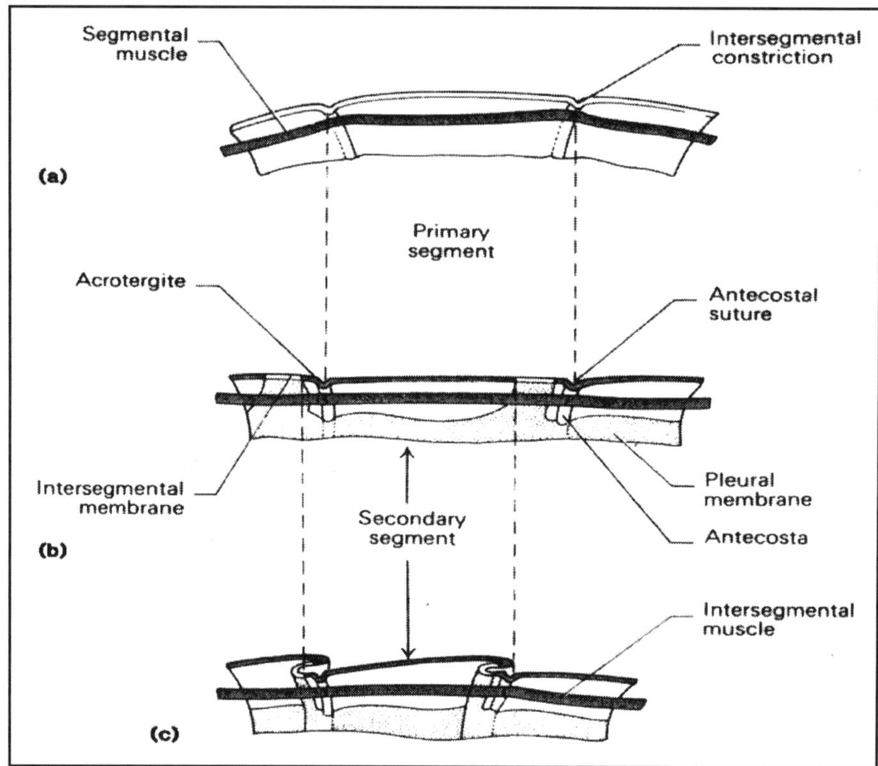

Fig. 17. Types of segmentation (tergites only). (a) Primary. (b) and (c) Secondary (adjacent secondary segments overlapping in c). After *Snodgrass*, 1935.

Antennal club: The enlarged distal segments of a clubbed antenna (Fig. 15).

Antennal fossa: A cavity or depression in which the antennae are located.

Antennal groove: A groove in the head capsule into which the basal segment of the antenna fits.

Antennal lobes: Lobes in the insect brain that receive both sensory and motor axons from the antennae (Fig. 54).

Antennal suture: A suture that surrounds the antennal base and delineates the antennal sclerite (Fig. 21).

Antennifer: Articular point of antenna located in the antennal socket (Fig. 19).

Antennule: The first antennae of Crustacea.

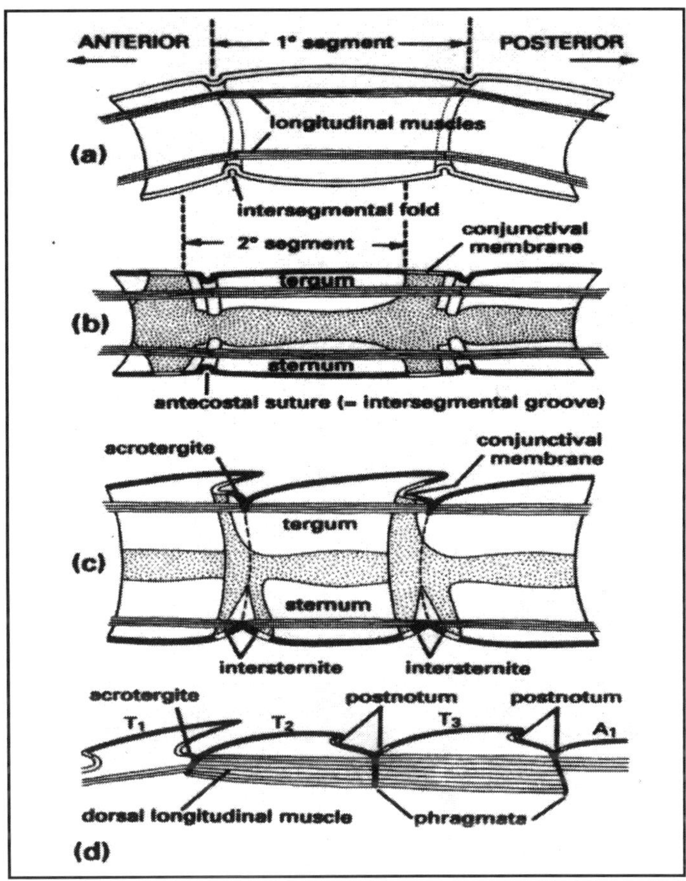

Fig. 18. Types of body segmentation: (a) primary segmentation, as seen in soft-bodied larvae of some insects; (b) simple secondary segmentation; (c) more derived secondary segmentation; (d) longitudinal section of dorsum of the thorax of winged insects, in which the acrotergites of the second and third segments have enlarged to become the postnota. After *Snodgrass*, 1935.

Antenodal cross veins: Cross veins along the costal border of the wing, between the base of the wings and the nodus, extending from the costa to the radius (Odonata).

Antenomere: A subdivision of the antenna.

Antepenultimate: The second from the last.

Antepygidial: Bristle one or more large bristles on the apical margin of the seventh (next to the last) tergum (Siphonaptera).

Anterior cross vein: The r-m cross vein (Diptera).

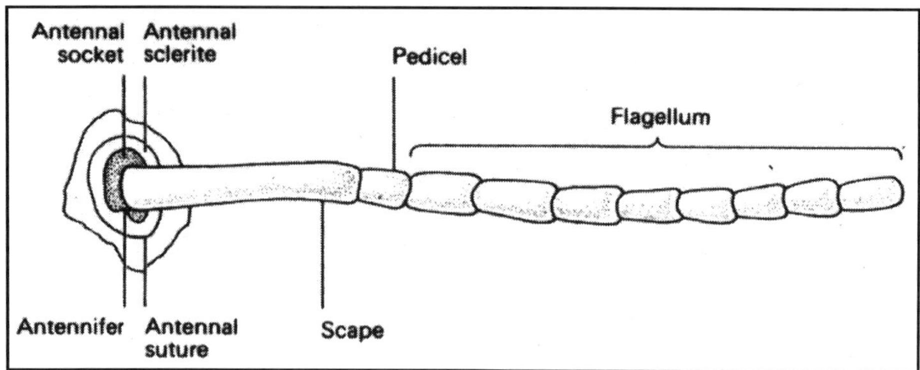

Fig. 19. Generalized insect antennae.

Anterior imaginal ring: Ring of embryonic cells located at the posterior end of the larval foregut. *See also* posterior imaginal ring.

Anterior intestine: The portion of the hindgut between the Malpighian tubules and the rectum.

Anterior notal process: Anterior wing articulation point on the notum (Fig. 11).

Anterior: Front; in front of; at or towards the front.

Anterodorsal: In the front and at the top or upper side.

Anteromesal: In the front and along the midline of the body.

Anteroventral: In the front and underneath or on the lower side.

Anther: The male part of a flower which produces pollen.

Anthophilous: Flower-loving.

Anthoxanthins: Whitish or yellow pigments found in insects and common in scale insects.

Anthraquinones: Red and orange pigments common in scale insects.

Anthropocentric: Human-centered point of view.

Anthropocentrism: Regarding humans or human objectives as the principal focus of a system or situation.

Anthropogenic: Caused by humans.

Anthropomorphism: The attribution of human characteristic to nonhumans or interpreting behavior patterns observed in animals in term of human characteristics.

Anthropophilic: Associated with humans.

Anti-: Against.

Antibiosis: A property of an organism that adversely affects the well-being of another organism that consumes it. In case of host plant resistance, it is a plant characteristics that affect insects in a negative manner (e.g., increase mortality, reduce fecundity); involves activities of both host plant and insect. The substance that prevents, inhibits, or destroys life is produced by one organism and has an adverse or deleterious physiological effect on another organism resulting from feeding on a resistant host. A host plant may show resistance against insect attack by having, for example, a thick cuticle, hairy leaves, toxic sap, etc. In antibiosis one organism exerts a toxic or negative effect on another organism. For example, Penicillin is an antibiotic that exerts a negative effect on certain bacteria. In IPM, the term also describes a type of host-plant resistance affecting the physiology of the herbivore; a mechanism of plant resistance to pests whereby harmful effects occur to the insect through feeding on the plant.

Antibiotic: Substance that prevents, inhibits, or destroys life. These are the chemicals produced by microbes that inhibit, or are toxic, to other microbes.

Antibody: A serum globulin which is produced in the blood of an immunized animal in response to the introduction of a foreign antigen (cf. Antigen, Antiserum, Serum globulin, Serology).

Anticoagulant: A substance that inhibits, delays, or suppresses blood clotting resulting in internal hemorrhaging; this is the mode of action of a major class of rodenticides.

Antidote: A practical immediate treatment that alleviates or reverses the physiological damage caused by toxins, including first aid. A remedy or treatment given to reverse the effects of poisoning.

Anti-feedant: A chemical possessing the property of inhibiting or stopping the feeding of certain insect pests. A control method often used for clothes moth larvae and termites.

Anti-frothing agent: Material added to prevent frothing of the liquid in the spray tank.

Antigen: A substance capable of including the formation of antibodies when introduced into the blood stream of animals (cf. Antibody, Precipitin reaction, Serology).

Antimetabolite: Substances or chemicals resembling essential nutrients that interfere with metabolism inhibit normal physiology, usually with detrimental results. These substances are chemical analogs that block normal absorption and function of nutrients.

Antiserum: Blood serum containing specific antibodies (cf. Antibody, Precipitin reaction).

Antixenosis (non preference): Plant characteristics that drive insects away from a particular host; involves activities of both host and insect. It is a type of plant resistance that affects herbivore behavior because of being unsuitable to a feeding insect.

Ant-lion: It is a predatory larva of a winged insect resembling a dragonfly. It digs a pit in sandy soil and lies with its large jaws partly exposed as it waits for passing insects particularly ants.

Anus: The posterior opening of the alimentary system or the digestive tract (Fig. 50). Opening of the alimentary canal to the exterior of the body of honey bee including its sting glands is also known by the same name *Apis* which has medicinal value especially in cases of rheumatism.

Aorta: The anterior subdivision of the dorsal blood vessel, lacking ostia and alary muscles. It is a blood-containing tube that extends forward from the heart and is open anteriorly (Fig.20). The forepart of the dorsal blood vessel of insects. In contrast to the hind part (heart), the aorta is without valves. It discharges into the head region.

Apenustic: A respiratory system without functional spiracles. *See also* **Oligopneustic; Polypneustic.**

Aphaniptera: An alternative name for the insect order Siphonaptera which comprises the fleas.

Aphicide: An insecticide that kills aphids.

Aphidophagous. Aphid eating;-predator that feeds largely on aphids e.g. lacewing larva.

Aphins: Red, orange, and yellow pigments found only in aphids.

Aphrodisiac: A chemical substance that stimulates sexual desire.

Apical cell: A cell near the wing tip.

Apical cross vein: A cross vein near the apex of the wing (Plecoptera, apc; Homoptera).

Apical margin: The outer edge of an insect wing (Fig. 21).

Apical: At or towards the apex (Fig. 320). At the tip or end; or outermost part; towards the apex or distal often applied to the growing points on plants.

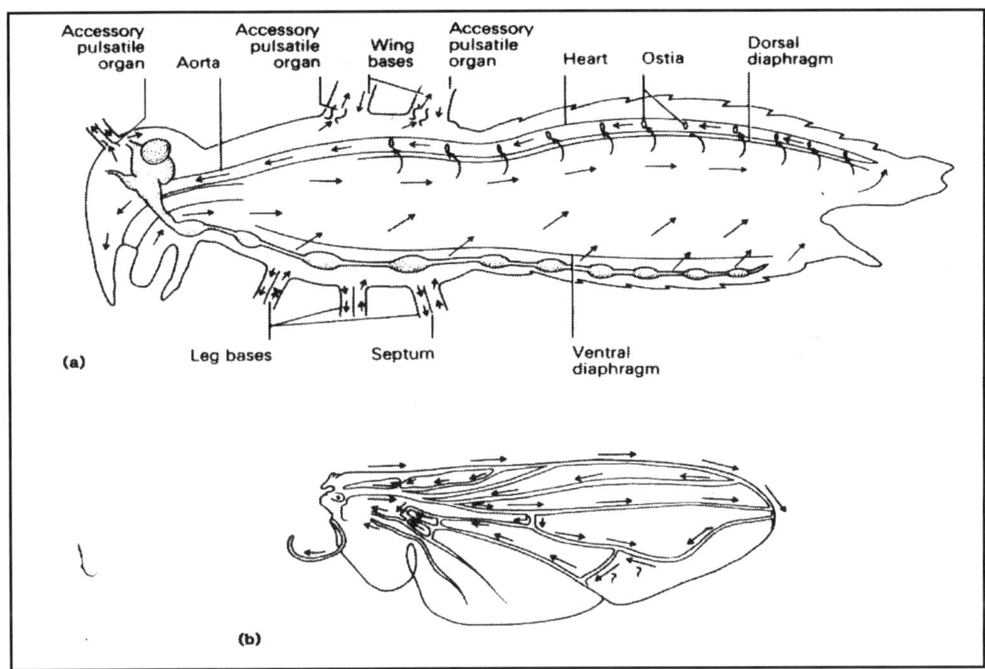

Fig. 20. Patterns of circulation. (a) Insect with a fully developed circulatory system. (b) Pattern of circulation in a house fly wing. *After West, 1951.*

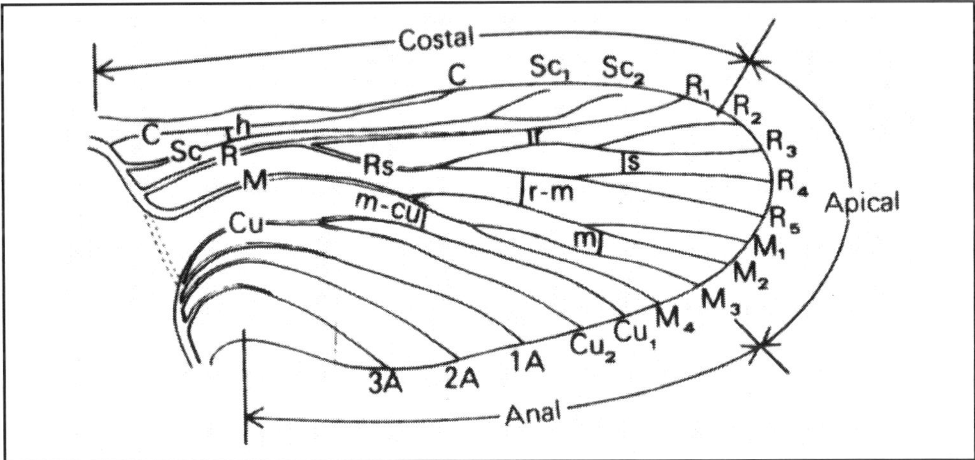

Fig. 21. Hypothetical primitive pattern of wing venation. Longitudinal veins: A = anal; C = costa; Cu = cubitus; M = media; R = radius sector; Sc = subcosta. Cross veins: h = humeral; m = medial; m-cu = mediocubital; r = radial; r-m = radiomedial; s = sectorial. Costal, apical, and anal margins are indicated. After *Comstock,* 1918.

Apiculture: Beekeeping; the keeping of honeybees or the management of honey bees. The commercial rearing of honeybees for bee products or for pollination purposes.

Apneustic: Insect larva (usually aquatic Diptera) without functional spiracles, but still possessing an internal tracheal system. The closed type of ventilatory system, without functional spiracles.

Apocritan: Belonging to the suborder of Hymenoptera (Apocrita) in which the first abdominal segment is fused to the thorax. *See also* **Propodeum.**

Apod larva: A larva lacking true legs. Legless; without feet; larva without thoracic legs.

Apode (adj. apodous)**:** An organism without legs.

Apodeme: An invagination of the body wall that forms a rigid process which provides a point or areas for muscle attachment and strengthens the exoskeleton and for strengthening the body wall. An internal inflection or projection of the integument or exoskeleton, tendon-like to which muscles are attached (Fig. 22).

Apolysial space: The space between the old and the new cuticle that forms during **apolysis,** prior to **ecdysis.**

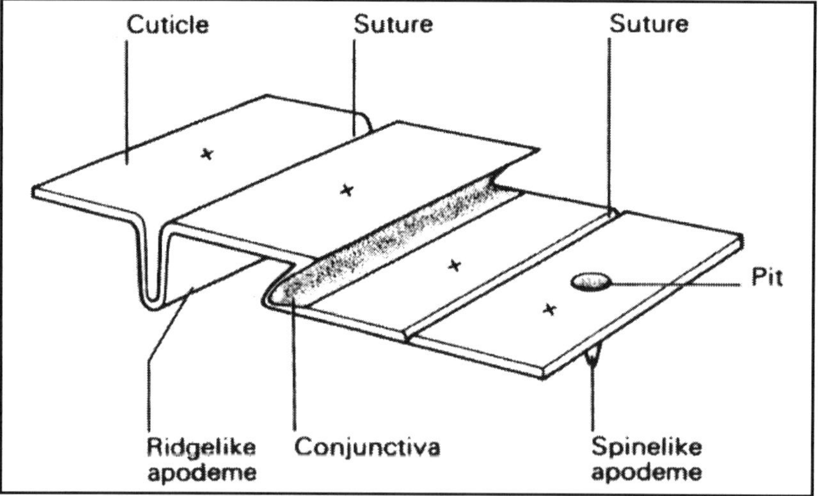

Fig. 22. Sclerites (denoted by x), sutures, and internal integumentary processes.

Apolysis: The separation of the old cuticle from the epidermis during molting; an early step in the molting cycle. It is because of the retraction of the epidermal cells from the inner surface of the endocuticle that is the first step in molting.

Apophysis (pl., apophyses): A tubercular or elongate process of the body wall or exoskeleton, either external or internal. This is spine like inflection (apodeme) of the insect integument to which muscles attach.

Aporous: Without a pore.

Aposematic colouration: These are conspicuous or warning colourations such as that of dangerous or poisonous insects (or animals).

Aposematic: Warning of unpalatability, particularly using colour. Possession of vivid coloration that identifies an animal as having distasteful or unpleasant properties. It is a kind of communication system that is based on warning signals; conspicuous insect coloration used to frighten away enemies or to warn predators that it is unpalatable/ poisonous.

Aposematism: A communication system based on warning signals.

Apparency: Obviousness (e.g. of a plant to an insect herbivore).

Apparent resistance (ecological resistance, pseudoresistance)**:** Temporary host-plant resistance characteristics that rely heavily on environmental conditions; includes host evasion, indued resistance, and host escape.

Appendage: A movable projecting part of a metazoan body having an active function. (L. *ad* = to + *pendere* = to hang); an internal limb or similar structure arising from segments of the head, thorax or abdomen of insect.

Appendicular ovipositor: The true ovipositor formed from appendages of segments 8 and 9. *See also* **Substitutional ovipositor.**

Appendiculate marginal cell: With the vein forming the posterior side of the marginal cell extending a short distance beyond the apex of the cell (Hymenoptera).

Appendix dorsalis: The medial caudal appendage arising from the epiproct, lying above the anus; present in apterygotes, most mayflies and some fossil insects.

Appendix: A supplementary or additional piece or part of the homopteran wing.

Appetitive behavior: Behavior that increases the probability of exposure to a releasing stimulus or searching behaviour of variable pattern, seeking an appropriate stimulus.

Applied control: The measures adopted by human beings to control pests.

Applied entomology: The application of pure entomology to the control of insect pests.

Apposition eye: A type of compound eye occurring in diurnal insects, in which each ommatidium is surrounded by a shield of pigment.

Apposition image: Image formed when the light reaches the rhabdom of an individual eye unit only via the dioptric apparatus of the same eye unit (Fig. 23).

Approved common name: Generic name of a pesticide (e.g., carbaryl) or common name of an insect as approved by the Entomological Society of America (e.g., European corn borer).

Approved product: Proprietary brand of pesticide officially approved by the Ministry of Agriculture, Fisheries and Food, UK.

Apterae: Wingless insects; usually refers to wingless forms of Aphididae.

Apterous neotenic: See **Ergatoid.**

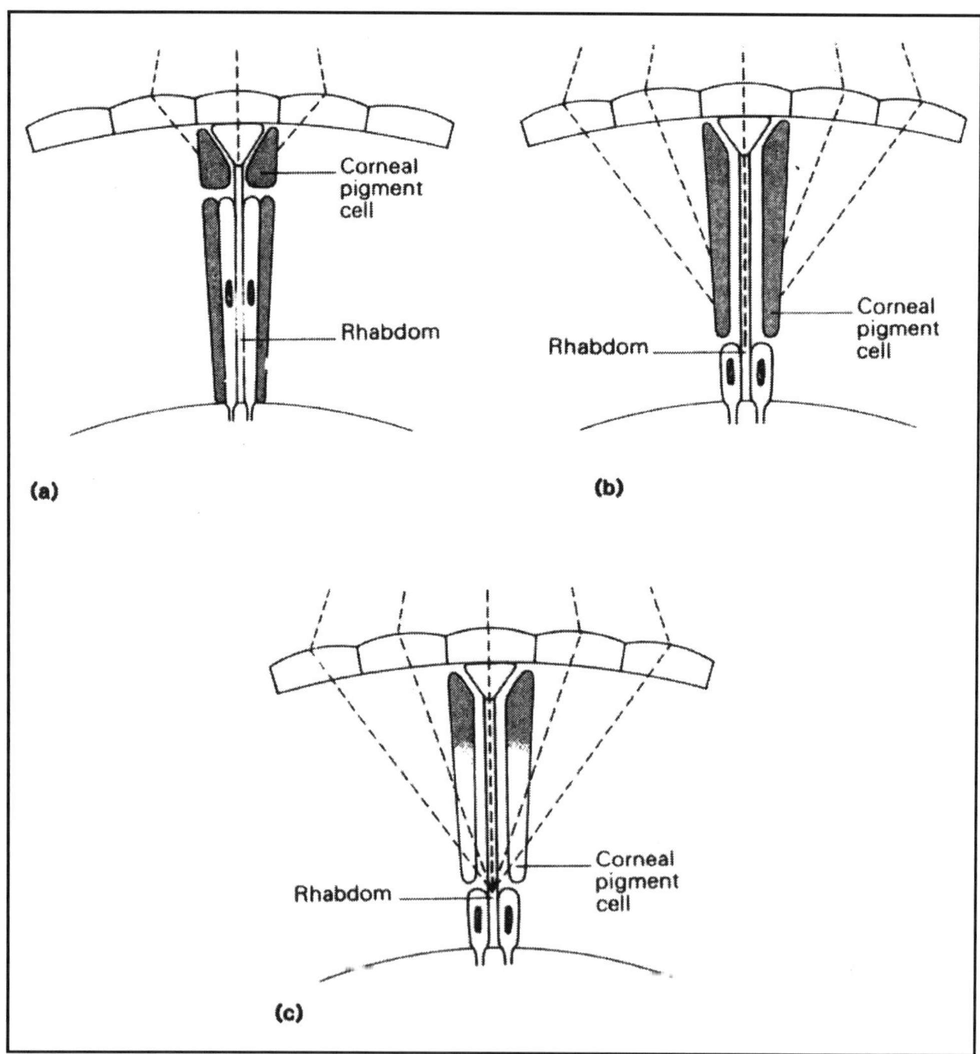

Fig. 23. Image formations in compound eyes. (a) Apposition type. (b) Superposition type, light adapted. (c) Superposition type, dark adapted. Dashed lines represent rays of light originating from a common point.

Apterous: Wingless; without wings; often used in reference to aphids. The opposite is macropterous, etc.

Apterygota: The subclass of Insecta that comprises the primitively wingless insects which do not undergo metamorphosis. A wingless insect of a group belonging to the subclass Apterygota believed never to have possessed wings in its past history. cf. pterygota.

Apterygote: Belonging to the subclass Apterygota.

Aquaculture: Water cultivation of plants or breeding (rearing) of animals.

Aquatic: Living in water.

Arachnida: The arthropod class that includes the spiders, scorpions, ticks, and mites, closely related to insects and characterized by having a cephalothorax and an abdomen, and six pairs of appendages including the chelicerae, pedipalps, and four pairs of walking legs.

Arachnophobia: Fear of arachnids (spiders and relatives).

Araneae: The order of arthropods within the class Arachnida that comprises the spiders.

Arboreal: Belonging to tree or tree inhabiting habit of animals *e.g.,* monkeys (L. *arbor* = tree).

Arbovirus: A virus transmitted by an arthropod vector like mosquitoes or other blood-sucking arthropods.

Arch-: Primitive; original.

Archedictyon: Irregular network of veins found between the principal wing veins in many fossil insect groups; similar networks in the wings of contemporary forms (e.g., dragonflies) are probably remnants of the archedictyon.

Archenteron: The primitive digestive cavity of a metazoan embryo formed by gastrulation. (Gr. *archo* = rule + *enteron* = intestine) *e.g.,* in Coelenterates.

Archeognatha, order: (*archeo,* ancient; *gnath,* jaw) jumping bristletails.

Archetype: A hypothetical ancestral type arrived at by the elimination of specialized characters (cf. Phylogeny).

Arcuate: Bent like a bow, or arched.

Arculus: A basal cross vein between the radius and the cubitus (Odonata).

Area-wide IPM: IPM tactics implemented across large geographic regions to suppress pest populations at the regional rather than field scale; requires cooperation among multiple agencies.

Areole: An accessory cell (see also *basal areole*).

Areolet: A small cell in the wing; in the Ichneumonidae, the small submarginal cell opposite the second recurrent vein.

Arista: A specialized large antennal bristle or process, usually located on the dorsal edge of the apical antennal segment in the certain Diptera; represents the reduced antennal flagellum

Aristate: Bristlelike; with an arista; aristate antenna.

Aroliar pad: Pretarsal ventral adhesive pad (Fig. 24).

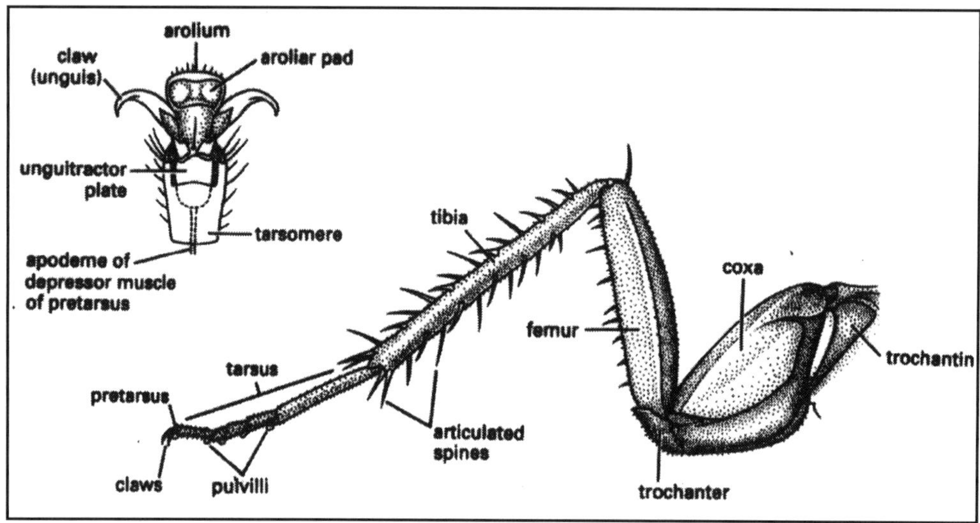

Fig. 24. The hind leg of a cockroach, *Periplaneta Americana* (Blattodea: Blattidae), with enlargement of ventral surface of pretarsus and last tarsomere.

Arolium (pl. arolia): An invertebrate animal with jointed appendages; a member of the phylum Arthropoda.

Arolium (pl. arolia)**:** Pretarsal pad-like or sac-like structure(s) lying between the claws; a bulbous, lobelike or padlike structure at the tip or apex of the last tarsal segment, between the claws (Orthoptera); a padlike structure at the base of each tarsal claw (Hemiptera) (Fig. 25).

Aromatic odour: Flavour like that of a compound belonging to the closed-chain class of organic compounds, or benezene derivatives.

Arrhenotokous parthenogenesis: The production of haploid male offspring from unfertilized eggs.

Artenkreis (Rensch)**:** Superspecies.

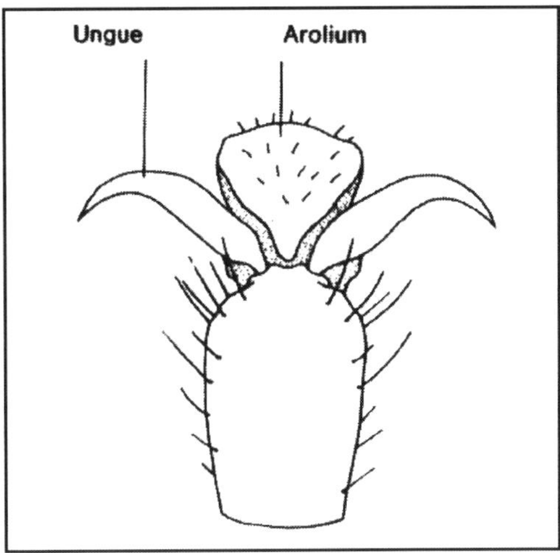

Fig. 25. Tarsal structure: grasshopper tarsus.

Arthrodial membrane: Soft, elastic cuticle between body segments. (Fig. 26).

Arthropleona: A sub-order of springtails (Collembola) having a distinctly segmented body of elongate form. cf. Symphypleona animals to which insects and other.

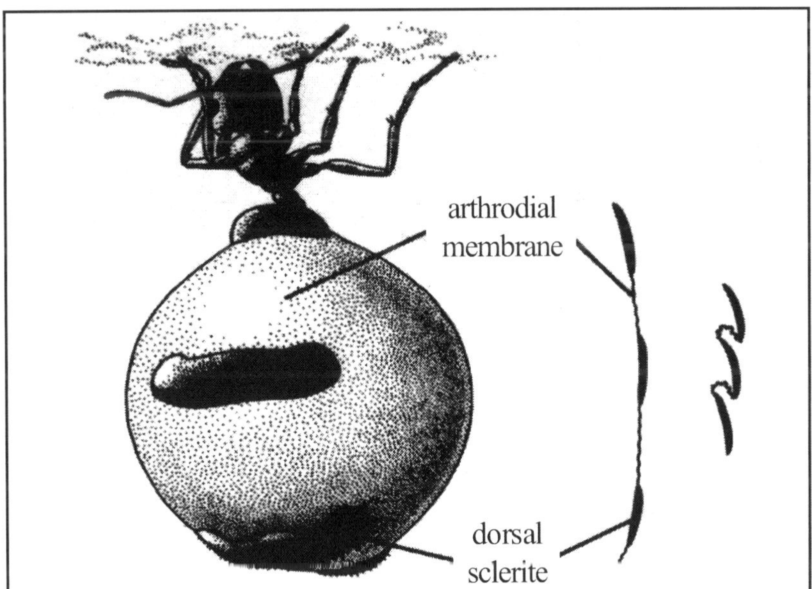

Fig. 26. A specialized worker, or replete, of the honeypot ant *Camponotus inflatus* (Hymenoptera: Formicidae), which holds honey in its distensible abdomen and acts as a food store for the colony. The arthrodial membrane between tergal plates is depicted to the right in its unfolded and folded conditions.

Arthropod phobia: The persistent fear of infestation by arthropods.

Arthropoda: The largest phylum of animal world to which insects and other groups belong; characterized by paired jointed appendages, bilateral symmetry, metameric segmentation and chitinous exoskeleton. The class Insecta is the largest and most important group of arthropods.

Arthropods: Animals belonging to phylum Arthropoda. They possess jointed legs (*arthros* means joint and *podos* means foot).

Arthropodins: A group of soluble proteins found in the insect cuticle.

Articular sclerites: Separate, small movable plates that lie between the body and a wing.

Articulation: A joint, as between two segments or structures.

Artificial classification: Classification based upon characters of convenience without relation to phylogenetic significance; classification based upon characters erroneously presumed to indicate phylogenetic relationship; also classification based on a single arbitrarily chosen criterion, instead of an evaluation of the totality of characters (cf. Classification, Phylogeny, Natural classification).

Artificial control: See applied control.

Ascospore: A spore produced by sexual reproduction in a fungus of the class Ascomycetes.

Asexual reproduction: Not involving the fusion of the nuclei of different gametes.

Asexual: Reproduction without the fusion of gametes, including budding and spore formation; parthenogenic (animals), vegetative (plants); not related to sex; not involving any kind of gametes or their fusion. *e.g.* Building in *Hydra.*

Aspirator: A sucking device with which insects may be picked up by suction. This is used for small insects.

Assembling scents: Chemicals (pheromones) that, when released, result in the formation of a group of individuals of the same species.

Association neurons: Found within ganglia, that may synapse with one another or with sensory or motor neurons; also called interneurons.

Associative learning: Acquisition of the capacity to associate a stimulus with a reward or punishment or learning based on the presence of reward or punishment; includes classical conditioning, instrumental learning, and shock-avoidance conditioning.

Asymmetrical: Not alike on the two sides; without symmetry.

Asymptote: The point in the growth of a population at which numerical stability is reached.

Asynchronous flight muscles: Flight muscles in which contraction is not synchronized with the reception of nervous stimuli.

Asynchronous muscles: A muscles that contracts many times per nerve impulse, as in many flight muscles and those controlling the cicada tymbal.

Athermopause: When factors other than temperature such as hygiene, nutritional, etc. are involved in dormancy.

Atlas: In taxonomy, a method of presenting taxonomic materials primarily by means of comparative illustrations rather than by comparative descriptions (cf. Monograph).

Atomization: Process of breaking liquid into a fine spray.

Atomiser: To break up into fine droplets (as applied to spray liquids). Device for breaking up a liquid stream into very fine droplets by a stream of air.

Atrial aperture: The outside opening of a chambered or atriate spiracle (Fig. 27 a and b).

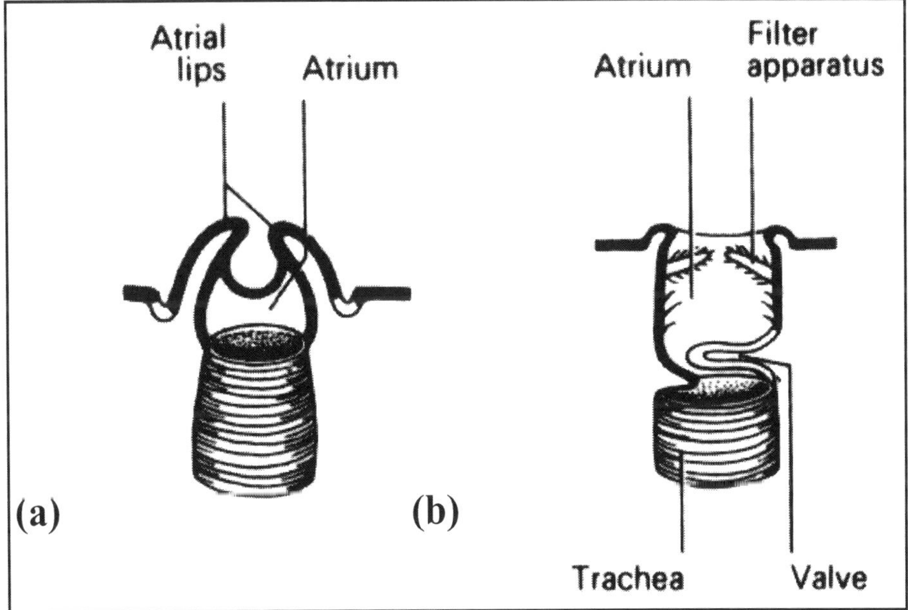

Fig. 27. Types pf spiracle. (a) Atriate spiracle with lip closure mechanism. (b) Atriate spiracle with filter apparatus and valve closure mechanism.

Atrium (pl., atria): A spiracular chamber housing the opening to a trachea; a chamber just inside a body opening especially inside a tubular conducting system, such as the tracheal system. (Fig. 28).

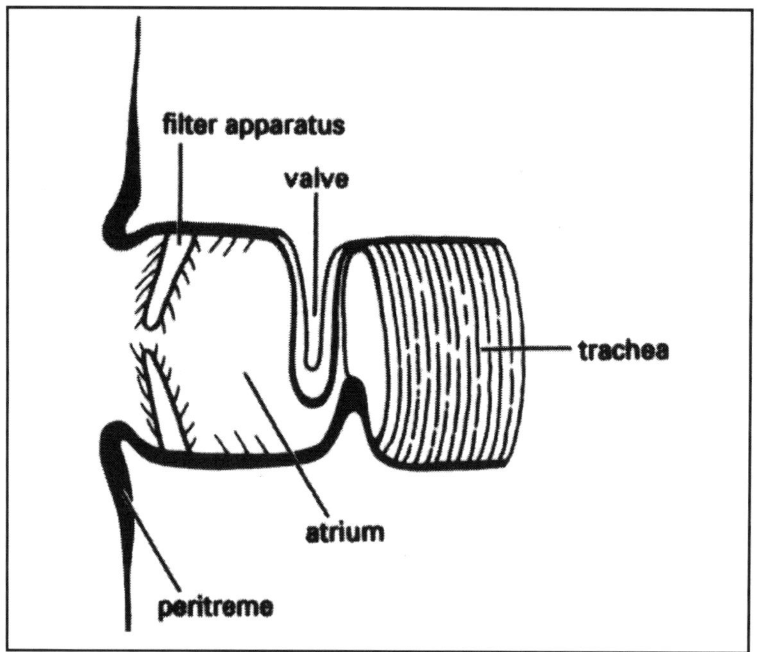

Fig. 28. An atriate spiracle with closing valve at inner end of atrium.

Atrophied: Reduced in size; rudimentary; vestigial.

Atropine: An anticholinergic alkaloid drug used as an antidote for organophosphate and carbamate insecticide poisoning.

Attapulgite: A magnesium silicate clay that is used as a carrier for the insecticide.

Attenuated: Very slender and gradually tapering distally.

Attractant: A chemical or physical source which induces insects to move towards it. cf. repellent; chemical substances with an odour that attracts certain insects or devices used to lure insects or other mobile pests to areas where they can be trapped or killed; for example, a light attracts moths at night. Attractants are based on feeding, oviposition, or mating behavior of insects. Several proprietory lures are manufactured.

Auditory organs: Organs for hearing.

Auger arrangement: It is spiral mechanism formed about a turning-shaft for moving the material.

Augmentation: Mass culture and local release of parasites, predators or pathogens to provide short term biological control of pest species; any biological control pratice designed to increase the number or effectiveness of existing natural enemies. This involves the release of massed reared parasitoids or predators that are released to augment existing parasitoid or predator populations or to reintroduce parasitoids or predators.

Augmentative biocontrol: The release of beneficial organisms, or biological control agents, to establish or augment a natural population.

Auricle: A small lobe or earlike structure (Hymenopetra).

Autecology: Study of individual organism or species in relation to habitat. This is study of ecology of a single species.

Authority citation: The custom of citing the name of the author of a scientific name or name combination [*e.g., X-us* Jones, *X-us albus* Jones, *Y-us albus* (Jones)]; **Double authority citation** includes the name of the author of the specific trivial name and the author of the currently accepted combination if different from the original combination [*e.g., Y-us albus* (Jones) Smith].

Auto-: Self; same.

Autoamputation: The spontaneous release of a leg that is being grasped by a predator; characteristic of nymphal walking sticks.

Autochthonous: Aboriginal; native; indigenous; formed where found (opp. allochthonous).

Autocidal control: Use of induced sterility for reducing pest population size. This is a type of control where insects are used for self-destruction, chiefly by release of sterile individuals.

Autocide: Self- destruction; control of an insect pest population by the sterile male technique.

Autogenous: Able to carry out vitellogenesis without an exogenous source of protein. In blood-feeding insects, the ability to produce eggs without taking blood; compare with anautogenous.

Automimic: A condition of **Batesian mimicy** in which palatable members of a species are defended by their resemblance to members of the same species that are chemically unpalatable.

Autosome: One of the chromosomes other than a sex chromosome; autosomes usually occur in identical numbers in the two chromosome sets of the same species (cf. Chromosome, Sex chromosome).

Autotomy: The shedding of appendages(s), notably for defence. Also see autoamputation.

Autotrophic: Organisms able to synthesize their food from inorganic materials using energy from an outside source. Most green plants are completely autotrophic, making organic foods from carbon dioxide, water, minerals, using the energy of sunlight which is in some way fixed by the chlorophyll.

Auxiliary vein: The subcosta (Diptera).

Auxin: A plant hormone that stimulates growth.

Available name: A name proposed in compliance with Art. 25 of the International Rules of Zoological Nomenclature (cf. Valid name).

Avicide: A pesticide toxic to birds.

Avirulent: Pathognes that do not cause disease because of the loss of virulence, or of a disease-causing characteristic or change in host susceptibility.

Avoidance: Growing crops in area that is largely free of pest species.

Axilla (pl., axillae): A triangular or rounded sclerite lateral of the scutellum, and usually just caudal of the base of the front wing (Hymenoptera).

Axillary area: An area at the wing base bearing the wing articulation. (Fig. 14).

Axillary cell: A cell in the anal area of the wing (Diptera, Hymenoptera).

Axillary plates: Two (anterior and posterior) articulating plates which are fused with the veins in an odonate wing; the anterior supports the costal vein, the posterior, the remaining veins, in Ephemeroptera there is only a posterior plate.

Axillary sclerite: A small sclerite at the wing base, articulating with the thorax.

Axillary sclerites: Three or four sclerites, which together with the **humeral plate** and **tegula** comprise that articular sclerites of the neopteran wing base.

Axillary vein: A cell in the anal area of the hind wing (Hymenoptera, *Symphyta*).

Axon: A fiber of a nerve cell that transmits nerve impulses away from the cell body; *see also* **Dendrite.** (Gr. *Axen* = axle) (Fig. 29).

Axonic poison: A chemical that interrupts normal transmission of impulses along the axon of a neuron.

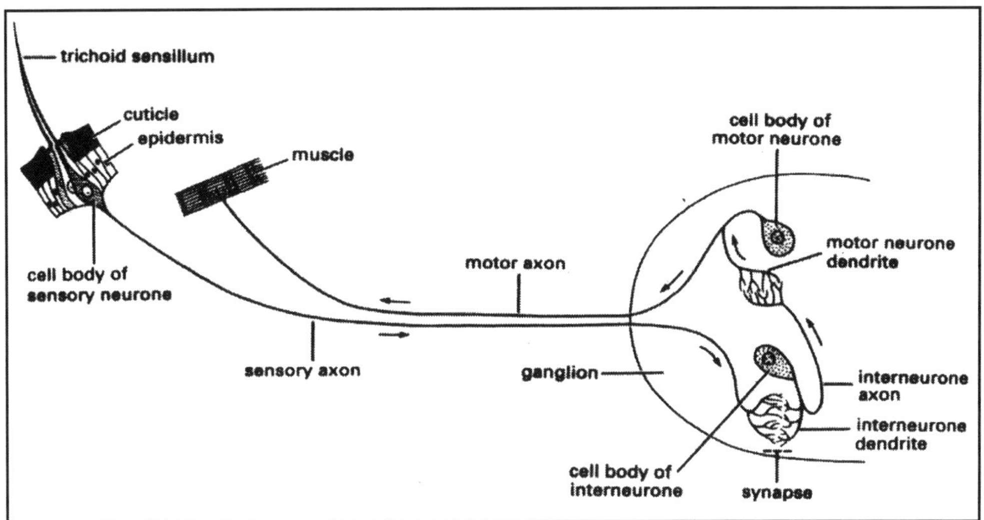

Fig. 29. Diagram of simple reflex mechanism of an insect. the arrows show the paths of nerve impulses along nerve fibres (axons and dendrites).

Azadirachtin: The chemically active principle of neem tree which act as an insect feeding inhibitor and growth regulator.

B

***Bacillus thuringiensis* (Bt):** Soil-inhabiting bacterium that produces an insecticide effective against larval stages of many species of Lepidoptera, although some strains are effective against various beetle, mosquito, and blackfly larvae.

Backcross: A cross between a hybrid and one of its parents; a cross between a heterozygote and a homozygote (cf. Hybridization, Heterozygous, Homozygous).

Bacterial insecticide: Bacteria pathogenic to insects (e.g., Bt or *B. popilliae*). Applied using application techniques also used for chemical pesticides.

Bacterium (pl. bacteria): Microscopic, prokaryotic, single-celled organisms having cell walls but lacking membrane-bound organelles. Important in decomposition, nitrogen fixation and other symbiosis, and as pathogens. A significant component of soil food webs.

Bait (B): A pesticide formulation that combines an edible or attractive substance with a small amount of toxic active ingredient (usually about 5%). Baits containing poisonous chemical and food stuffs are primarily offered to pests as food for control of animal pests like mollusks, cutworms, and rodents.

Balanced polymorphism: The occurrence of several different forms resulting from variation in selection pressures acting on the same species.

Ballooning: A dispersal method for early instars of lepidopterous larvae, some spiders, and mites. The organism produces a silken thread that is caught by the wind resulting in the organism being transported by air movement.

Band application: Treatment of a band of soil in row-crops, usually covering plant rows, with either sprays or granules. This is a method of applying pesticides or fertilizer to defined, limited, continuous area, such as in or along the row, rather than over an entire field.

Band: A transverse marking broader than a line.

Basad: Toward the base.

Basal anal cell: An anal cell near the wing base; a cell at the wing between 1A and 2A (Plecoptera).

Basal areole: A small cell at the base of the wing; the cell at the base of the wing between Sc and R (Lepidoptera).

Basal cell: A cell near the base of the wing, bordered at least in part by the unbranched portions of the longitudinal veins; in the Diptera, one of the two cells proximal to the anterior cross vein and the discal cell; Hymenoptera.

Basal vein: A vein in about the middle of the front, extending from the median vein to the subcostal or cubital vein (Hymenoptera).

Basal: Proximal; at or towards either the base or the main body, or closer to point of attachment (of an appendage) (Fig. 31).

Basalare (*pl.* basalaria)**:** A small sclerite, one of the epipleurites that lies anterior to the pleural wing process; an attachment for the direct flight muscles (Fig. 11); a small sclerite in the upper part of the pleuron that articulates with the axillary sclerites.

Basement membrane: The noncellular innermost layer of the insect integument (Fig. 24) or the noncellular layer on the hemocoel- side of the gut epithelium (Fig. 32); or a noncellelar membrane underlying the epidermal cells of the body wall; or a noncellular sheath separating the epidermal cells from the hemolymph or any such layer associated with an epithelium.

Basisternum: Portion of the sternum lying between the anterior presternum and posterior sternellum (Fig. 33 b and c) or that part of the sternum anterior to the sternacostal suture.

Batesian mimic: A palatable and otherwise harmless species that gains protection from predators by virtue of its close resemblance to an unpalatable or harmful species.

Batesian mimicry: A form of similarity in the coloration of two or more species, one of which is unpalatable or distasteful, the others of which are edible; a mimetic system in which a palatable species obtains protection from predation by resembling an unpalatable species; *see also* **Mullerian mimicry;** resemblance of a palatable species to one that is unpalatable or has effective defenses.

Beak: The protruding mouth-part structures of a sucking insect; proboscis (Fig. 34).

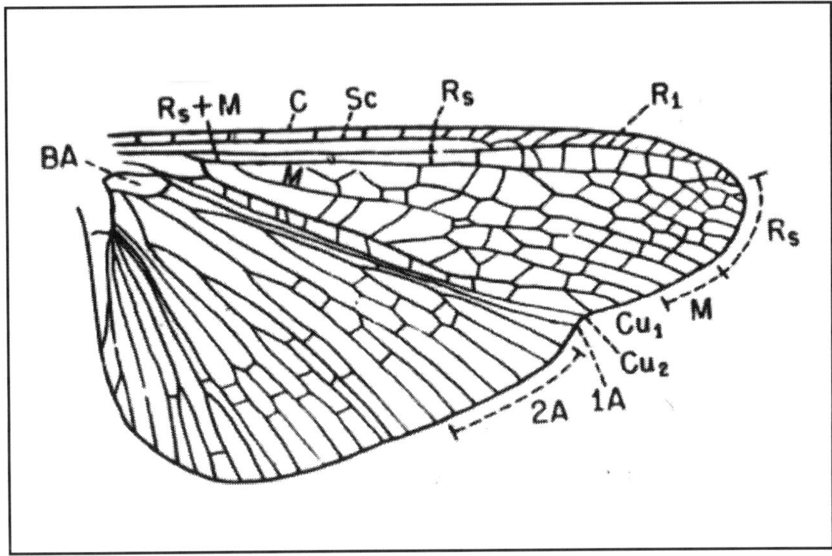

Fig. 30. Wing of Plecoptera. *Pteronarcys* (Pteronarcidae) showing basal anal cell (BA).

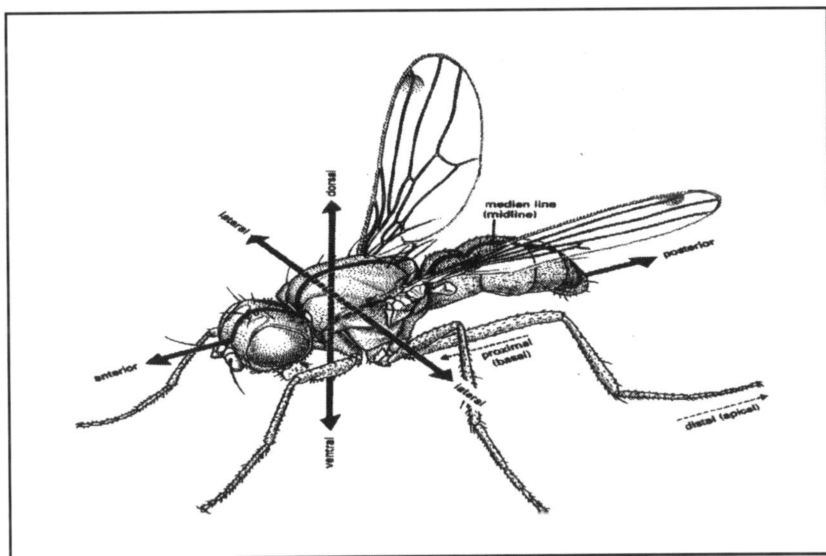

Fig. 31. The major body axes and the relationship of parts of the appendages to the body, shown for a sepsid fly.

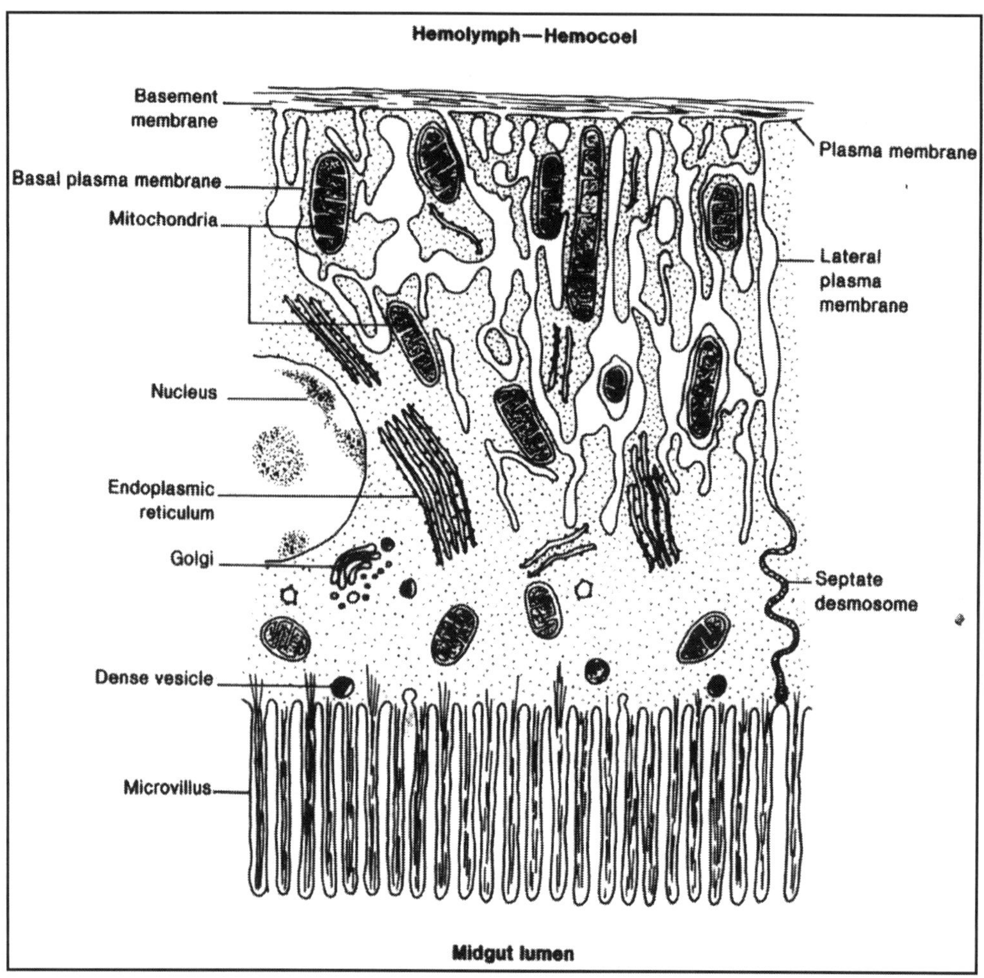

Fig. 32. Generalized midgut epithelial cell. After *Berridge,* 1970.

Bed: (i) A raised ridge of soil for planting crops above the furrows which are located on each side. (ii) An area where seedlings or transplants are grown for later transplanting into the field.

Behaviour: Any response or action of an organism.

Behavioural resistance: The ability of an insect population to change its behavior in order to avoid insecticides or other injurious factors; a mechanism in insects of resistance to insecticides which involves a behavioural change whereby contact with the insecticide is avoided.

Beneficial: Organism that provides biological control by killing, or competing with, pests or benefits mankind in way or the other such as natural enemies of insect bests, honeybees, lac insects, silkworm, pollinators etc.

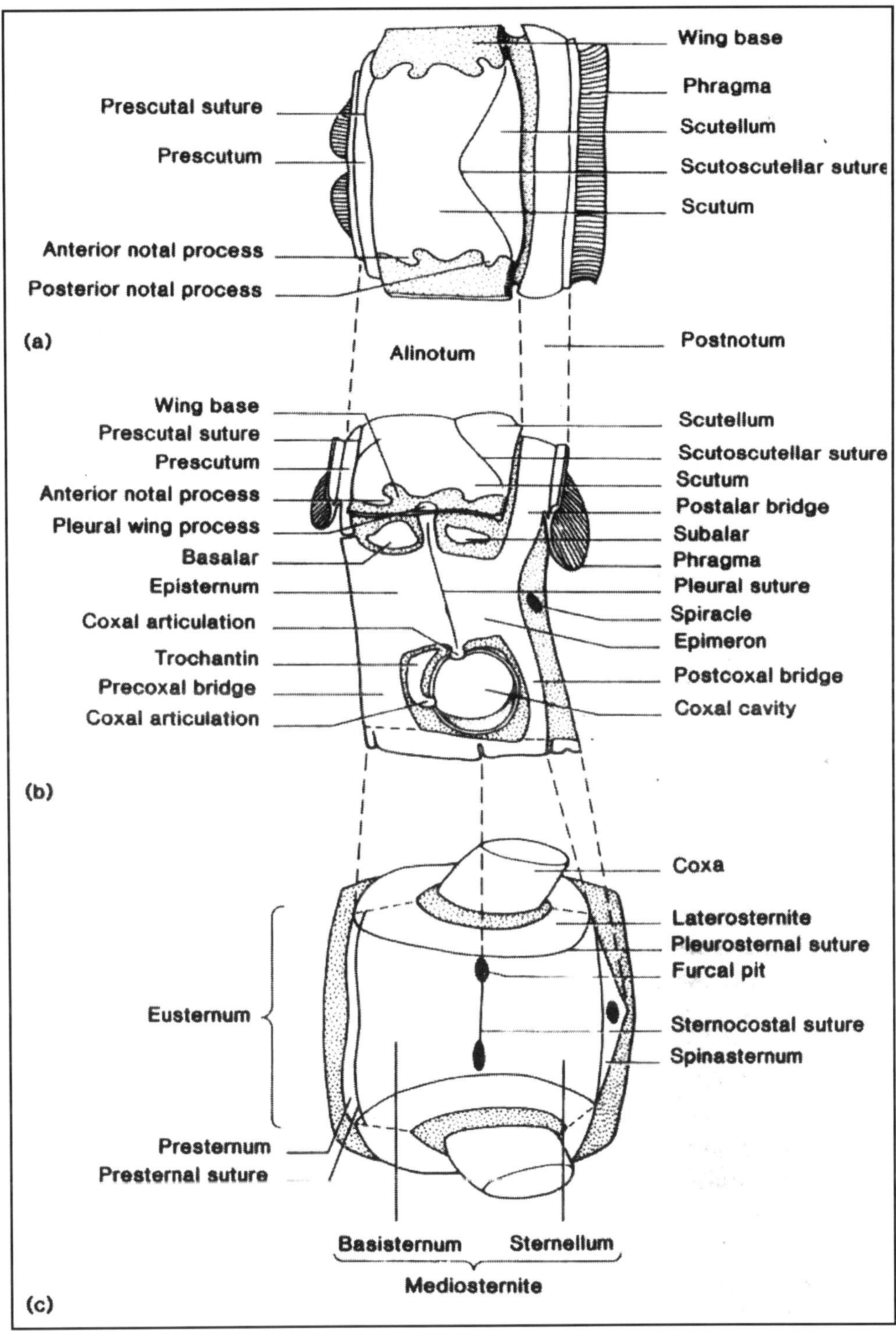

Fig. 33. Generalized wing- bearing thoracic segment. (a) Notum. (b) Pleuron. (c) Sternum.

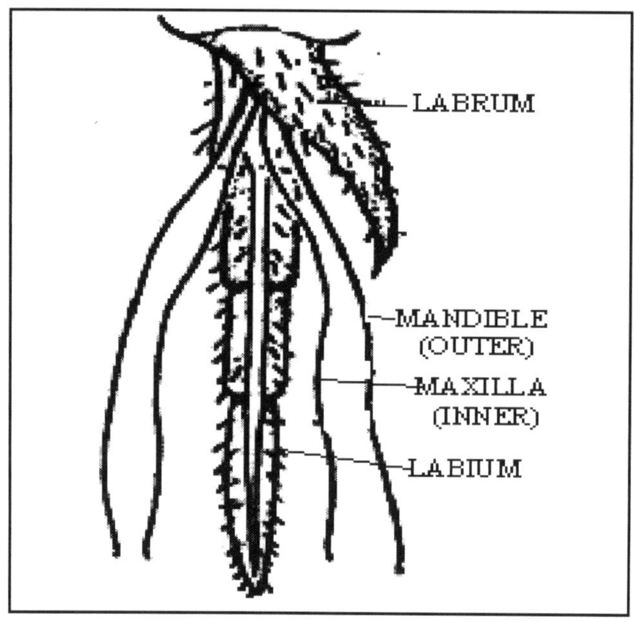

Fig. 34. Piercing and sucking type of mouthparts of a plant

Benthic. Living on the lake bottom or sea bed.

Benthos: The bottom sediments of aquatic habitats and/or the organisms that live there.

Benzyl benzoate: A repellent for use against ticks and chiggers.

Berlese funnel: A sampling device that uses heat to drive insects from a sample of soil, vegetation, or litter (Fig. 35).

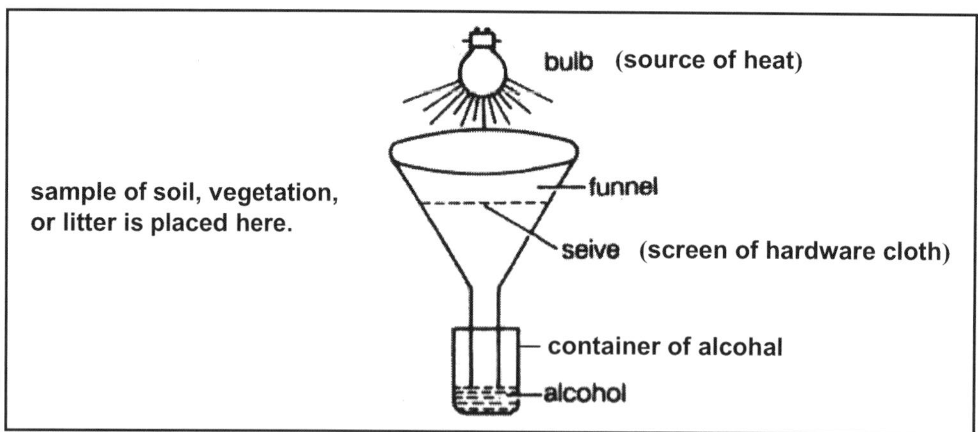

Fig. 35. Berlese

Beta diversity: The organisms that occur within a specified region, resulting from the exchange of species among habitats.

Beta taxonomy: The level of taxonomy concerned with the arranging of species into a natural system of lesser and higher categories (cf. Alpha taxonomy, Gamma taxonomy).

Bi-: Two.

Bibliographical reference: For nomenclatural purpose, the citation of the name of the author and date of publication for a scientific name; a **Full bibliographical reference** includes, in addition, the citation of the exact place of publication of a scientific name (i.e., title of book or journal, volume, page, etc.).

Biennial: A plant with a life cycle that takes two years. In the first year it grows vegetatively, in the second it flowers, produces fruits and seeds, senesces, and dies.

Bifid: Forked, or divided into two parts.

Bilateral symmetry: Symmetry in which the body or part can be divided in a median plane into identical right and left halves.

Bilobed: Divided into two lobes.

Binary nomenclature: Synonymous with, and to be replaced by, the term *binominal nomenclature,* by decision of the International Commission on Zoological Nomenclature (cf. Binominal nomenclature, Binomial nomenclature).

Binomen: The two parts, genus and epithet that make up an organism's scientific name.

Binomial: Having two names, as in a scientific name, composed of a genus name and a trivial name or specific epithet; e.g., *Helicoverpa armigera* (Chickpea pod borer).

Binominal, Latin: The two-part formal name of organisms that includes the genus and specific epithet; e.g., *Bemisia tabaci* (Tobacco while fly).

Binomial nomenclature: The system of classification of organisms, as advanced by Carolus Linnaeus.

Binomnial nomenclature: The system of nomenclature first standardized by Linnaeus and now generally referred to as binominal nomenclature (cf. Binomial nomenclature, Binary nomenclature); the system of naming organisms with two names, generic and specific; the system of nomenclature adopted by the International congress of Zoology, by which the scientific name of an animal is designated by both a generic and specific trivial name (cf. Binary nomenclature, Binomial nomenclature).

Binomial system: The system of dual Latin names for plants and animals devised by the Swedish naturalist Linnaeus and now almost universally applied to living organisms.

Bio-: Life; living.

Bioassay or biological assay: The use of living organisms in experiments to measure the effect of biotic or abiotic stresses or toxins.

Biochemical resistance: The ability of an insect population to develop enzymes that detoxify an insecticide or other injurious material before it can reach its site of action.

Biocide: A general poison or toxicant or toxin that kills all organisms under a comparable range of exposure.

Biodiversity: The variety and variability within and among living organisms and the communities in which they occur.

Bioeconomics: The relationship between pest numbers and economic losses.

Biofix: An identifiable phenological event or stage in the life cycle that represents the point for starting degree-day based models.

Biogenic salts: Salts which are vital to the life of an organism.

Biogeochemical cycle: The cyclic movement of chemical elements of the biosphere between the organism and the environment is termed biogeochemical cycle.

Biological character: A taxonomy attribute of a living organism (in contrast to the museum specimen), hence usually a character which is not strictly morphological (cf. Taxonomic character).

Biological control (also **biocontrol):** The human use of selected living organisms to control populations of pest species (plant or animal); the employment of any biotic or biological agent such as natural enemies that include herbivores, predators, parasitoids, parasites, and disease organisms, to control pest populations.

Biological monitoring: Using plants or animals to detect changes in the environment.

Biological magnification: When a small organism feeds on several microorganism containing residues of a pesticides, the pesticide may get deposited in its body. A predatory animal which feeds on several such small animals ingests a large quantity of pesticide. Thus, large predators and the human beings at the top of food chain consume quantity of the pesticide much larger than that was present in the microorganism. This increase in chemical concentration as it passes through food chain is called biological magnification.

Biological race: Noninterbreeding sympatric populations, which differ in biology but not, or scarcely, in morphology; supposedly prevented from interbreeding by preference for different food plants or other hosts (cf. Sibling species).

Biological species: A group of interbreeding organisms reproductively isolated from other closely similar groups.

Biological transfer: The movement of a disease organism from one host to another by one or more vectors in which there is a biological cycle of disease.

Bioluminescence: Light produced by living organisms; the production by an organism of cold light, commonly involving the action of an enzyme (luciferase) on a substrate (luciferin).

Biomagnification: The increase in concentration of a substance (usually a pesticide) in animal tissue as related to the animal's higher position in the food chain (Fig. 59); or the relative accumulation and concentration of a chemical in the tissues of organisms, that result in increasing tissue concentrations of the chemical in organisms that at successively higher levels in the food chain. An example is DDT.

Biomass: The total dry weight or volume of all living organisms in a given area.

Biomass pyramids: Biomass pyramids can be constructed on the same lines as pyramids of numbers to summarize the biomass structure of the communities.

Biome and ecotone: There have been several systems of classification of ecosystems, but one which has currently gained acceptance is that of biomass. Each biome is an ecosystem of a large area in which the nature of stable communities (also termed climax) are similar in appearance. However, it is often difficult to determine the boundaries of a biome and even to determine how many biomes there are. Biomes are recognized primarily by the dominant plants of climax communities but in many parts of the country, communities have been severely disturbed by man. Some communities which are considered to be transitional between two biomes are termed 'ecotone'.

Bionomics: The study of the habits, life histories and adaptations of living organisms.

Biosphere: The land, water, and atmospheric portions of our planet in which life exists.

Biota: The flora and fauna of a region (cf. Fauna, Flora).

Biotic: Living. The biotic elements of an ecosystem are the living organisms therein, including plants, animals, and microorganisms.

Biotic insecticide: Natural or introduced enemies of a pest including predators and parasites that are applied using standard pesticide application techniques.

Biotic isolation: The occurrence of organisms in isolation from others of their species. In many animals a breeding pair may isolate themselves, occupying a definite territory from which others of the species are excluded. Such biotic isolation provides each individual or pair with a living space from which it can obtain food or water or nutrients, more or less free from competition with other members of its species.

Biotic potential (alternative term, reproductive potential)**:** The maximum possible rate of increase of an organism in the absence of any limiting factors; a measure of the innate ability of a population to survive and reproduce. Or the maximum level of increase that a population is capable of under optimal conditions.

Biotic release: The sudden release of a population from its natural enemies, often resulting in a population explosion.

Biotype (race): A population of a pest species that differs from other populations of the species in its ability to attack a particular cultivar or a population of a species that differs genetically from another population with respect to host affiliation (also called "host race") it is a designation below the species level for organisms that are distinguished from other members of the same species by morphological, ecological (e.g., temperature or humidity requirements) or physiological characteristics (parasite susceptibility or host preference); a population or group of individuals composed of a single genotype (cf. Population, Genotype); a genetically distinct strain or sub - group of a species distinguished by some behavioural or physiological difference but indistinguishable morphologically. Usually applied to strains of insect species which can overcome resistant plants.

Biovoltine: Having two generations in one year; *see also* **Univoltine; Multivoltine.**

Bipectinate: Having branches on two sides like the teeth of a comb.

Bipolar neuron: Neuron with a cell body bearing an axon and a single, branched or unbranched dendrite (Fig. 36).

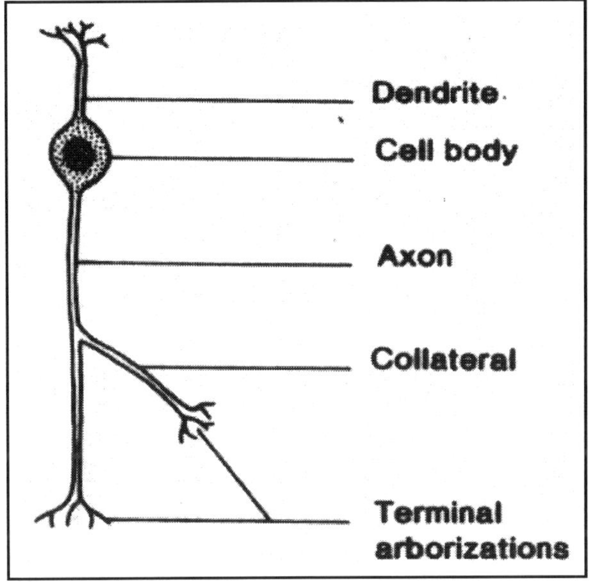

Fig. 36. Bipolar neuron.

Biramous appendages: Two-branched appendage; for example, characteristic of Crustacea.

Biramous: With two branches; consisting of an endopodite and an exopodite (Crustacea). It is an end branching or bifurcating into two *e.g.,* most of the appendages of practically all crustaceans are biramous.

Bisexual: A population composed of functional males and females; applied also to an individual possessing functional male and female reproductive organs (= hermaphrodite); or with males and females (insects).

Bituberculate: With two tubercles or swellings.

Bivalved: With two valves or parts; clamlike.

Bivoltine: Producing two generations per year.

Bivouac: An army ant camp during the mobile phase.

Blast: The common name for the sudden death of flowers, buds, or young fruit caused by disease.

Blasto: Embryo.

Blastocoel: An internal cavity surrounded by blastoderm.

Blastoderm: The first distinct layer of cells to form in morphogenesis or the one-cell thick layer of cells that surrounds the yolk of an egg early in embryological development; or the peripheral cell layer in the insect egg following cleavage; or the thin cellular layer that surrounds the yolk of an egg.

Blastokinesis: All displacements, rotations, or revolutions of the embryo in the egg.

Blastomeres: Cells composing the blastoderm.

Blastula: An early embryonic form that usually consists of a hollow sphere composed of a single layer of cells.

Blattaria, order: (blatta, cockroach) cockroaches (Fig. 37).

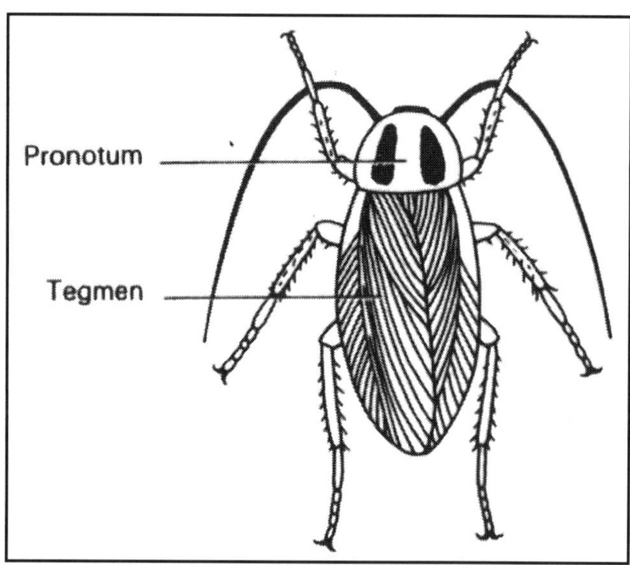

Fig. 37. Blattaria. A cockroach, *Blattella germanica* (Me.).

Blending characters: Characters which merge and do not show clear-cut Mendelian segregation (cf. Blending inheritance).

Blending inheritance: Inheritance (generally due to multiple factors) in which clear segregation is not evident in the F_2 generation (cf. Multiple factors).

Blight: Disease or injury that results in the sudden and conspicuous withering of leaves, fruit, twigs and branches.

Blister beetle: Those beetles of family Meloidae which cause blisters on human skin, by their secretions. *e.g.,Mylabris pustulata.*

Boll: The pod in which the lint and seeds of cotton develop.

Bombykol: The sex attractant pheromone of the female silkworm moth.

Book gill: The leaflike gills of a horsehoe crab (*Limulus* sp.).

Book lung: The respiratory cavity of some spiders that contains a series of leaf like folds.

Boom (spray)**:** Horizontal (or vertical) light frame carrying several spray nozzles; a horizontal support arm or pipe that allows for mounting applicator nozzles for dispersing liquid mixtures of agricultural chemicals.

Bordeaux mixture: A mixture of copper sulphate and calcium hydroxide applied for the control a range of plant pathogens.

Borer: The maker of burrows in dead or living tissue.

Borrow pit: A pit formed by an excavation, where earth has been "borrowed" for use elsewhere.

Boss: A smooth lateral prominence at the base of the chelicera (spiders).

Botanical insecticide: An insecticide produced from a plant or plant product (e.g., NSKE, tobacco decoction, pyrethrum etc). It is secondary plant substances having a toxic, repellent, deterrent, or hormonal effect on insects.

Botanicals: Pesticides derived from plants, such as pyrethrum, rotenone (derris), ryania, and nicotine.

Boundary layer: With respect to insects, the layer of air that extends from ground level upward through increasing wind speeds to a height where wind speed and flight speed are equal.

Brace vein: A slanting cross vein; in Odonata, a slanting cross vein just behind the proximal end of the stigma.

Brachypterous: Having very short wings that do not cover the abdomen; having short underdeveloped wings that do not cover the abdomen and are generally incapable of flight.

Brain hormone (AH, activation hormone or PTTH, prothoracicotropic hormone)**:** The hormone that stimulates secretory activity of the prothoracic gland; also called *prothoracic hormone* or *ecdysiotropin.* This is a hormone which is secreted by neurosecretory cells of the insect brain which activates prothoracic glands.

Brain: In insects, the supraoesphageal ganglion of the nervous system, comprising protocerbrum, deutrocerebeum and tritocerebrum (Fig. 38);

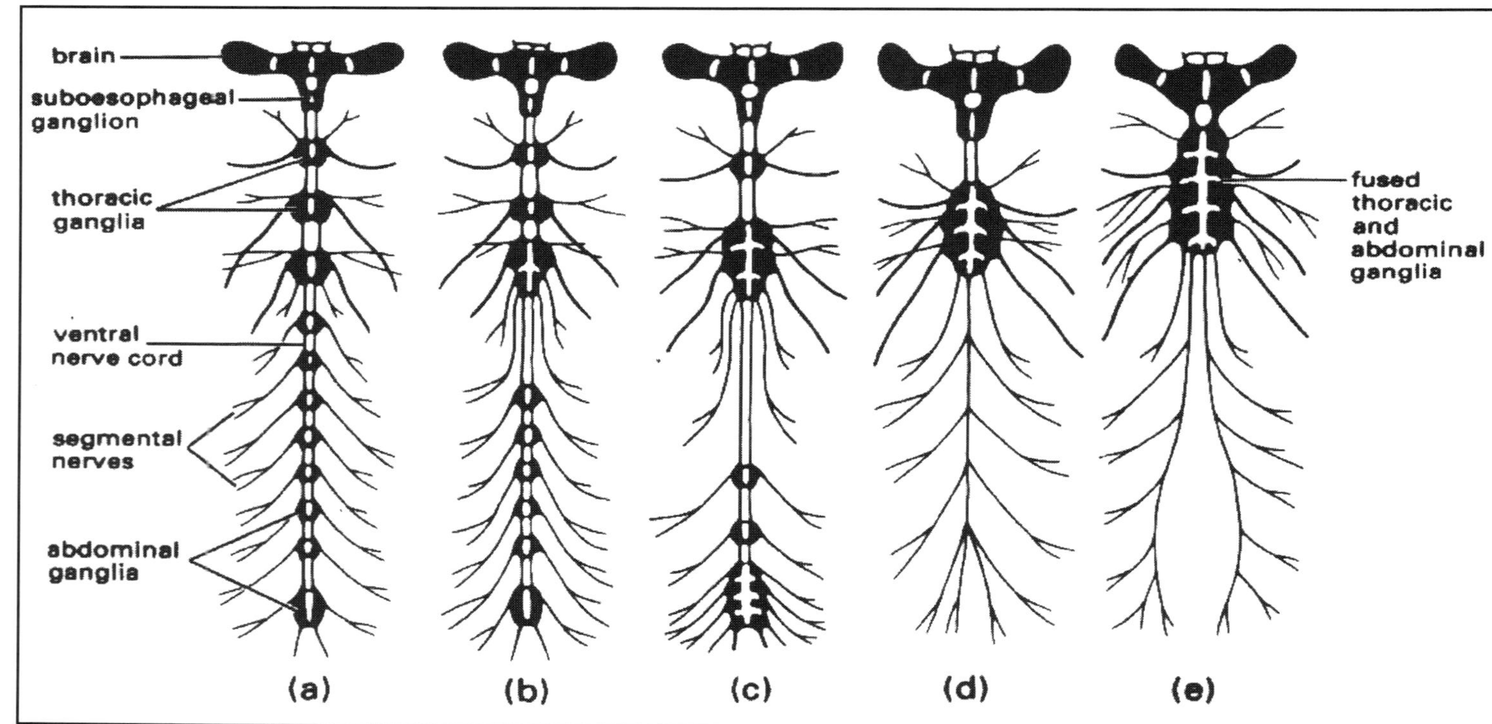

Fig. 38. The central nervous system of various insects showing the diversity of arrangement of ganglia in the ventral nerve cord. Varying degrees of fusion of ganglia occur from the least in the most specialized: (a) three separate thoracic and eight abdominal ganglia, as in *Dictyopterus* (Coleoptera: Lycidae) and *Pulex* (Siphonatera: Pulicidae); (b) three thoracic and six abdominal ganglia, as in *Blatta* (Blattodea: Blattidae) and *Chironomous* (Diptera: Chironomidae); (c) two thoracic and considerable abdominal fusion of ganglia, as in *Crabro* and *Eucera* (Hymenoptera: Sphecidae and Anthophorinae); (d) highly fused with one thoracic and no abdominal ganglia, as in *Musca, Calliphora* and *Lucilia* (Diptera: Muscidae, Calliphoridae); (e) extreme fusion with no separate suboesophageal ganglion, as in *Hydrometra* (Hemiptera: Hydrometridae) and *Rhizotrogus* (Scarabaeidae).

the anterior composite ganglion located dorsal to the foregut in the insect head.

Branchial: Pertaining to the mouth or cheek. (L. *bucca* = cheek).

Brand name: The name that manufacturers use for commercial purposes.

Breaking: The separation of the phases from emulsion.

Brevicomin: An aggregation pheromone of the bark beetle *Dendroctonus brevicomis.*

Broad spectrum (as applied to insecticides): Having an effect on a wide range of insects; a pesticide that is effective against many pest species, as opposed to a narrow spectrum (see selective) pesticide that is toxic to a limited number of pest species.

Broad spectrum insecticide: An insecticide with a broad range of targets, usually acting generally on the insect nervous system.

Broad spectrum: Broadcast: Application of fertilizer, pesticide, or seed over the entire soil surface, rather than in rows or bands.

Brood care: Parental care of offspring.

Brood: The individuals that hatch from the eggs laid by one mother, individuals that hatch at about the same time and normally mature at about the same time or the progeny of any animal. Cohorts of progeny hatched or born at the same time and maturing at about the same rate and live together in a defined and limited area.

Bubonic plague: A bacterial disease of rodents and man caused by *Pasteurella pestis* and transmitted chiefly by the oriental rat flea.

Bucca (pl., bucculae): One of two ridges on the underside of the head, on each side of the beak (Hemiptera).

Buccal cavity: The opening enclosed by the mouthparts, leading to the true mouth and the pharynx.

Bud: see **Imaginal disc.**

Budworm: Common name in the USA for various tortricid larvae e.g. *Cydia pulverula.*

Buffer: (i) Any chemical that prevents sudden changes in pH. (ii) Area of field where no pesticide is applied to prevent worker exposure or contamination of adjacent areas.

Bursa copulatrix: A saclike vagina. It is a pouch on the median oviduct of the female that receives the aedeagus of the male; the female **genital chamber**

if functioning as a copulatory pouch in Lepidoptera, the primary receptacle for sperm (Fig. 39 a and b); a specialized part of the female insect reproductive system into which sperm is received from the male during copulation.

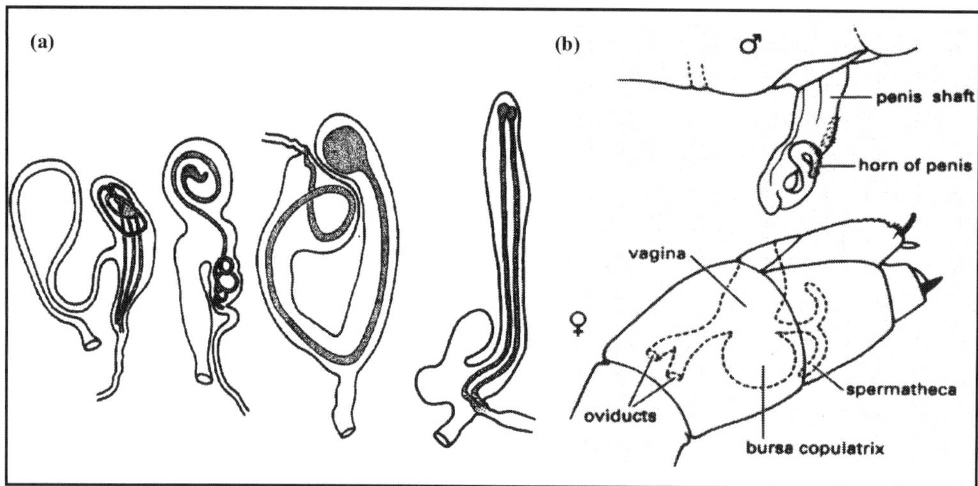

Fig. 39. (a) Spermatophore lying within the bursae of the female reproductive tracts of moth species from different genera (Lepidoptera: Noctuidae). The sperm leave via the narrow end of each spermatophore, which has been deposited so that its opening lies opposite the 'seminal duct' leading to the spermatheca (not drawn). The bursa on the far right contains two spermatophores, indicating that the female has remated; (b) Illustration of bursa copulatrix and related organs of female and male damesflies (*Calopteryx* sp.)

Bursicon: A hormone secreted by certain cells of the nervous system that plays a role in the control of cuticular hardening and darkening. It is neurosecretory hormone that stimulates sclerotization of the cuticle following ecdysis; also called *tanning hormone* or it is the hormone controlling tanning and expansion, produced by neurosecretory cells of the brain; A neuropeptide hormone that controls hardening and darkening of the cuticle after **ecdysis.**

C

Caecae: Saclike or tubular structures open at only one end (blind), often opening from the mid-intestine (Fig. 41).

Caecum (pl. caeca)**:** A blind-ending tube or sac or pouch sac of the alimentary canal (L. *caecus*=blind) *e.g.*, haepatic caecae in the alimentary canal of insects (Fig. 41).

Calamistrum: One or two rows of curved spines on the metatarsus of the hind legs (spiders; fig. 608, clm).

Calibration: The adujustment of application equipment so that a defined quantity of chemical or other material is applied per unit time or area.

Calling: A vergin female moth releasing sex pheromones to attract males for the purpose of mating.

Calliphorin: A protein produced in the fat body and stored in the haemolymph of larval Calliphoridae (Diptera).

Callus (pl. calli): (i) A rounded swelling. (ii) Plant tissue that grows over an injury site, such as wound or graft, as a protective response. (iii) Plant tissue grown in tissue culture that remains undifferentiated but retains the potential to differentiate into specific tissues.

Calypter (pl. calypteres): One of two small lobes at the base of the wing, located just above the halter (Diptera; fig. 408, cal). (Also called *squama*).

Calyx (pl. calyces)**:** A cup-like expansion, especially of the oviduct into which the ovaries open. Cavity joined with the lateral oviduct and into which the pedicels of the ovarioles open or the bulbous junction of the pedicels of the ovarioles (Fig. 5).

Camera lucida: A device enabling one to make accurate drawings of objects seen through a microscope; when it is attached to the eyepiece of the microscope, the observer can see the object under the microscope and his drawing paper at the same time.

Camouflage: Crypsis, in which an organism is indistinguishable from its background.

Campaniform sensilla (sensilla campaniformia): Type of mechanoreceptor consisting of a pit covered with a cuticular dome; a mechanoreceptor that detects stress on the cuticle, comprising a dome of thin cuticle overlying one neurone per sensillum, located especially on joints; a sense organ consisting of a dome-shaped cuticular area into which the sensory cell process is inserted like the clapper of a bell or a sense organ consisting of a dome-shaped portion of the cuticle with associated sensory neuron; perceives stresses in the cuticle.

Campodeiform larva: A larva shaped like the thysanuran *Campodea*, that is, elongate, flattened, with well-developed legs and antennae, and usually active.

Canker: A localized dry or dead area on a woody part of a plant such as the stem, branch, or trunk, often resulting from pathogen infection.

Cannibalism: The phenomenon of feeding on individuals of the same species, e.g., praying mantids, *Helicoverpa* spp., earwigs, etc.

Cannibalistic: Feeding on other individuals of the same species.

Canopy: The foliage of plants.

Cantharidin: Chemical found in the body of adult Meloidae and Oedemeridae; mostly found in the elytra, with urticating properties and a reputation as an aphrodisiac; an irritating substance produced by blister beetles; used medicinally in keratolytic preparations or a defence allomone of blister beetles (also known as "Spanish fly") or secretion of the accessory glands of the genitical tract of blister beetles (Meloidae); has medicinal and commercial value (L. *Cantharis*=beetle).

Cantharophily: Plant pollination by beetles.

Cap cell: The outermost cell of a sense organ such as a chordotonal organ.

Capillary: A minute tubular vessel with walls composed of a single layer of thin cells, through which diffusion may occur; commonly in a connecting network between arteries and veins. (L. *capillus*=hair).

Capitate: Head-shaped or with an apical knoblike enlargement; capitate antennae.

Capitulum: A small head; in particular the false head of certain mites and ticks on which the mouthparts are borne.

Carapace: A hard dorsal covering consisting of fused dorsal sclerites (Crustacea) or dorsal plate that covers the cephalothorax in many crustaceans (Fig. 40).

Carbamate (insecticide)**:** A group name applied to insecticides which are based on the carbamic acid nucleus; an organic compound derived from carbamic acid, NH_2COOH; includes many important synthetic insecticides (e.g., carbaryl).

Carbohydrate: A compound of carbon, hydrogen, and oxygen that follows the general formula $(CH_2O)_n$.

Carbon-based defenses: Allelochemicals that make food plants less nutritious.

Carcinogen: A chemical or biological agent or material which produces, accelerates or increases frequency of cancers.

Carcinogenic: Cancer inducing.

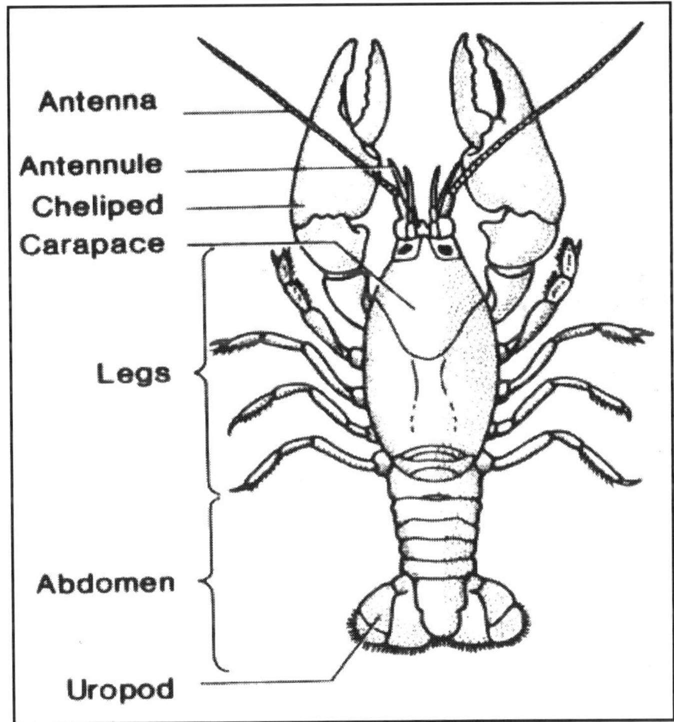

Fig. 40. Crayfish (Crustacea) showing carapace.

Cardenolides: Secondary plant substances in milkweed plants that may render milkweed-feeding insects unpalatable to vertebrate predators.

Cardia: A structure composed of both foregut and midgut tissue and in which the foregut-midgut junction is located; see also proventriculus and stomodael valve (Fig. 41).

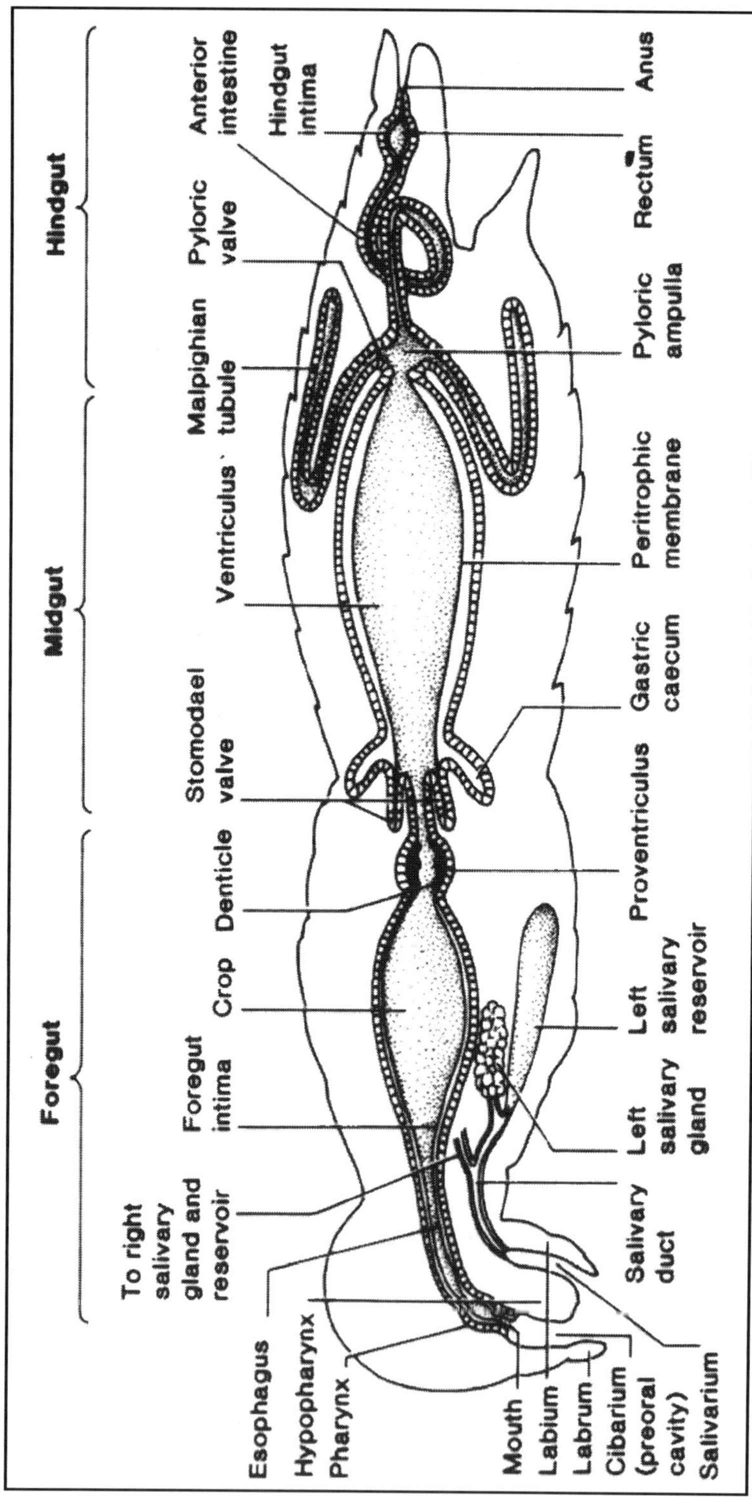

Fig. 41. Generalized insect alimentary system.

Cardiac glycosides: Chemicals from milkweed plants that render insects that feed on these plants unpalatable to predators.

Cardiac valve: The valve at the junction of the foregut and the midgut (Fig. 41).

Cardial epithelium: The most anterior tissue of the midgut.

Cardiopeptide: A neuropeptide hormone that stimulates the dorsal vessel ('heart') causing haemolymph movement.

Cardo (pl., cardines): The basal segment or division of a maxilla; one of two small laterobasal sclerites in the millipede gnathochilarium; proximal segment of the maxilla a basal sclerite forming part of the maxilla in the mouthparts of insects (Fig. 54).

Carina (pl., carinae): A ridge or keel.

Carinate: Ridged or keeled.

Carnivore: An organism that eats living animals. It is zoophagous or animal-eating animal.

Carnivorous: Eating or living on flesh of other animals. (L. *flesh+voro*=eat) *e.g.,* lion.

Carotenoids: Insect pigments of red and yellow, derived from plant tissues.

Carrier: (i) A substance added to a chemical compound to improve the uniformity of application. Carriers can be organic (e.g., walnut shells, bark) or mineral (e.g., sulfur, lime, clay). (ii) The water or air used to transport a pesticide from the application machine to the target. (iii) An asymptomatic organism that is harboring an infectious disease agent.

Carrier: Material serving as diluent and vehicle for the active ingredients; usually in dusts.

Carrying capacity: The maximum density of a population of a given species that an ecosystem can sustain at a specific time (Fig. 42).

Caryo: Nucleus.

Caste (s): A form or type of adult in a social insect (termites); morphologically and functionally different types within a colony of social insects as in Isoptera (Fig. 43) and some Hymenoptera; morphologically distinctive groups of individuals within a single species of social insect, usually differing in behavior.

Catalogue: An index to taxonomic literature arranged by taxonomic categories so as to provide ready reference to at least the most important taxonomic and nomenclatural references to the category involved (cf. Check list).

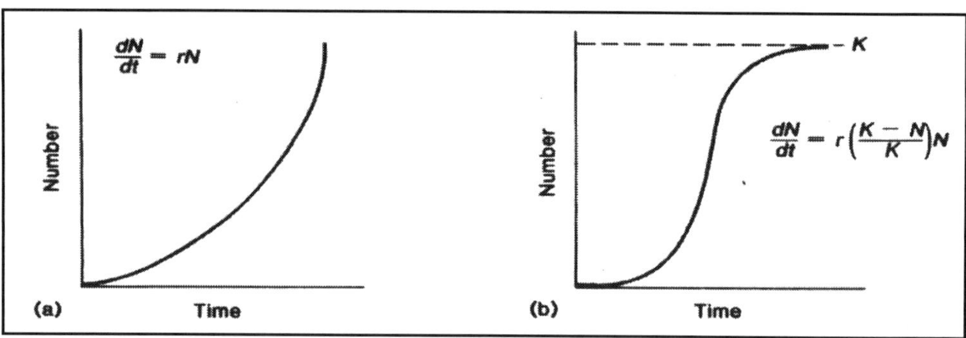

Fig. 42. Population growth curves. (a) Exponential population growth. (b) Logistic population growth.

Catastrophic and facultative factors: Howard and Fiske (1911) coined these terms. When the factors act in such a way that a constant proportion of population is destroyed irrespective of the number of individuals it contains, they are designated as catastrophic. On the other hand, facultative factors are those which involve the destruction of a percentage population which increases as the number of individuals per unit area increases. Smith (1935), however, named these factors as density-independent and density-dependent factors, respectively.

Category I pesticide: Pesticides that are highly toxic, with LD_{50} between 0 and 50 mg/kg body weight, and are labeled with the signal words DANGER-POISON in large, bold letters with a skull and crossbones logo.

Category II pesticide: Moderately toxic pesticides with LD_{50} between 50 and 500 mg/kg body weight, that are labeled with the signal word WARNING in large, bold letters.

Category III pesticide: Pesticides with low order acute oral toxicity and **LD50** between 500 and 5000 mg/kg body weight, that are labeled with the signal word CAUTION in large, bold letters.

Category IV pesticide: Pesticides with very low acute oral toxicity and LD_{50} greater than 5000 mg/kg body weight. These present minimal hazards to humans when handled properly, although some are flammable and so must be stored as such.

Caterpillar: Worms that are the immature stage of Lepidoptera; a polypod or an eruciform larva; the larva of a butterfly, moth, skipper, sawfly, or scorpionfly (Fig. 44).

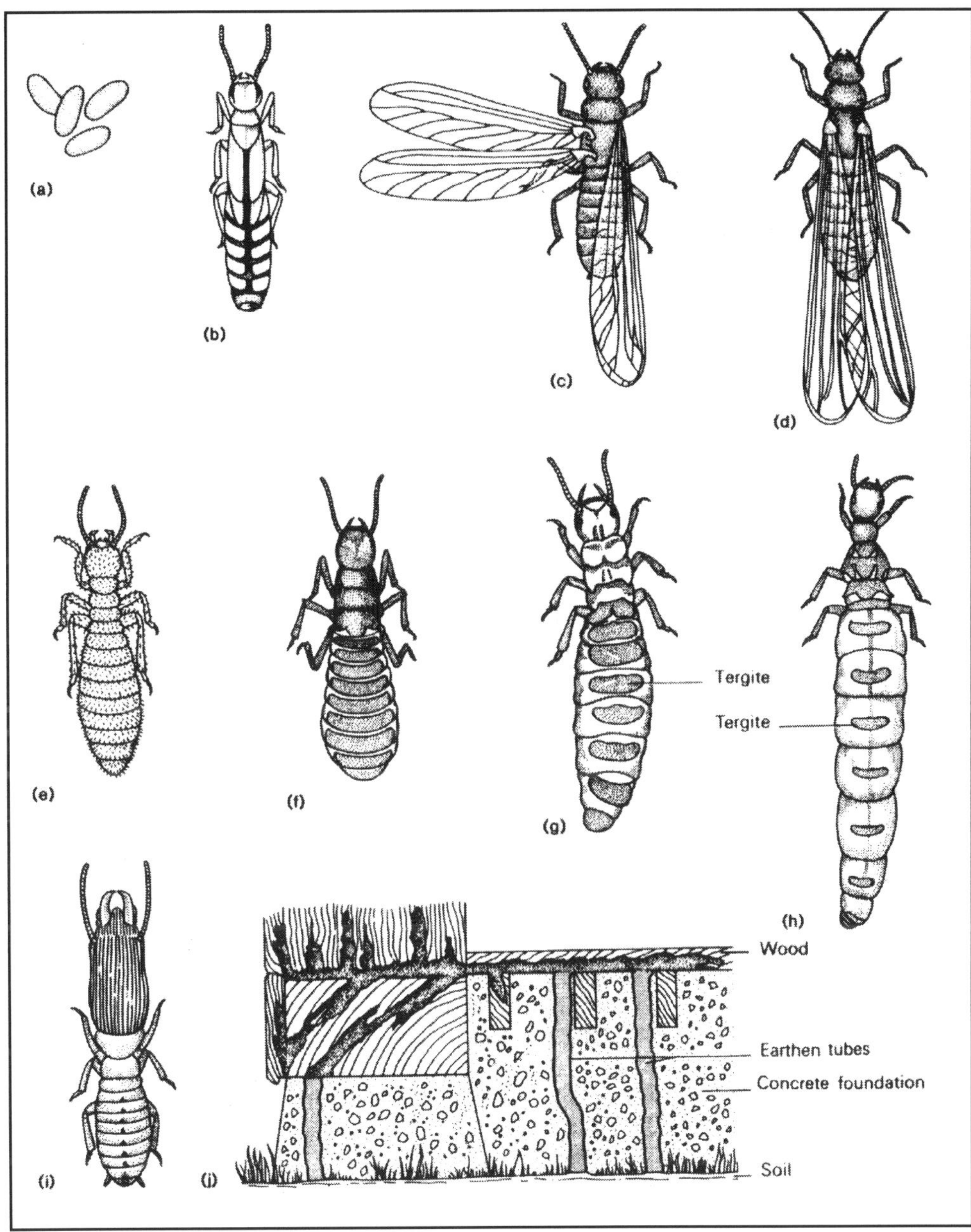

Fig. 43. Subterranean termite, *Reticulitermes.* (a) Eggs. (b) Nymph. (c) Female winged form. (d) Male winged form. (e) Worker. (f) King. (g) Secondary reproductive (abdomen distended with enlarged ovaries). (h) Queen (abdomen distended with enlarged ovaries). (i) Soldier. (j) Earthen tubes from soil across surface of concrete foundation to wooden structures. After *Boreal Labs Key Card.*

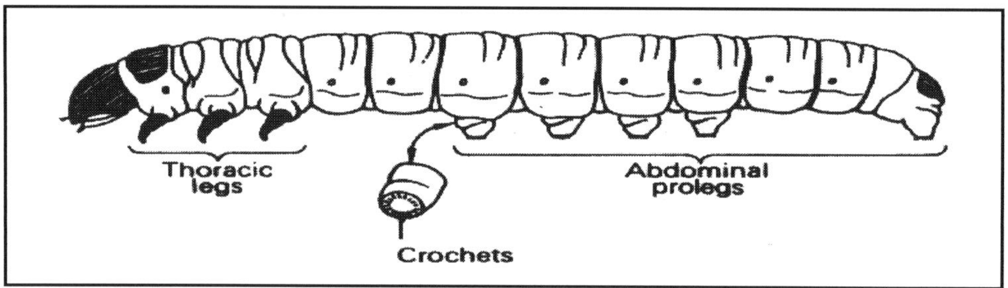

Fig.44. Caterpillar.

Catfacing: Fruit deformation caused by insect feeding during fruit development; injury resulting from the feeding of piercing-sucking insects on developing fruit; a result of uneven growth and deformation.

Cationic surfactant: Material in which surface activity is determined by the basic part of a compound.

Caudal: Toward the tail, or toward the posterior part of the body or refers to a tail or tail- like structure at the end of the abdomen.

Caudal filament: A threadlike process at the posterior end of the abdomen; One of two or three terminal filaments.

Caudal lamella: One of two or three terminal gills.

Caudal visceral system: The part of the visceral nervous system associated with the posterior segments of the abdomen.

Causal agent: The biotic or abiotic factor(s) responsible for inducing disease.

Cauterization: Destruction of tissue by burning, usually with an electrical current.

Caution: A signal word that appears on pesticides labels to inform users that the pesticide has a low toxicity as defined by FIFRA. See *Category III pesticide.*

Cavernicolous (*also* **troglodytic; troglobiont**): Living in caves.

Cecidology: The study of plant galls.

Cecidozoa: Gall-including animals.

Cell (of the wing)**:** A thin, membranous area surrounded by veins; a space on an insect wing delineated by various combinations of longitudinal and cross veins and the wing margins; a space in the wing membrane partly (an open cell) or completely (a closed cell) surrounded by veins.

Cell: A functional unit of living matter bounded by a membrane and containing one or more nuclei or nuclear materials; a unit mass of protoplasm, surrounded by a cell membrane and containing one or more nuclei or nuclear material; a small mass of living matter usually containing a nucleus and cytoplasm; the fundamental unit of structure and function in plants and animals. (L. *Cella*=small room).

Cell body: The swollen portion of a neuron that contains the nucleus and is responsible for metabolic and maintenance functions of the neuron.

Cellular rearing: Rearing of individual eggs.

Cellulase. Enzyme capable of breaking down or digesting plant cellulose.

Cellulose: An enzyme that catalyzes the breakdown of cellulose.

Cement layer: The outermost layer of the epicuticle (Fig. 84), often absent.

Cenchri. Special lobes on the metanotum of adult sawflies (Hymenopetera, Symphyta) which engage on rough areas of the undersides of the forewings to hold them in place.

Centipede: Arthropods of the class Chilopoda have one pair of legs per body segment and a pair of venomous claws on the first segment. Centipedes live in detritus and the soil and are not usually considered pests.

Central body: Neuropilar mass connecting the two lobes of the protocerebrum (Fig. 54).

Central nervous system (CNS): In insects, the central series of ganglia extending for the length of the body (Fig. 38); *see also* **Brain;** the chain of paired segmental ganglia connected by paired cords; the location of the integrative activities of the insect nervous system (Fig. 38).

Central place foragers: Insects like ants and bees that return to their nests or hives after foraging.

Cephalad: Toward the head or anterior side.

Cephalic: On or attached to the head; anterior.

Cephalopharangeal. Refers to a piece of skeleton in some larval Diptera; held withdrawn into the thoracic segments but the mouth-hooks protrude during feeding.

Cephalopharyngeal skeleton: Reduced, invaginated mouthparts that provide articular points for the mouthparts of higher dipteran larvae (Fig. 55).

Cephalothorax: A body region consisting of head and thoracic segments; the fused head and thorax of arachnids and crustaceans; a body region consisting of head and thoracic segments (Crustacea and Arachinda) (Fig. 56).

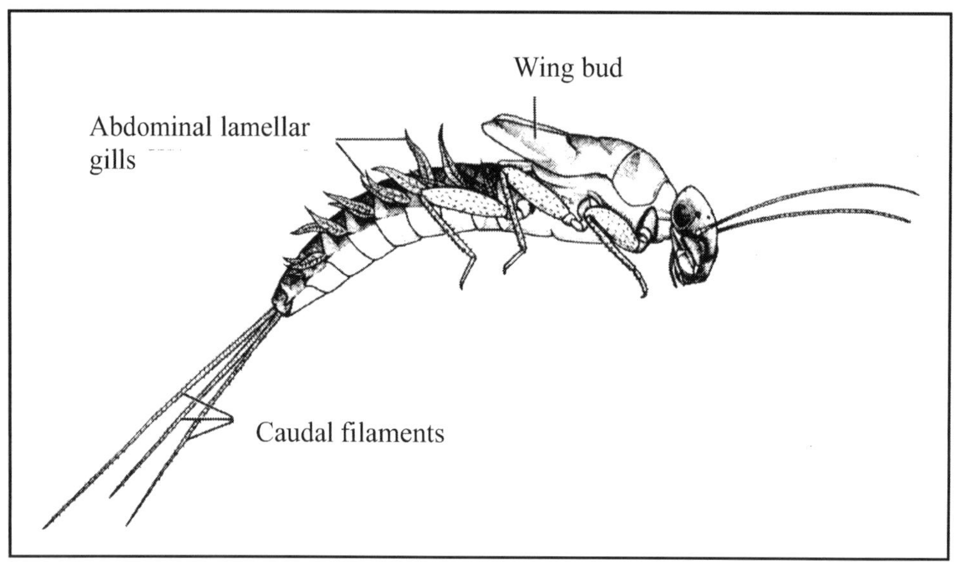

Fig. 45. Leptophlebiid mayfly nymph showing caudal filaments.

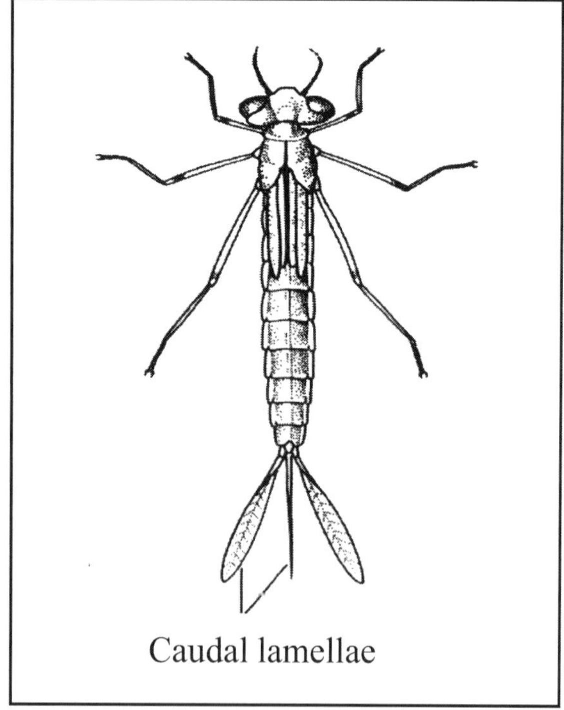

Fig. 46. Lestid damselfly larva showing caudal lamellae.

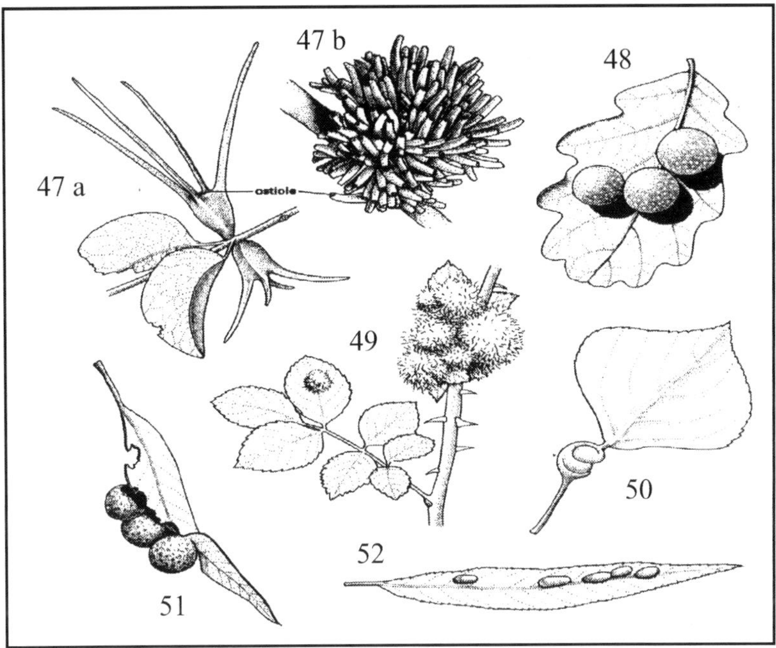

Fig. 47 - 52. A variety of insect-induced galls: (47a) two coccoid galls, each formed by a female of *Apiomorpha munita* (Hemiptera: Eriococcidae) on the stem *Eucalyptus melliodora;* (47b) a cluster of galls, each containing a male of *A. munita* on *E. melliodora;* (48) three oak cynipid galls formed by *Cynips quercusfolii* (Hymenoptera: Cynipidae) on a leaf of *Quercus* sp.; (49) rose bedeguar galls formed by *Diplolepis rosae* (Hymenoptera: Cynipidae) on *Rosa* sp.; (50) a leaf petiole of Lombardy poplar, *Populus nigra,* galled by the aphid *Pemphigus spirothecae* (Hemiptera: Aphididae); (51) three psyllid galls, each formed by a nymph of *Glycaspis* sp. (Hemiptera: Psyllidae) on a eucalypt leaf; (52) willow bean galls of the sawfly *Pontania proxima* (Hymenoptera: Tenthredinidae) on a leaf of *Salix* sp.

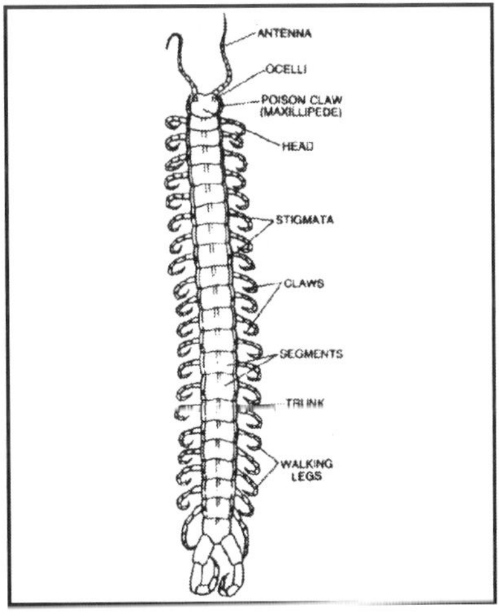

Fig. 53. A Centipede (dorsal view).

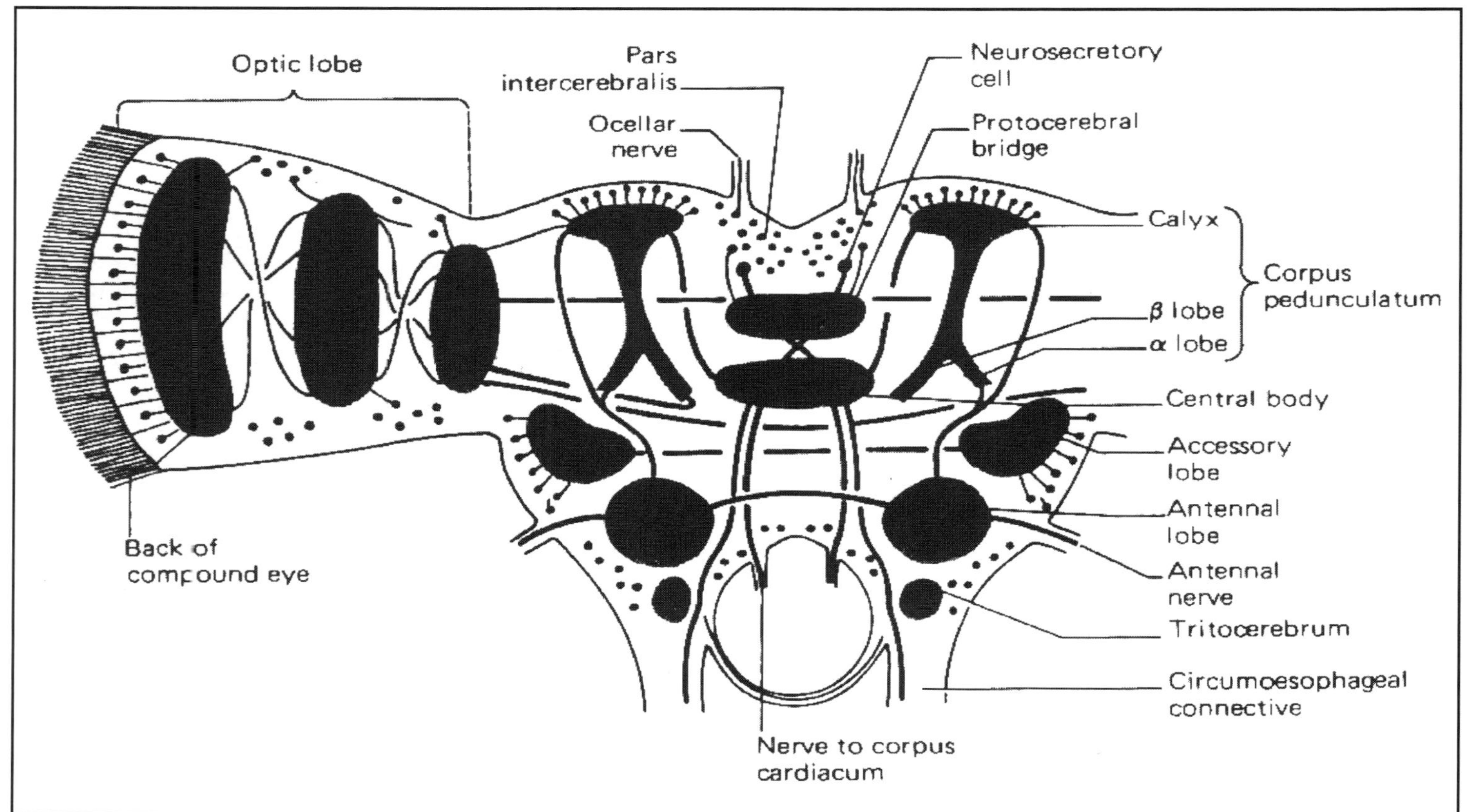

Fig. 54. Diagram showing major neuropilar regions (shaded) of the brain and some connections between these regions. After *Chapman,* 1971.

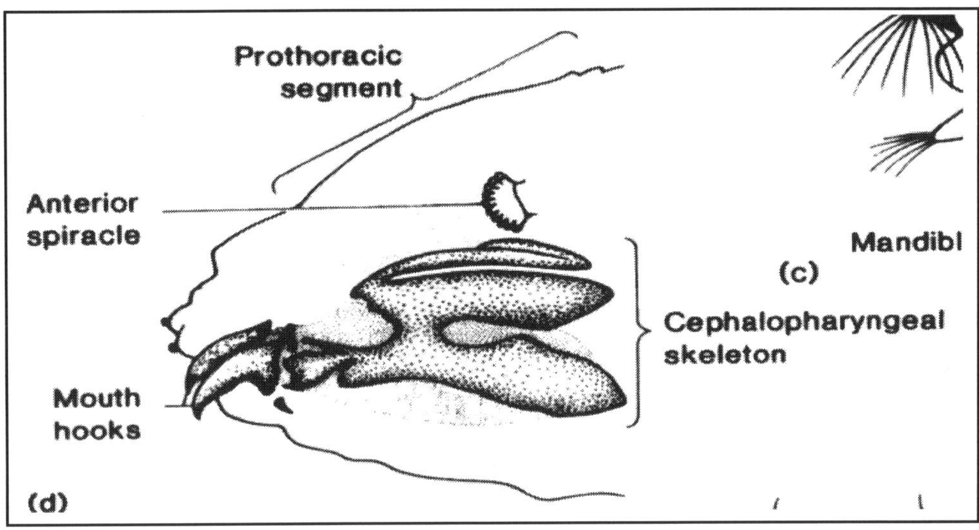

Fig. 55. Example of mouthparts from immature form larva of a true fly (maggot): Cephalopharyngeal skeleton.

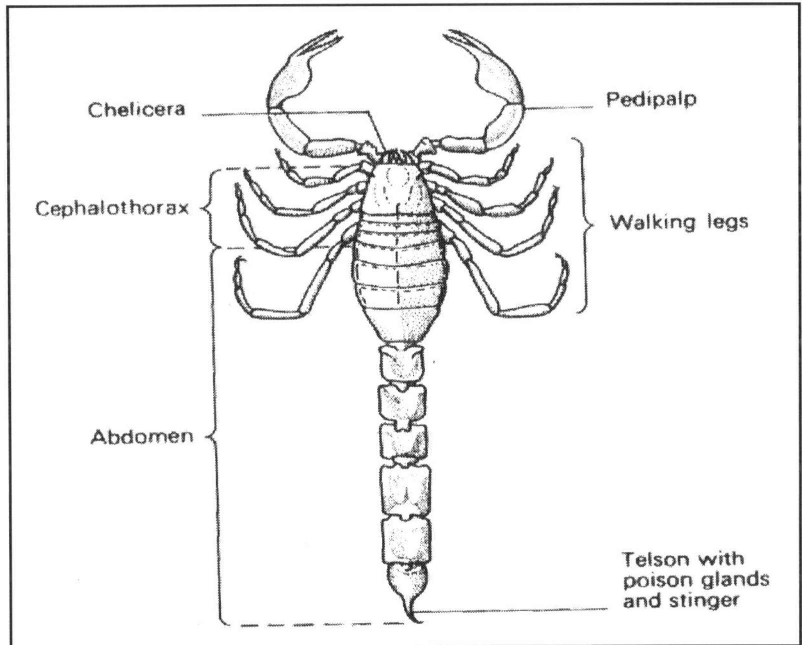

Fig. 56. Scorpion, illustrating the major chelicerate features.

Cercaria: A larval form of flukes (Trematode) produced asexually from redia (another larval form), parasitic on snails which act as intermediate host to transfer the causative organism to the vertebrate host *e.g. Cercaria of liver fluke of sheep.*

Cerci: Paired appendages at the distal end of the abdomen, usually many segmented; most obvious in more primitive insects (Fig. 1).

Cercus (pl. cerci)**:** One of a pair of appendages originating from abdominal segment 11 but usually visible as if on segment 10; an antennalike sensory appendage arising from the posterior end of the abdomen (Fig. 1).

Certified applicator: Person licensed by the EPA and or state regulators to apply restricted-use pesticides.

Certified seed: Seed that has its varietal status and freedom from weeds and other pests documented by a government agency.

Certified stock: Plant propagation material that has been documented by State Departments of Agriculture or other government agencies as free from certain diseases or other pests.

Cerumen. A mixture of wax and plant resin used by some bees for nest construction.

Cervical: Pertaining to the neck or cervix.

Cervical sclerite: A sclerite located in the lateral part of the cervix, between the head and the prothorax; lateral plates found on the membranous region connecting the head with the thorax (Fig. 1).

Cervix: The membranous neck region of an insect connecting the head with the thorax (Fig. 1).

Chaetotaxy: The arrangement and nomenculture of the bristles on the exoskeleton, both adults and larvae; the arrangement of setae in constant patterns, useful in systematics.

Character: A trait; See Taxonomic character.

Character gradient: See Cline.

Check list: usually a skeleton classification of a group listed by taxonomic categories for quick reference and as an aid in the arrangement of collections (cf. Catalogue).

Check: (i) An earthen barrier that keeps water in a field or orchard. (ii) Units in an experiment that receive either a standard treatment that will result in a predictable response or no treatment.

Cheek: The lateral part of the head between the compound eye and the mouth (see gena).

Cheironym: A manuscript name (q.v.).

Chelate: Pincerlike, having two opposable claws.

Chelicera (plural, **chelicerae**): One of the major elements in the mouthparts of spiders and related arthropods; not jawlike, but in the form of fangs, pincers, or piercing organs; one of the most anterior pair of preoral appendages of arachnids such as spiders, scorpions, mites and ticks. (Gr. *Chele* = claw + *keras* = horn); the pincer-like first pair of head appendages found in Arachnida and Merostomata; the inner portions of the mouthparts of mites; modified to piercing stylets in plant feeding mites.

Cheliped: A leg terminating in an enlarged pincerlike structure (Crustacea).

Chemical control: The reduction of pest populations by chemical pesticides.

Chemical control: The use of chemicals to kill, deter or in any way suppress pest populations.

Chemical ecology: The study of adaptation of animals or preference of particular organism to a particular chemical substance.

Chemical name: The name that specifies the chemical structure of a compound such as the active ingredients of a pesticide; e.g., malathion is 0, 0-dimethyl-(1, 2-dicarb-ethoxyethyl) phosphorodithioate; naming follows rules of the International Union of Pure and Applied Chemistry.

Chemigation: The delivery of pesticides in irrigation water; use of irrigation systems for dispensing pesticides.

Chemoreception: The perception of atoms and molecules as stimuli; sensory perception of chemical substances; includes both taste and smell.

Chemoreceptor: A sense organ involved in the perception of tastes and /or odors; a sense organ modified for the reception of chemical stimuli.

Chemosterilant: A chemical which has the property of sterilizing an insect but which does not kill it; a chemical that induces reproductive sterility or a chemical that effectively sterilizes organisms and so prevents reproduction.

Chemotaxis: Orientation with respect to a chemical gradient.

Chilopoda: The arthropod class that includes the centipedes; characterized by mandibulate mouthparts and one pair of appendages per body segment, including a pair of venomous claws on the first segment.

Chisel: A plough with penetrating points used to turn up moist subsoil while leaving dry soil on top, allows deep cultivation, breaks hardpans.

Chitin inhibitor: An insecticide that disrupts chitin production.

Chitin synthesis inhibitor: An insecticide that prevents chitin formation.

Chitin: A complex polysaccharide that is a primary structural component of insect cuticle and nematode egg shells, providing mechanical strength and protection; a high-molecular-weight polymer of N-acetyl-D-glucosamine that resembles cellulose and constitutes one-fourth to one-half of the dry weight of the exo-and endocuticle; a nitrogenous polysaccharide with the formula $(C_{32}H_{54}N_4O_{21})_r$, occurring in the cuticle of arthopods (Fig. 57). A polymer of the nitrogenous polysaccharide n-acetyl-D-glucosamine that occurs in the exoskeleton of arthropods; or the tough, insoluble polysaccharide making up a major part of the insect procuticle.

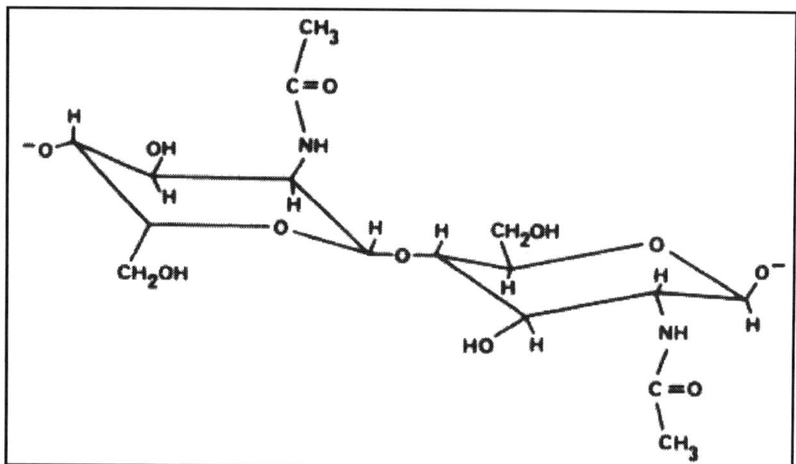

Fig. 57. Structure of part of a chitin chain, showing two linked units of *N*-acetyl-D-glucosamine.

Chlorinated hydrocarbon: A category of organic compound containing chlorine, hydrogen, carbon and occasionally oxygen and sulfur; they have long residual life and are active against a wide range of pests; includes the first widely used synthetic organic insecticides i.e. include endrin, DDT, and toxaphene.

Chlorosis: A symptom of disease, insect injury, herbicide injury, or physiological disorder in plants, wherein tissues that are normally dark green turn yellow or light green; abnormal condition characterized by the absence of green pigment in plant due to magnesium / ion deficiency, can also be due to lack of light or genetic factor inhibiting chlorophyll synthesis.

Chlorotic: Affected by chlorosis.

Cholesterol: A fatlike steroid alcohol, $C_{27}H_{45}OH$, which is an important precursor in the formation of other steroids like ecdysone.

Cholinesterase (acetylcholinesterase): An enzyme that breaks down acetylcholine as part of normal signal transmission across a nerve synapse; The latter is formed at nerve connections (synapses) each time a nerve impulse is transmitted some insecticides and nematicides deactivate cholinesterase; shortened term for acetylcholinesterase.

Chordotonal organ: A sense organ, the cellular elements of which form an elongate structure attached at both ends to the body wall; or an elongate sense organ attached to the inner surface of the body wall and sensitive to stretching and to vibrations; sense organs (mechanoreceptors) that perceive vibrations, comprising one to several elongate cells called **scolopidia**; examples include the **tympanal organ, subgenual organ** and **Johnston's organ.**

Chordotonal sensillum: A mechanoreceptor that consists of a bundle of bipolar nerve cells stretched between two surfaces of the integument.

Choriogenesis: Formation of the shell (chorion) of the egg.

Chorion: Eggshell secreted by the ovarian follicle cells (Fig. 58); or the outer covering of an insect egg, which may be multilayered, including exo- and endochorion and wax layer (Fig. 58).

Chorology: The study of the geographical distribution of organisms.

Chrom-, -chrome: Color; pigment.

Chromosomal puffs: Bulges at various points along a chromosome associated with localized uncoiling of the DNA molecule; taken to indicate mRNA transcription associated with the expression of a specific gene or genes.

Chromosome: One of the deeply staining chromatin bodies, formed in the nucleus of a cell during mitosis, which carries the genetic factors (cf. Gene).

Chronic poisoning: Illness or death from long-time exposure to low levels of a toxicant.

Chronic toxicity: A measure that indicates the effects of multiple exposures to a toxin over time; the toxicity of a pesticide (to higher animals) when administered in small sub-lethal doses over a long period of time.

Chrysalis (pl., chrysalids or chrysalides): The pupa of a butterfly that lacks a protective cover; or the pupal stage in lepidopterans that do not construct cocoons.

Cibarial pump: See cibarium (Fig. 60).

Cibarium: The dorsal food pouch, lying between the hypopharynx and the inner wall of the clypeus, often with muscular pump (Fig. 60 a, b and c); the portion of the preoral cavity between the hypopharynx and the labrum.

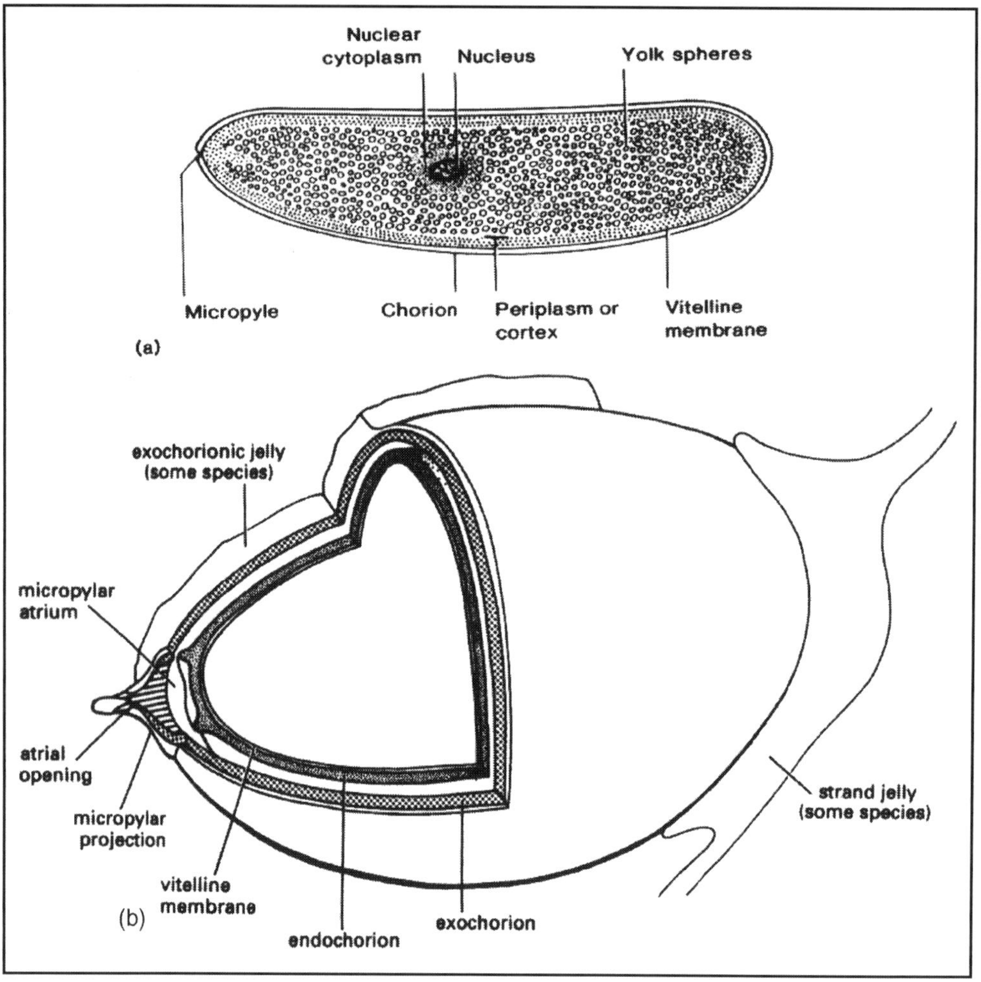

Fig. 58. (a) Sagittal section of a typical insect egg. (b) The generalized structure of a libelluloid dragonfly egg (Odonata: Corduliidae, Libellulidae).

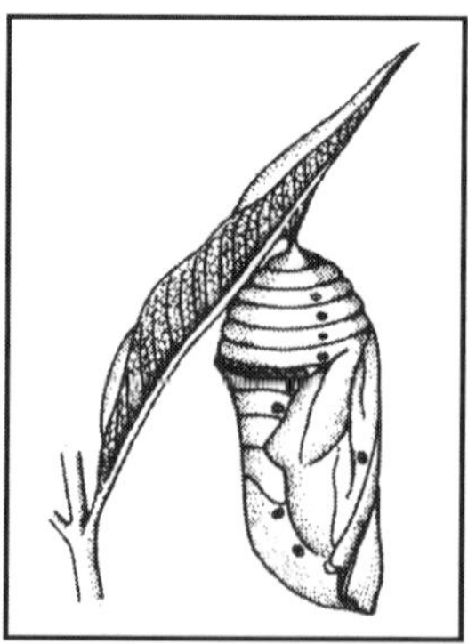

Fig. 59. Pupa (chrysalis).

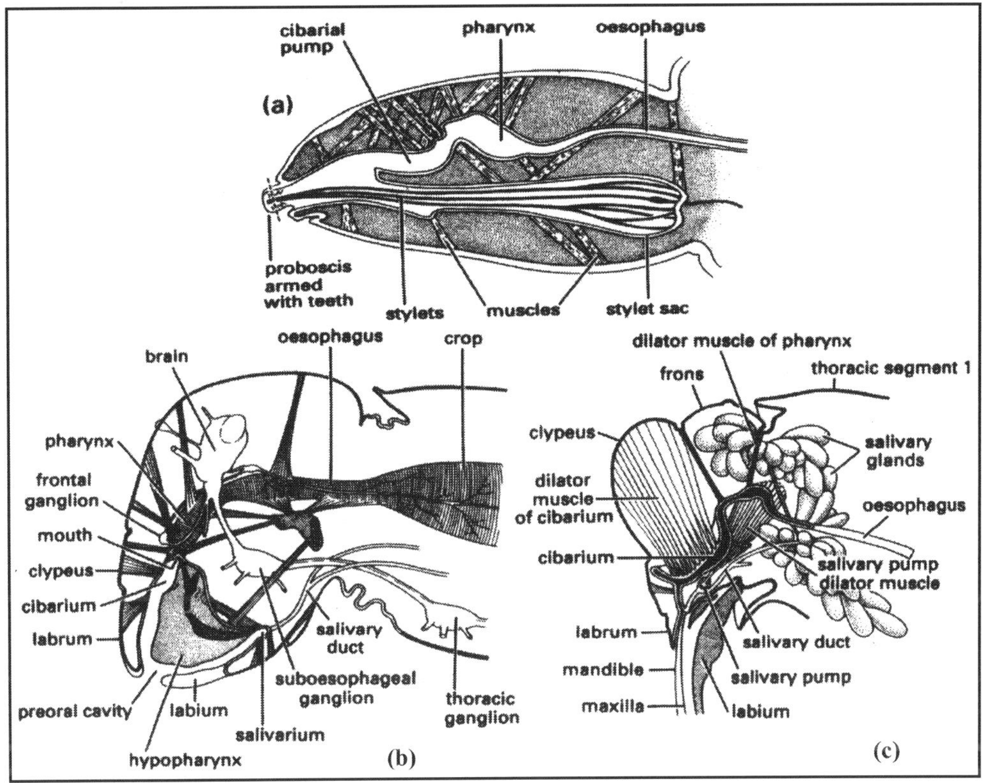

Fig. 60. (a) Head and mouthparts of a sucking louse, *Pediculus* (Phthiraptera: Anoplura: Pediculidae); Preoral and anterior foregut morphology in (b) a generalized orthopteroid insect and (c) a xylem-feeding cicada. After *Snodgrass*, 1935.

Cilium (pl. cilia): Cytoplasmic fine thread like, projecting from all surfaces, help the organism by their movement in locomotion/in feeding; distinguished from flagellum by the size-present in one class of Protozoa (Ciliata) and in most metazoans.

Circadian rhythm: An endogenous rhythm involving a response at about 24-hour intervals; behavioral or physiological events that occur at approximately 24-hours intervals.

Circulative (virus): Plant and animal viruses that persist, multiply, or accumulate in their vectors; a virus that circulates within the body of an insect before being introduced into a new host.

Circumesophgeal connectives: Two large nerves, one on each side of the esophagous that run between the brain and the subesophageal ganglion.

Clade: A group of organisms proposed to be monophyletic, that is all descendants of one common ancestor.

Cladistics: A classification system in which **clades** are the only permissible groupings; a school of systematic thought grounded in the belief that classification can and should reflect phylogeny.

Cladogram: A diagram based on cladistic analysis of a group of organisms; a diagram illustrating the branching sequence of purported relationships of organisms, based on the distribution of shared derived features (synapomorphies).

Claspers: A pair of processes at the end of the abdomen of male insects which serve to clasp the female during copulation.

Class: A category of the classification scheme that includes one or more orders (e.g., Insecta); a subdivision of a phylum or subphylum, containing a group of related orders.

Classical conditioning: See associative learning.

Classification: The ordering of organisms into a hierarchy of categories; the definition, ranking, and arrangement of taxonomic categories and taxonomic entities (cf. Taxonomy, Systematics, Horizontal classification, Vertical classification, Artificial classification, Natural classification).

Claval furrow: A flexion line on the wing that separates the **clavus** from the **remigium** (Fig. 14).

Claval suture: The suture of the front wings separating the clavus from the corium (Hemiptera).

Claval vein: A vein in the clavus (Hemiptera, Homoptera).

Clavate: Club-shaped or enlarged at the tip; clavate antennae (Fig. 15).

Clavus. A club; antennal club, usually the terminal three segments; also the ablong or triangular anal area of the hemelytron (forewing) in the Hemiptera and Homoptera; an area of the wing delimited by the claval furrow and the posterior margin (Fig. 15).

Claw (*strictly*, **pretarsal claw; unguis**): A hooked structure on the distal end of the pretarsus, usually paired; more generally, any hooked structure.

Claw tuff: A dense tuff of hairs below the claws (spiders).

Clay: Very fine soil particles that are less than 0.002 mm in diameter. Clay soils have little air space and drain poorly. Clays may be used as carriers in pesticide formulations.

Cleavage center: Location where cleavage and migration of cleavage nuclei begin.

Cleavage: The divisions of the zygote and subsequent daughter cells that lead to the formation of the blastoderm (Fig. 61. a and b).

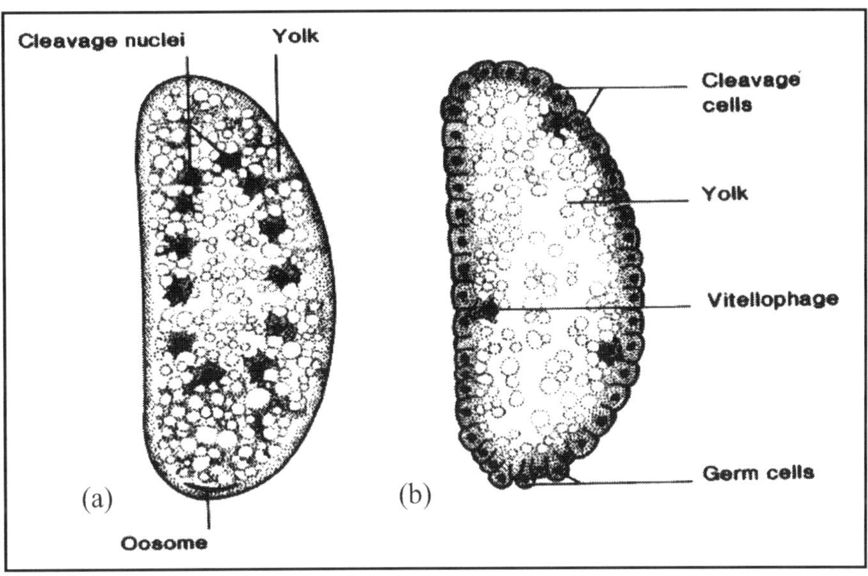

Fig. 61. (a) Cleavage. (b) Blastodeum forming and germ cells differentiating.

Cleft: Split or forked.

Cleptoparasite: A "thief parasite," one that consumes the food stored by another insect in a nest.

Cleptoparasitism: A process in which members of one species take over the nest of another species for their own use.

Climate: It is the average course or condition of the weather in a locality over a period of several years. It tends to remain much the same or change very slowly over a period of a many years. The influence of environmental factors discussed earlier relates to macroclimate or climate as measured by meteorologist. The climate in which an organism lives is very much different from the gross environmental conditions, this immediate environment is the microclimate and relates to the habitats of organisms living in the soil, within crevices of barks of trees, inside sheaths of grasses, within burrows, etc.

Climatic release: Release of climatic restraints, such as a period of favorable weather or entry into a favorable region, resulting in population increase.

Climatograph: A polygonal diagram resulting from plotting temperature means against relative humidity.

Climax community: A complex or organized community of plant, animal, and microbial populations which has a high probability of perpetuating itself on a specified set of land or water areas. The word 'climax' carries the implication that previous communicates on these areas were not so self perpetuating, but instead prepared the biological or physical conditions which allowed the invasion or prolonged regeneration by the climax species.

Cline: A gradual and nearly continuous change of a character in a series of continuous populations; a character gradient (cf. Subspecies).

Cloaca: The terminal portion of the digestive excretory, and reproductive ducts *e.g.* in frog.

Cloacal. Refers to the rectum or cloaca at the termination of the intestine; a common chamber into which opens both rectum and genital ducts (some Coleoptera and Lepidoptera).

Clone: A group of identical individuals propagated vegetatively from a single plant; or Descendants of a single parent or cell that are generally assumed to be genetically uniform because they result from reproduction that does not involve sex and recombination.

Clone: All the offspring derived by asexual reproduction from a single sexually produced individual.

Close - loading system: Apparatus that allows transfer of pesticide from a mixing tank to an application tank by direct connections such as hoses, pipes, and couplings and so prevent worker exposure to the mixture.

Closed cell: A wing cell bounded on all sides by veins and not reaching the wing margin; an area of the wing membrane bounded entirely by veins; *see also* **Open cell.**

Closed coxal cavity: One bounded posteriorly by a sclerite of the same thoracic segment (front coxal cavities, Coleoptera, or one completely surrounded by sternal sclerites and not touched by any pleural sclerites (middle coxal cavities, Coleoptera).

Closed- mixing system: A procedure that allows a pesticide to be moved from its original container, diluted with water, and transferred to an application tank using equipment that is sealed to prevent worker exposure to the pesticide.

Closed tracheal system: A gas exchange system comprising tracheae and tracheoles but lacking spiracles and therefore (Figs 62); *see also* **Open tracheal system.**

Clubbed: With the distal part (or segments) enlarged; clubbed antennae (Fig.15).

Clypeus: A sclerite on the lower part of the face, between the frons and the labium (Fig. 63); or lobelike structure located on the anterior part of the insect head just beneath the epistomal suture; the part of the insect head to which the labrum is attached anteriorly (Fig. 63); it lies below the frons, with which it may be fused in a **frontoclypeus** or separated by a suture.

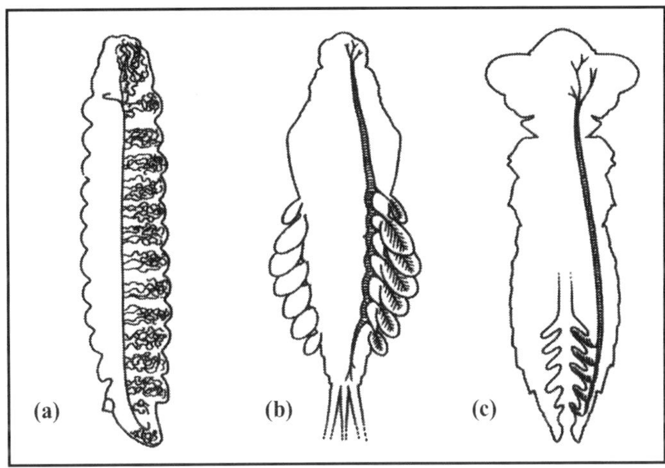

Fig. 62. Some basic variations in the closed tracheal systems of insects: (a) entirely closed tracheal system with cutaneous gas exchange, as in most endoparasitic larvae; (b) closed tracheal system with abdominal tracheal gills, as in mayfly nymphs; (c) closed tracheal system with rectal tracheal gills, as in dragonfly nymphs.

Co-: With; together.

Coadapted: See coevolved.

Coarctate: Compacted and oval in form, concealed by a hardened, thickened, cylindrical case; coarctate pupa.

Coarctate larva: A larva somewhat similar to a dipterous puparium, in which the skin of the preceding instar is also called a pseudopupa.

Coarctate pupa: A pupa enclosed inside a hardened shell formed by the previous larval skin (Diptera) or a type of pupa typical of cyclorrhaphous Diptera, in which the next-to-last larval skin forms a hard barrel-shaped case, the puparium. The puparium spilts at emergence, and the adult fly pushes away the top using an eversible head sac, the ptilinum; Pupal form in which the last larval skin is retained as an extra covering; occurs particularly in higher Diptera.

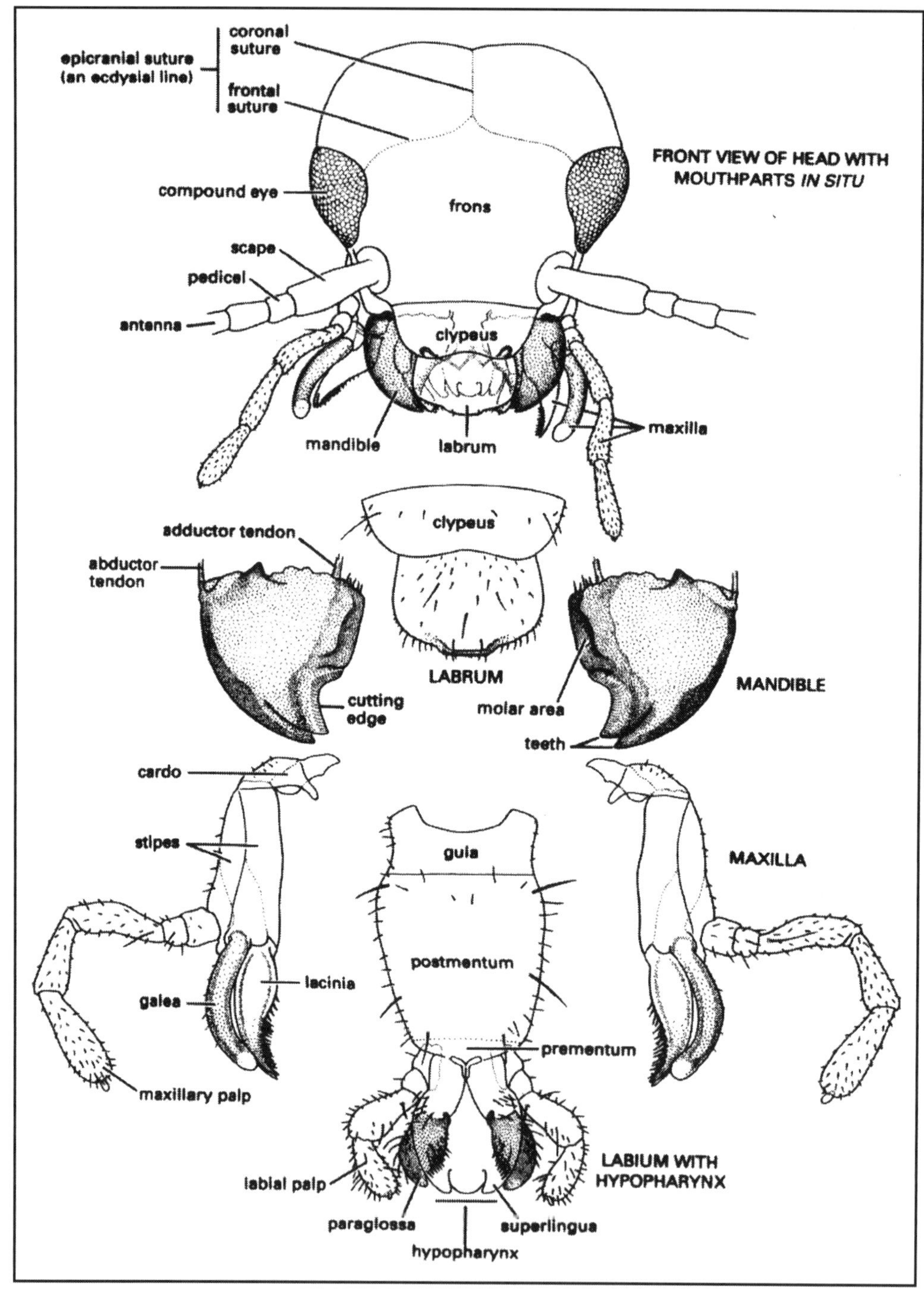

Fig. 63. Frontal view of the head and dissected mouthparts of an adult of the European earwing, *Forficula auricularia* (Dermaptera: Forficulidae). Note that the head is prognathous and thus a gular plate, or gul, occurs in the ventral neck region.

Coccidiasis: Disease caused in rabbits/fowls by the parasitic sporozoan; causative organism is *Eimeria* sp. *e.g. E. stiedae* in domestic rabbits and *E. tenella* and *E. mitis* in fowls.

Cochineal: Red dye produced from the dried bodies of certain scale insects.

Cocoon: A silken covering that is produced by an insect larva to protect it during the pupal stage; or a silken case or silk-bound debris in which a pupa develops; A protective case or covering about a mass of eggs, a larva or pupa, or even an adult animal *e.g.* cocoon of silkworm.

Cocoonase: An enzyme that digests the silken cocoon in saturniid and bombyliid moths.

Codlemone: The synthetic sex pheromone of the codling moth.

Coefficient of difference: Difference of means divided by sum of standard deviations

$$C.D. = \frac{M_B - M_A}{S.D._A + S.D._B}$$

Coefficient of variability: The standard deviation as percentage of the mean.

$$C.V. = \frac{S.D. \times 100}{M}$$

Coel-: Cavity; chamber.

Coelom: A body cavity which originates from mesoderm *e.g.* in Coelenterata.

Coelomic sacs: Small cavities in the mesoderm of a developing insect (Fig. 64).

Coenospecies: All the ecospecies so related that they may exchange genes among themselves to a limited extend through hybridization (cf. Ecospecies, Introgressive hybridization).

Coevolution: The simultaneous evolution of traits in two or more species because of the selection pressures resulting from the mutual interactions of those species; or an evolutionary change in a trait of individuals of one population in response to trait of individuals of a second population, followed by an evolutionary response of the second population to a change in the first; evolutionary interactions between two organisms, such as plants and pollinators, hosts and parasites; the degree of specificity and reciprocity varies; *see also* **Guild coevolution; Specific coevolution; Phyletic tracking.**

Coevolved: Closely associated species that have acted as agents of natural selection on one another.

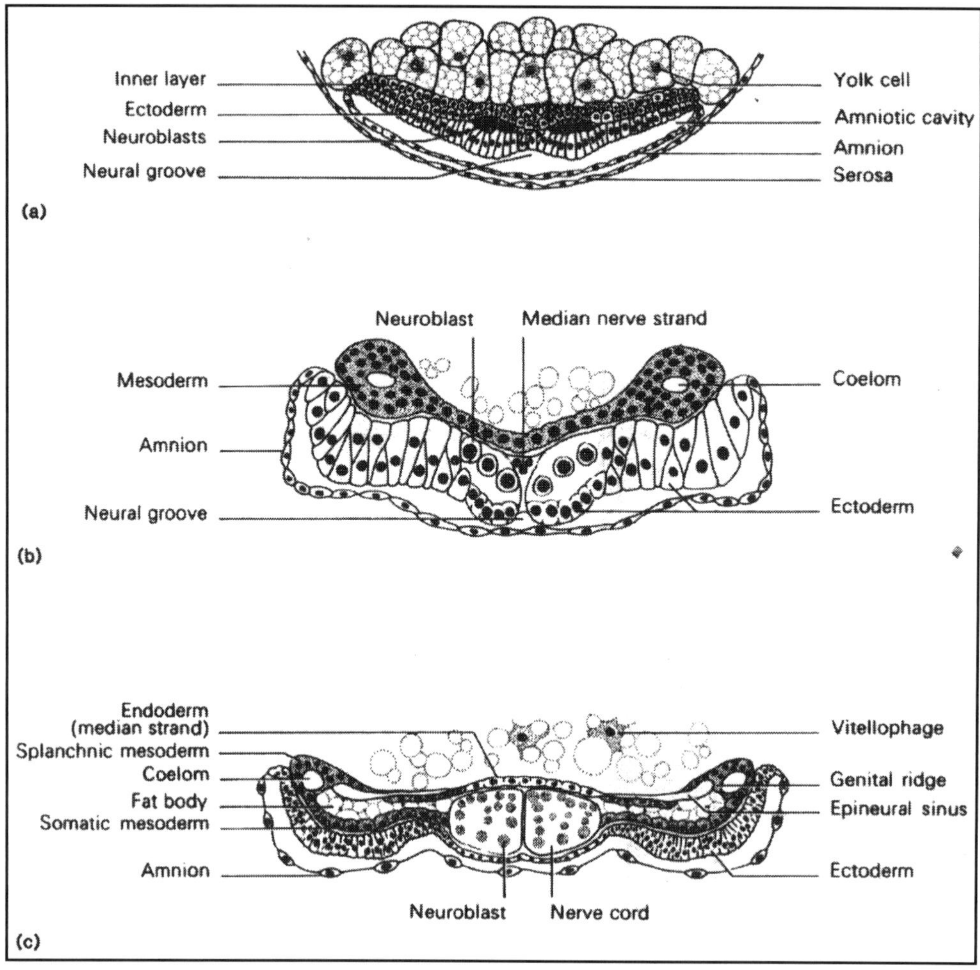

Fig. 64. Cross section of an insect embryo at successive stages of differentiation (chorion and most of yolk omitted); serosa omitted in (b) and (c). After *Johannsen and Butt,* 1941.

Cohesion: Mutual attraction of atoms or molecules; for example, the cohesion of water molecules.

Cohort: Group of individuals of a single species of the same age.

Cold-blooded: Animals whose body temperature keeps on changing according to the atomospheric temperature changes; have to device ways and means to escape the extremes of atmospheric temperatures *e.g.* insects, pisces, amphibians and reptiles.

Coleoptera: The order of insects comprising the beetles (including weevils) (Fig. 65).

Coleopteran, order: (*coleo-,* sheath; *ptera,* wing) beetles and weevils (Fig. 65).

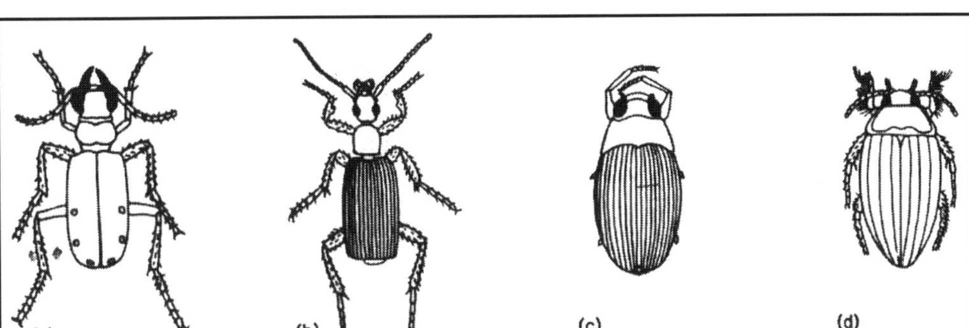

Fig. 65. Coleopteran beetles: (a) Tiger beetle (Cicindellidae). (b) Ground beetle (Carabidae). (c) Whirligig beetle (Gyrinidae). (d) Predaceous diving beetle (Dytiscidae).

Collateral: Axon branch.

Collembola: The order of insects comprising the springtails (Fig. 66).

Collembolan, order: (*coll,* glue; *embol,* a wedge) springtails (Fig. 66).

Colleterial gland: An accessory gland of the female that produces the ootheca; accessory glands of the female internal genitalia that produce secretions used to cement eggs to the substrate; gland that secretes adhesive materials.

Colloidal formulation: Solution in which the particle is less than 6μm in diameter, and the particles stay indefinitely dispersed.

Collophore: Extendable tubelike structure found on the ventral side of the first abdominal segment of collembolans; or a nongenital structure found on the venter of the first abdominal segment in Collembola (Fig. 66).

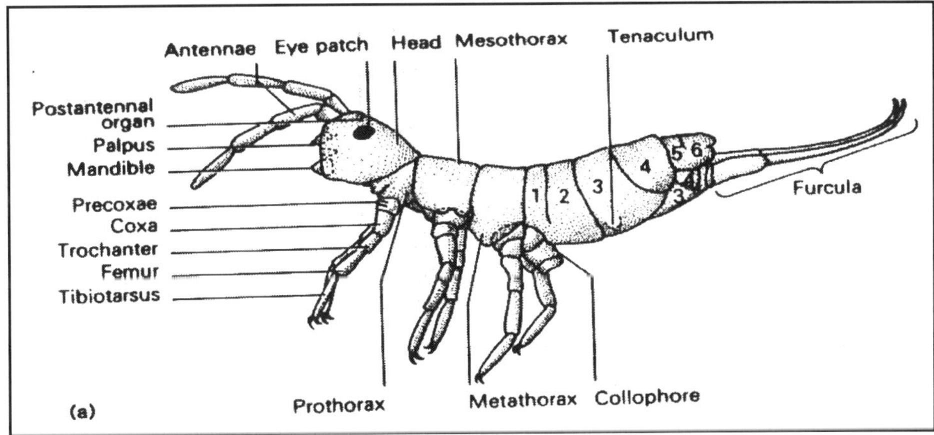

Fig. 66. Collembola, springtails.

Colloquial name: Common name = vernacular name (q.v.).

Collum: The tergite of the first body segment (Diplopoda).

Colon: The large intenstine; that part of the hingut between the ileum and the rectum; or a portion of large intestine extending between caecum to rectum (Gr. *Kolon* = large intestine).

Colony: A congregation of organisms of the same species living together; opposite of solitary. (L. *Colonus*=farmer).

Colony fission. The breaking up of an insect aggregation and dispersal of part to form a new colony.

Colulus: A slender pointed structure lying just anterior to the spinnerets (spiders).

Com-: Together.

Comb: A layer of nest cells crowded in a distinctive regular arrangement, found in social Hymenoptera(Fig. 67); the hexagonal wax structure made by honey bees in which youngs are reared and food is stored; serrated structure bearing a resemblance to a hair comb (pectinae antenna).

Combination cleavage: Total cleavage of the egg followed by peripheral cleavage in blastoderm formation, for example, in Collembola.

Commensalism: The association of 2 or more individuals of different species in which one or more are benefited. (L. *cum*=together + *mensa* = table); or two organisms living together and sharing food, both species usually benefiting from the association; a type of symbiosis; or a living together of two or more species, none of which is injured thereby, and at least one of which is benefited; or a symbiotic association in which one individual benefits and the other is neither helped nor harmed.

Commercial applicator: A person licensed by the state, in compliance with an EPA-approved state plan, to apply pesticides for other than private use. Must also be certified to apply restricted use pesticides.

Common chemical name: The formal name assigned to the active ingredient of a pesticide that is usually shorter than the chemical name. Each chemical has a single common name that is used in the scientific literature.

Common name: The well known name for an organism in colloquial use of the regional language.

Common (median) oviduct: In female insects, the tubes leading from the fused **lateral oviducts** to the **vagina** (Fig. 5).

Common oviduct: The duct that joins the lateral oviducts to the vagina (Fig. 5).

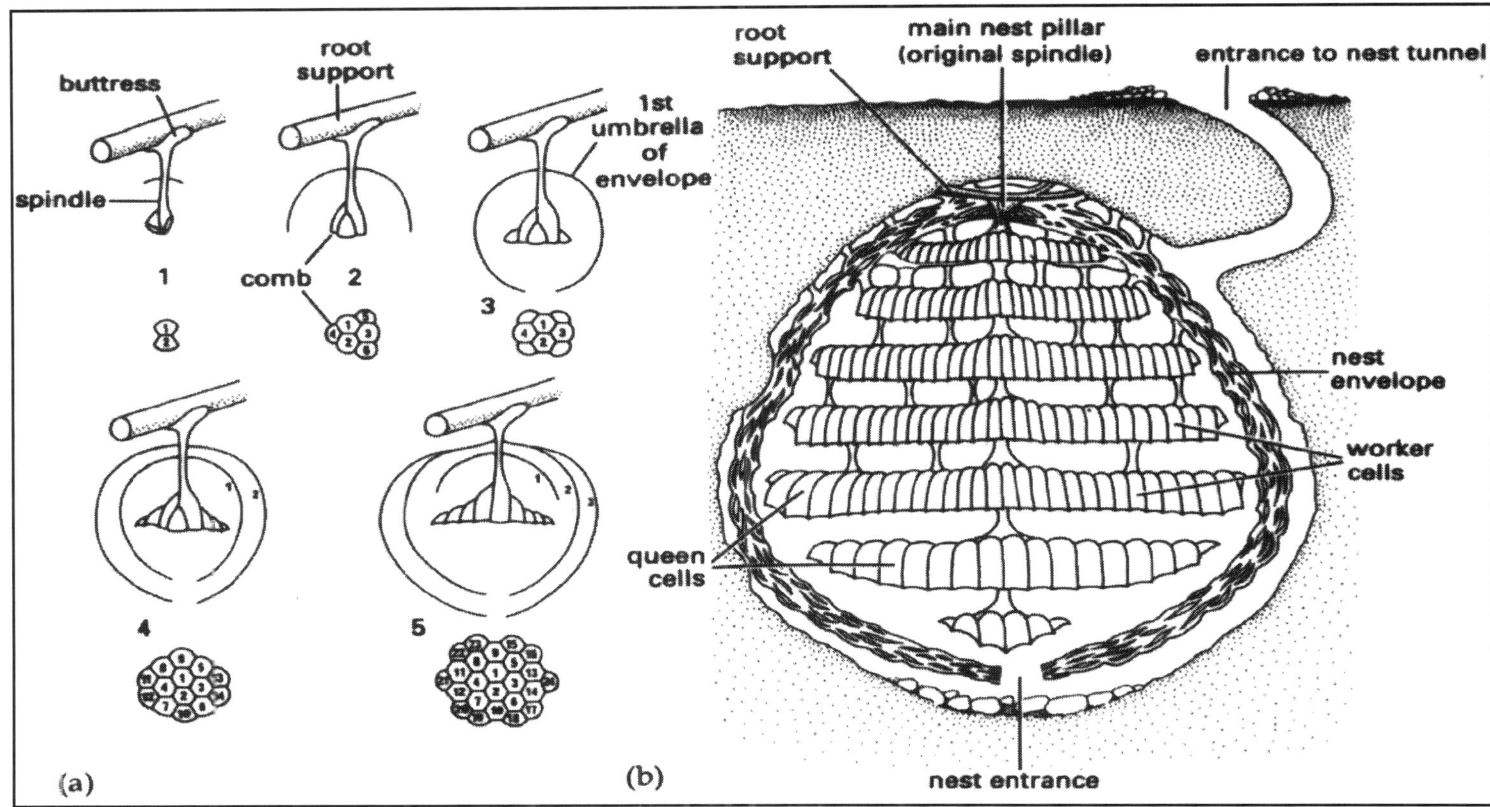

Fig. 67. The nest cf the European wasp, *Vespula vulgaris* (Hymenoptera: Vespidae): (a) five initial stages of nest construction by the queen (the embryonic phase of the colony's life); (b) a mature nest. After *Spradbery,* 1973.

Common salivary duct: The duct that joins the lateral salivary ducts to the salivarium (Fig. 41).

Communal behavior: Cooperation in nest construction but not brood rearing.

Communication: The influence of signals from one organism on the behavior and/or physiology of another organism; or the production of a signal by an individual that influences the behavior of another individual and that is mutually beneficial.

Community: The interacting assemblage of all organisms in a defined area. Sometimes also used to refer to all members of some particular grouping in a defined area, such as an insect community; or the collection of different species and types of plants and animals, in their respective niches, within the habitat. 1. Closed: the habitat is completely colonized by plants; no areas of bare soil; strong competition for space. 2. Open: habitat is not completely colonized by plants; bare areas of soil; competition for space is thus reduced; or the interacting "web" of populations in a given area.

Companion planting: Growing two or more plant species together so that the different species benefit each other.

Companionate planting: The intercropping of certain repellent plants with crop plants.

Compatibility: (as applied to pesticides)**:** The ability to mix different pesticides without physical or biological efficiency or increase in phytotoxicity. Refers to whether or not different pesticides can be mixed together for application. Safe mixtures are compatible. Unsafe mixtures are in compatible.

Compatible (as applied to pesticides)**:** Safe to mix together; pesticides are compatible if they can be mixed without their properties being altered.

Competency: The potential of a termite of one caste to become another, e.g. a worker to become a soldier.

Competition: The simultaneous demand by two or more organisms or species for the same, limited essential resource.

Competitive displacement: One species, having a competitive advantage, eventually eliminates another species that occupies the same or a closely similar niche; the replacement of an indigenous species by a recent arrival with similar ecological requirements.

Competitive exclusion principle: The concept that two species cannot long coexist if they have identical niches.

Competitive exclusion: See competitive displacement.

Complete metamorphosis: Striking changes between larva and adult, with an intervening pupal stage; metamorphosis in which the insect proceeds through four distinct developmental stages—egg, larva, pupa and adult. cf. incomplete metamorphosis.

Complex: A neutral term for a number of related taxonomic units, most commonly involving units in which the taxonomy is difficult or confusing (cf. Group, Neutral term).

Compound eye: A visual sensory organ composed of many individual elements, each of which is represented externally by a hexagonal facet; or an eye composed of many individual elements or ommatidia, each of which is represented externally by a facet; located on either side of the insect head, compound eyes are composed of a number of individual units, the ommatidia; an aggregation of **ommatidia,** each acting as a single facet of the eye (Fig. 68).

Compressed air sprayer: A sprayer with a 1-to 3-gallon capacity, a delivery tube, an air pump for pressure, and a strap so the sprayer can be carried.

Compressed: Flattened from side to side.

Concave veins: Longitudinal veins in troughs formed by wings folded in a pleated fashion.

Concentrate spraying: Direct application of the pesticide concentrate without dilution.

Concentrate: A liquid pesticide formulation that must be diluted prior to application.

Concentrated solution (c.s): Commercial pesticide preparation before dilution for use.

Concentration: Proportion of active ingredient in a pesticide preparation, before or after dilution.

Concurrence: The assemblage of organisms leads to overcrowding and then there is a keen struggle for the same requirements. It results in reduction of food, contamination of surroundings and mortality.

Conditional lethal: An allele that results in mortality of carriers under certain environmental conditions (e.g., cold temperatures).

Conditioning: *See* Associative learning.

Conducive soil: A soil in which particular plant pathogens are able to cause disease.

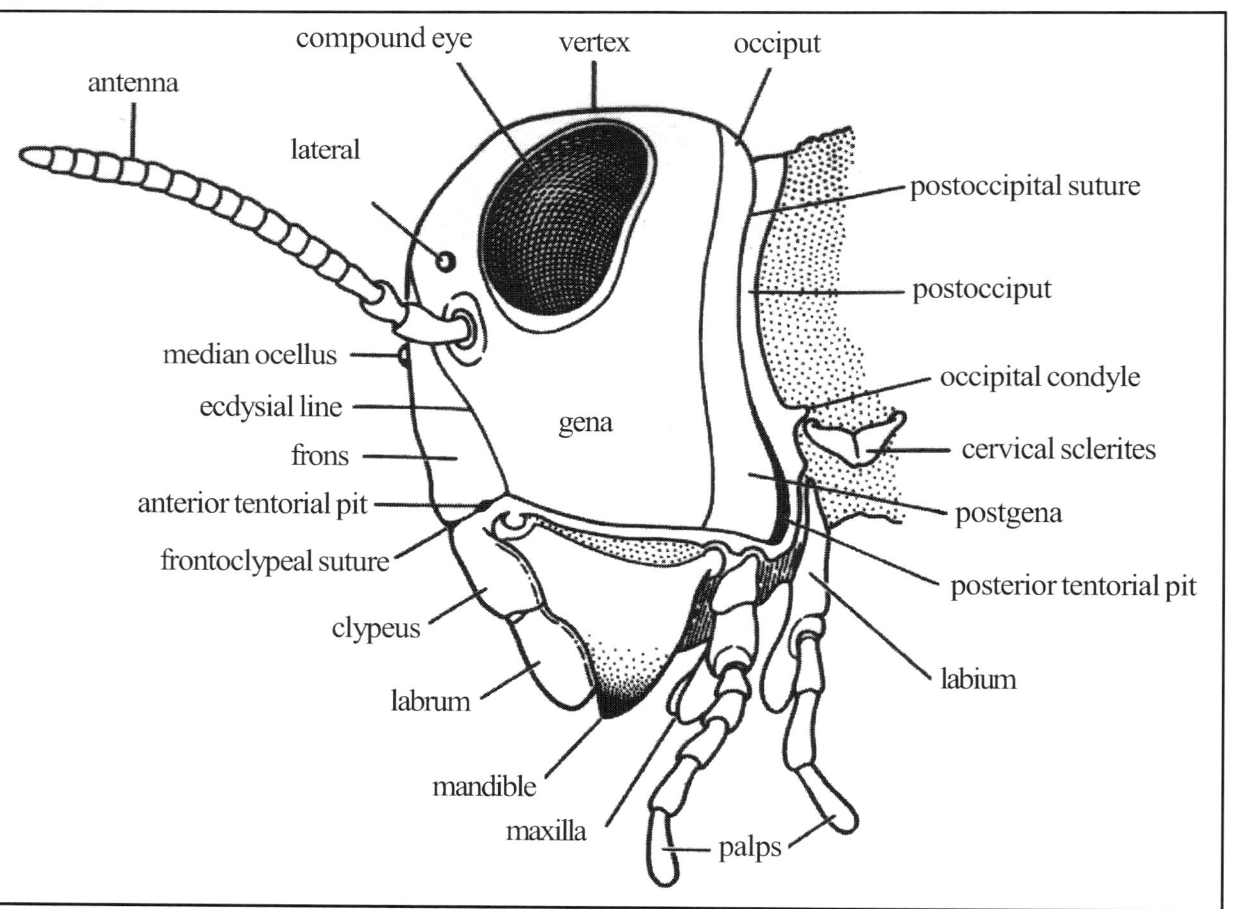

Fig. 63. Lateral view of the head of generalized pterygote insect showing position of mouth parts relative to the capsule: Hypognathous.

Condyle: A knoblike process forming an articulation.

Congeneric: A term applied to species of the same genus (cf. Genus).

Conidia: Asexual fungal spores.

Conidium (pl. conidia): Generally, any type of that is produced asexually and is nonmotile.

Conjunctiva, conjunctivial membrane: see **Intersegmental membrane.**

Conjunctiva: Soft membranous area separating sclerites in the insect skeleton.

Connate: Fused together or immovably united.

Connective: Anything that connects; more specifically, the paired longitudinal nerve cords that connect the ganglia of the central nervous system.

Conservation: A biological control practice that includes any activity designed to protect and maintain existing populations of natural enemies.

Conspecific: A term applied to individuals or populations of the same species (cf. Species).

Constitutive defence (s): Part of the normal chemical composition; Plant defenses present at all times, for example, physical defenses; *see also* **Induced defence.**

Constricted: narrowed.

Consumer: (i) A person who buys agricultural produce. (ii) The ecological term for all nonautotrophic organisms that must consume food.

Consummatory act: Behavior that follows appetitive behavior when an appetite releasing stimulus is encountered.

Contact herbicide: A herbicide that does not move within the plant and kills only that plant tissue on which it is actually sprayed.

Contact insecticide: One which is able to kill insects by contact with the external body surface.

Contact poison: A pesticide that is lethal when it touches or is absorbed into the body; will act as a stomach poison if eaten; or material killing pests by contact action, presumably by absorption through the cuticle; or an insecticide that enters the body when the insect walks or crawls over a treated surface; insecticide that is absorbed through the body wall; or toxic substances that are absorbed through the cuticle.

Contemporary species: Synchronic species (q.v.).

Contest competition: Competition involving aggressive interactions between individuals.

Contiguous: Touching each other.

Continuity: In nomenclature, the principle that continuity of usage should supersede priority of publication in determining which of two or more competing scientific names should be adopted for a particular taxonomic category (cf. law of priority).

Continuous variation: Variation in which individuals differ from each other by infinitely small steps, as variation in quality of expression of a character or group of characters (cf. Discontinuous variation).

Control (as applied to pest species): Reduction or maintenance of pest populations below the damage threshold; Keeping the pest population at such a level as not to allow it to act as a pest.

Control (noun)**:** Untreated subjects used for comparison with those given a particular crop protection treatment. In an experiment, an untreated plot used to determine population response in the absence of treatment; also known as a check. Keeping a pest population below a specified density.

Control (verb)**:** To reduce damage or pest density to a level below the economic threshold. 1. Legislative: the use of legislation to control the importation and to prevent any spread of a pest within a country. 2. Physical: the use of mechanical (hand picking, etc.) and physical methods (heat, cold, radiation, etc.) of controlling pests. 3. Cultural: regular farm operations designed to destroy pests. 4. Chemical: the use of chemical pesticides as smokes, gas, dusts, and sprays to poison pests. 5. Biological: the use of natural predators, parasites and disease organisms to reduce pest populations. 6. Integrated: the very carefully reasoned use of several different methods of pest control in conjunction with each other to control pests with a minimum disturbance to the natural situation.

Control action guideline: Used to define if pest control actions are necessary and when actions should be taken.

Control action threshold: See *economic threshold.*

Control, biological: Control of a pest by the use of living organisms.

Control, chemical: Control of a pest by the use of poisonous chemicals.

Control, cultural: Control of a pest by the use of cultural practices like rotation of crops, summer ploughing, flooding the field *etc.*

Control, integrated: Control of pest by combining several methods of control like biological, chemical, mechanical, all or some to have a better impact so that whatever is not covered by one may be covered by the other.

Control, mechnical: Control of a pest by the use of mechnical means *e.g.*, hand picking.

Controlled droplet application (CDA): Equipment that uses rotary atomizers (spinners) to produce spray droplets of uniform size.

Convergence: Morphological similarity in distantly related forms (cf. Adaptive convergence).

Convergent evolution: The process by which nonhomologous structures become similar in appearance due to similar selection pressure.

Convergent: Becoming closer distally.

Convex veins: Longitudinal veins located on crests resulting from wings folded in a pleated fashion.

Coprophage: A feeder on dung or excrement (Fig. 69).

Coprophagous: Feeding on fecal material.

Coprophilous: Living in dung (for example, dung beetles,) or decaying vegetable matter resembling excrement.

Copulation: The joining of male and female genital structures (Fig. 70); or sexual union (L. *copulo* = join together).

Corbicula (pl., corbiculae): A smooth area on the outer surface of the hind tibia, bordered on each side with a fringe of long curved hairs, which serves as a pollen basket (bees; cb); the pollen basket of bees (Fig. 71).

Corbiculum: Specialized structure on the legs of honey bees used to store pollen for transport to the hive; also called a *pollen basket* (Fig. 71).

Corium: A section of the heteropteran hemelytron (fore wing), differentiated from the clavus and membrane, usually leathery; the elongate, usually thickened, basal portion of the front wing (Hemiptera).

Cornea: The cuticlar part of an eye or ommatidium; see corneal lens; the cuticle covering the eye or ocellus; the outer transparent layer of the eye.

Corneagen cells: A layer of cells that secretes the corneal lens of a dorsal ocellus.

Corneagenous cell: One of the translucent cells beneath the **cornea** which secretes and supports the corneal lens.

Corneal lens: The outermost part of an ommatidium (Fig. 72) or dorsal ocellus (Fig. 72), derived from cuticle.

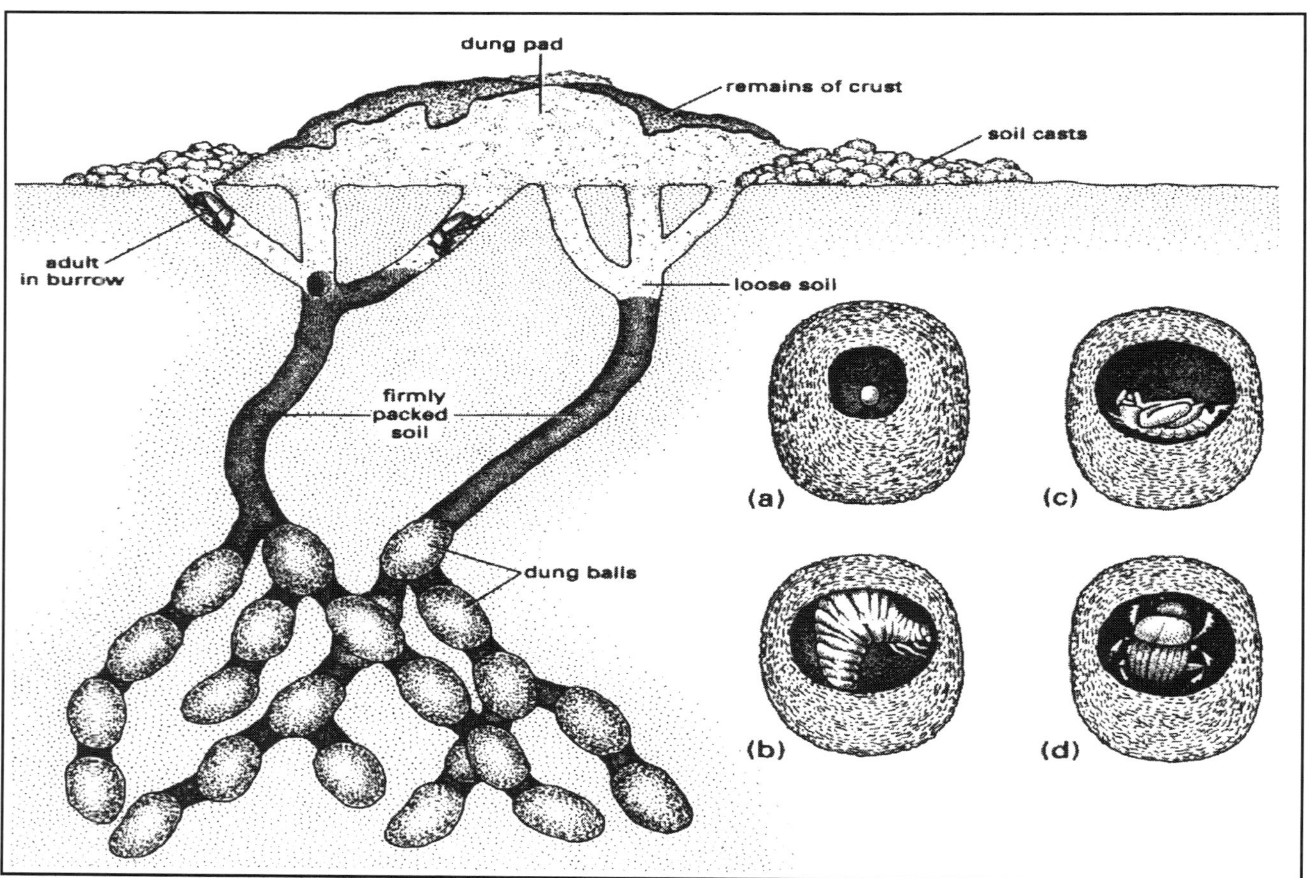

Fig. 69. A pair of dung beetles of *Onthophagus gazelle* (Coleoptera: Scarabaeidae) filling in the tunnels that they have excavated below a dung pad. The inset shows an individual dung ball within which beetle development takes place: (a) egg; (b) larva that feeds on the dung; (c) pupa; (d) adult just prior to emergence.

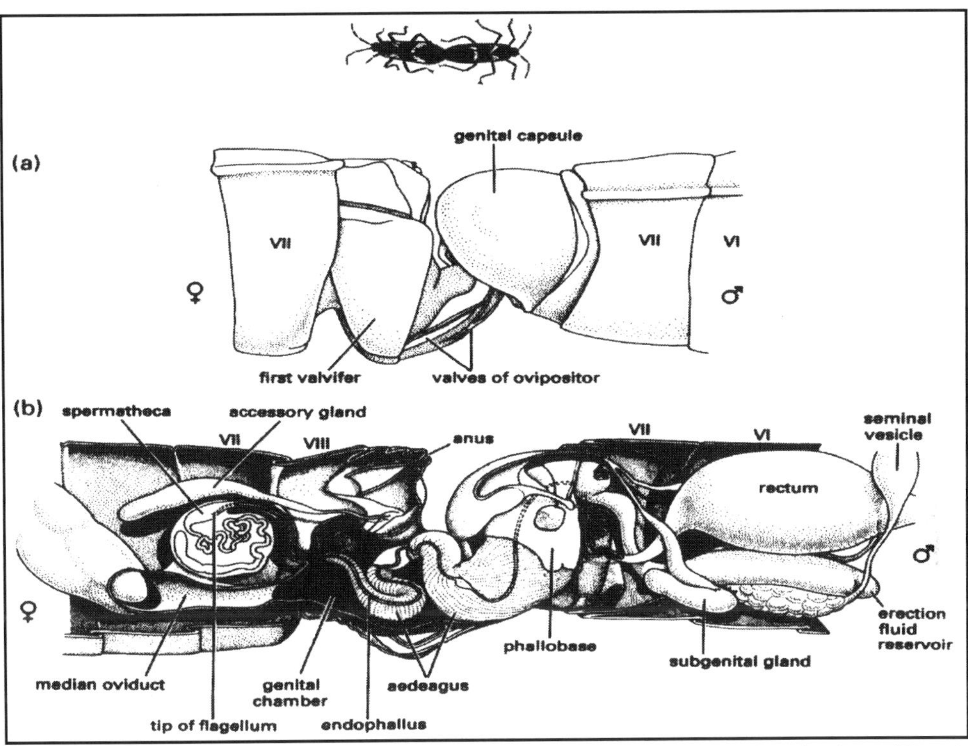

Fig. 70. Posterior ends of pair of copulating milkweed bugs, *Oncopeltus fasciatus* (Hemiptera: Lygaeidae). Mating commences with the pair facing in the same direction, then the male rotates his eight abdominal segment (90^0) and genital capsule (180^0), erects the aedeagus and gains entry to the female's genital chamber, before he swings around to face in the opposite direction. The bugs may copulate for several hours, during which they walk around with the female leading and the male walking backwards. (a) Lateral view of the terminal segments, showing the valves of the female's ovipositor in the male genital chamber; (b) longitudinal section showing internal structures of reproductive system, with the tip of the male's Aedeagus in the female's spermatheca.

Corneal pigment cells: Cells of an ommatidium, which contain pigments responsible for blocking the entry of light from an adjacent ommatidium (Fig. 72).

Cornicle (also **siphunculus**): A peg-like gland on the abdomen of aphids that secretes wax and pheromones; one of a pair of dorsal tubular structures on the posterior part of the abdomen of aphids; secretes an allomone and an alarm pheromone (Fig. 75).

Coronal suture: A longitudinal suture along the midline of the vertex, between the compound eyes.

Corpora allata (sing. Corpus allatum): Paired endocrine glands associated with the stomodeal ganglia behind the brain (Fig. 74), the source of

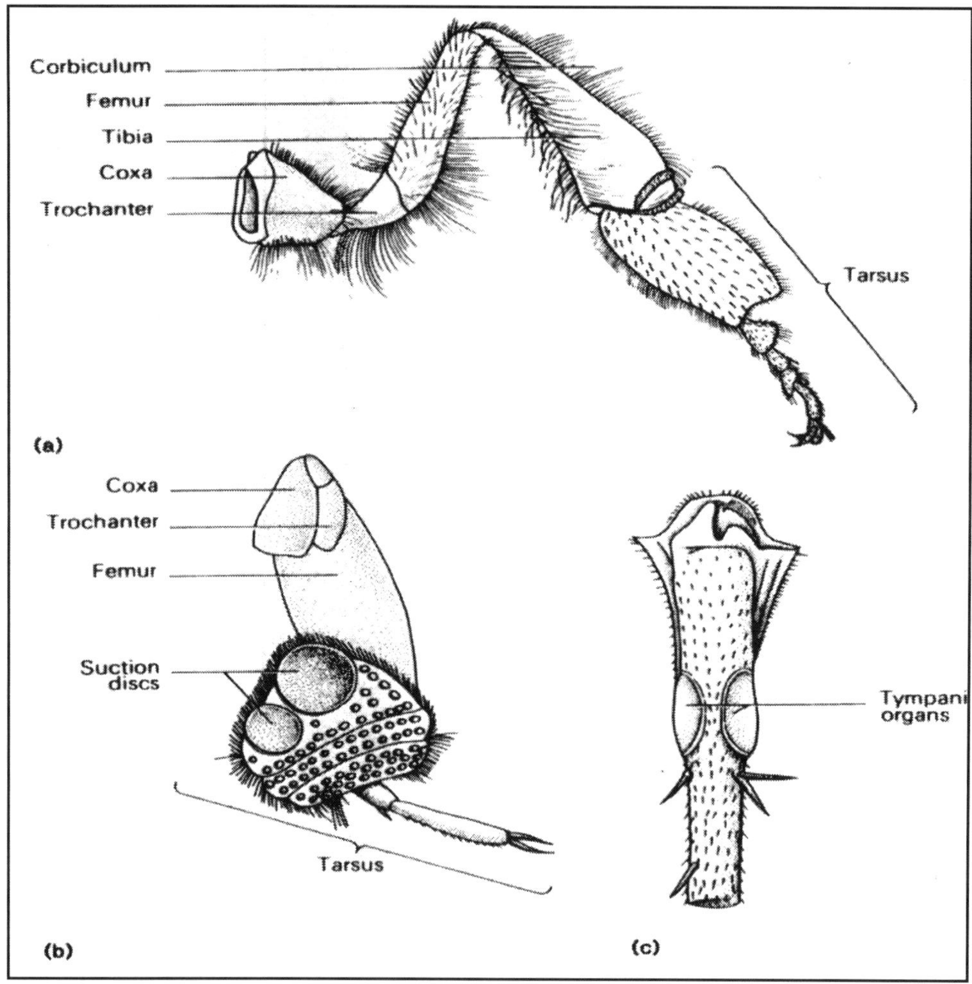

Fig. 71. Specialized structures borne on the legs. (a) Corbiculum or pollen basket on the hind tibia of the honey bee. (b) Suction discs on the fore tarsus of a male dividing beetle. (c) Tympanic organs on the tibia of a long-horned grasshopper.

juvenile hormone; a pair of glands associated with the brain of insects which secrete juvenile hormone.

Corpora cardiaca (sing. Corpus cardiacum): Paired glands lying close to the aorta and behind the brain (Fig. 74), acting as stores and produces of neurohormones.

Corpora pedunculata: Cell masses in the insect brain that contain association neurons and are considered to be the location of the "higher centers" controlling the most complex behavior; also called mushroom bodies (Fig. 54).

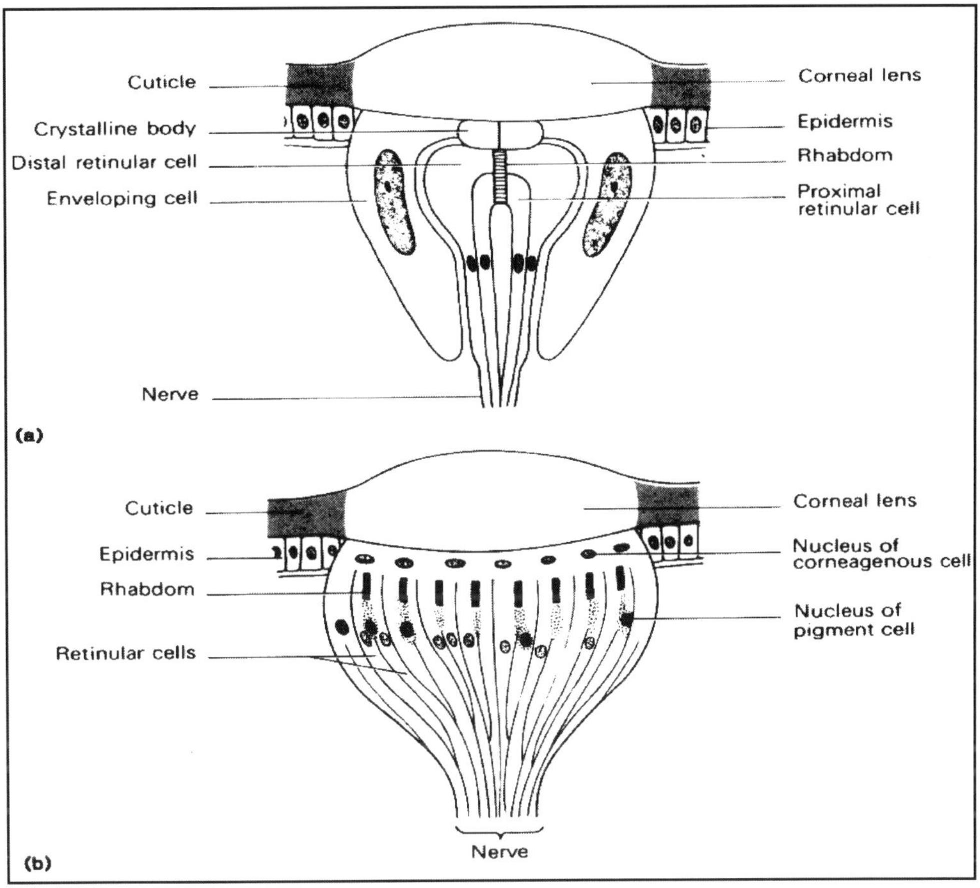

Fig. 72. (a) Section of a stemma. (b) Section of a dorsal ocellus.

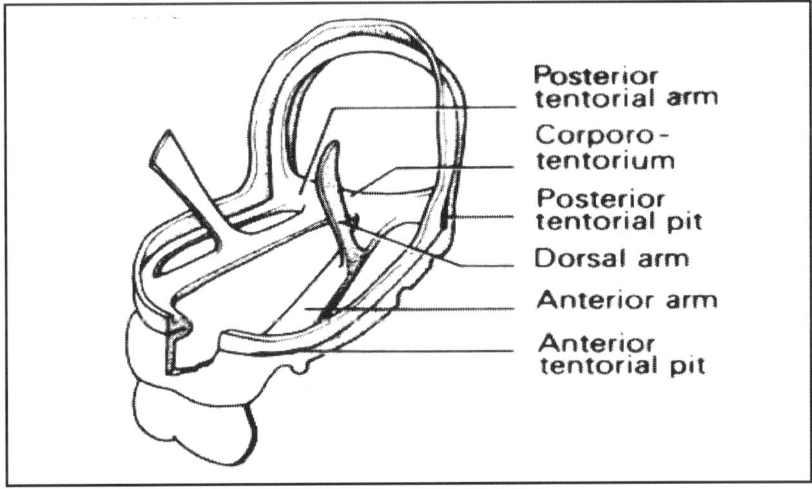

Fig. 73. Generalized insect head. Top of head capsule cut away to show tentorium.

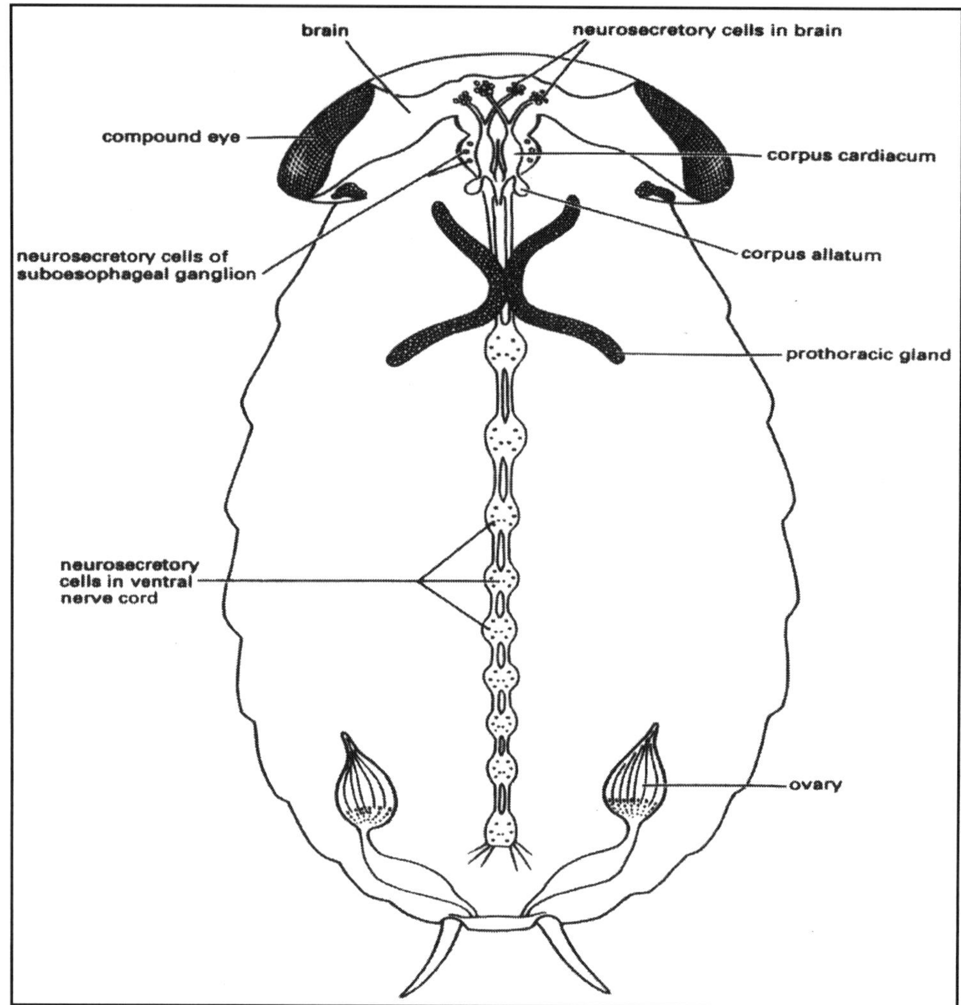

Fig. 74. The main endocrine centres in a generalized insect. After *Novak,* 1975.

Corporotentorium: Body of tentorium formed by the fussion of the anterior and posterior tentorial arms (Fig. 73).

Corpus allatum (pl. Corpara allata): One of a pair of small neuroendocrine glands just posterior to the brain that secrete juvenile hormone and store and release some brain hormones (Fig. 74 and 185).

Corpus cardiacum (pl. corpora cardiaca): One of a pair of small neuroendocrine glands adjacent to the corpora allata and posterior to the brain (Fig. 74); a small organ of nervous origin just behind the brain, associated with storage and release of PTTH and other hormones neurohaemal organs where some brain hormones are stored and released.

Corpus pedunculatum (plural, ***corpora pedunculata***): *See* Mushroom body (Fig. 54).

Cosmopolitan: A species occurring very widely throughout the major regions of the world.

Costa: Wing vein; the first longitudinal vein running along the anterior edge and terminating; before the apex; usually reduced in the hind wing; also any elevated ridge rounded at its crest; a longitudinal wing vein, usually forming the anterior margin of the wing (leading edge) (Fig. 81 & 83).

Costal area: The portion of the wing immediately behind the anterior margin.

Costal break: A point on the costa where the sclerotization is weak or lacking or the vein appears to be broken (Diptera).

Costal cell: The wing space beween the costa and the subcosta.

Costal fracture: A break or weakness in the costal margin in Heteroptera that divides the corium, separating the cuneus from the embolium (Fig. 82).

Costal margin: Anterior edge of the wing.

Cotyledon: A modified leaf that is formed within the seed, and the leaf is present on a seedling at germination; seed leaf. Two are present in dicotyledons; one is present in monocotyledons.

Cotype: Syntype (q.v.).

Counter shading: A form of coloration that tends to reduce the three-dimensional appearance of an animal by suppressing highlights.

Cover crop: Non-crop plantings used to protect soil during a fallow period, to add organic matter to soil when ploughed under, and in some cases to reduce pest and pathogen numbers.

Cover: Proportion of the surface area of the target plant on which the pesticide has been deposited.

Coverage: The degree to which pesticide applied reaches the area intended. See ground cover, plant cover, target cover.

Coxa (pl. coxae): The basal (proximal) segment of an arthopod appendage e.g. the basal segment of the leg (Fig. 71); the most basal segment of the insect leg, articulating with the thorax; The uppermost joint of insect legs by which they are attached to the insect body at the sternum of the thorax.

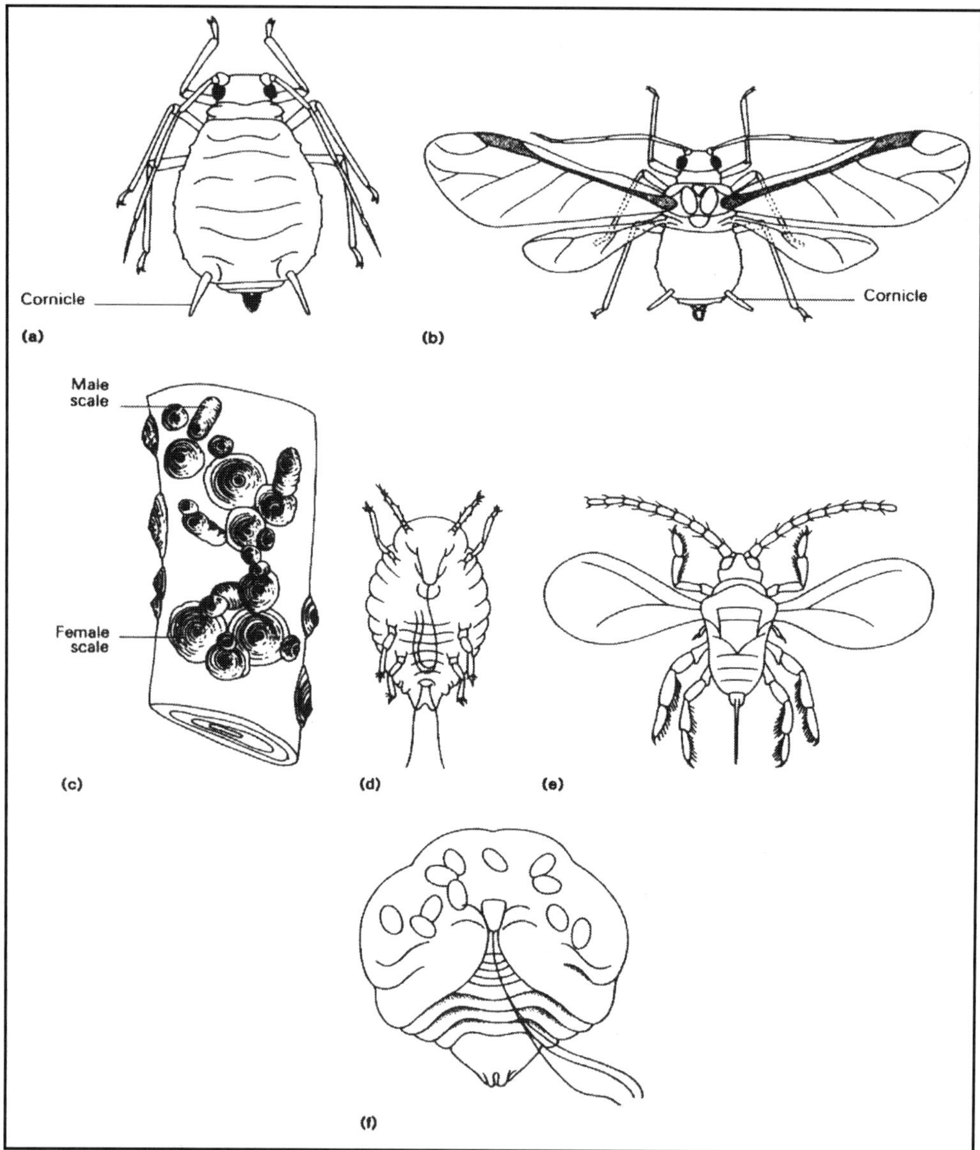

Fig. 75. Homoptera, Sternorrhyncha. (a) Aphid, *Aphis pomi* (Aphididae), wingless viviparous female. (b) Same species as (a) winged viviparous female. (c-f) San Jose scale, *Aspidiotus permiciosus* (Diaspididae): (c) Scales on a twig; (d) Nymph; (e) Adult male; (f) Adult female with scale removed.

Coxapleurite: One of three primitive, subcoxal sclerites (Fig. 76).

Coxopodite: Basal segment of a genital appendage.

Cranium: Head capsule; the hardened capsule of the head that bears the mouthparts, antennae, and eyes.

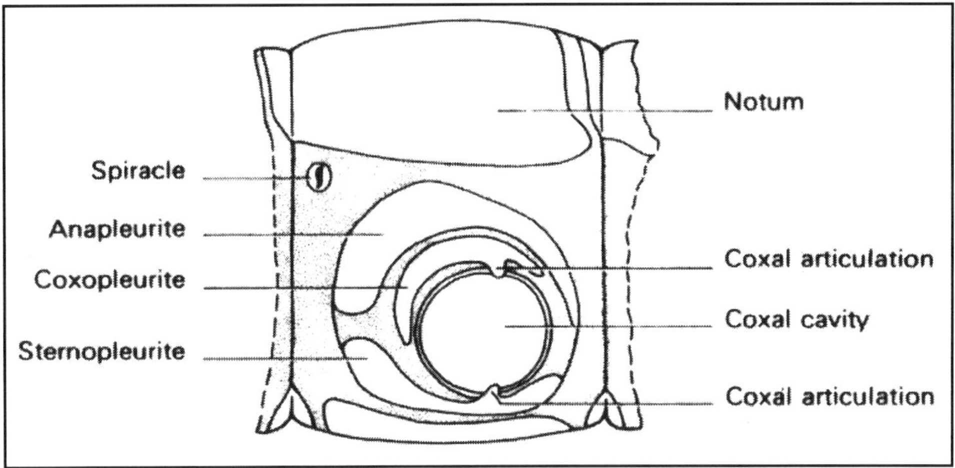

Fig. 76. Hypothetical primitive arrangement of subcoxal sclerite.

Crawler: The active first instar of some homopterans; immature mobile stage of scale insects and whiteflies which possesses legs and thus enables dispersal to take place.

Cremaster: A hooked or spine-like process at the posterior end of the pupa, often used for attachment (Lepidoptera); a spine at the tip of the abdomen of a chrysalis; (Lepidoptera); a structure, on the end of the abdomen, that bears small hooks by which some lepidopterous pupae attach to a substrate.

Crenulate: Wavy, or with small scallops.

Crepuscular: Animals that are active in the twilight, pre-dawn and at dusk in the evenings; active at low light intensities, dusk or dawn; *see also* **Diurnal; Nocturnal.**

Crepuscular period: The daily cycle of dark and light in between day and night and vice versa (i.e., dawn and dusk) is termed crepuscular period.

Crescentdric. Crescent-shaped

Cribellum: A sievelike structure lying just anterior to the spinnerets (spiders).

Crista acustica (*also* **crista acoustica**): The main chordotonal organ of the tibial tympanal organ of katydids (Orthoptera: Tettigoniidae) (Fig. 77).

Crochets (pronounced **croshays**): Hooked spines at the tip of the prolegs of lepidopterous larvae; curved hooks, spines or spinules on **prolegs;** minute spines which form a pattern on the extremity of false legs of lepidopterous caterpillars (Fig. 44).

Crop (as applied to insect anatomy): The distended part of the foregut of insects that receives ingested food from the oesophagus; the dilated

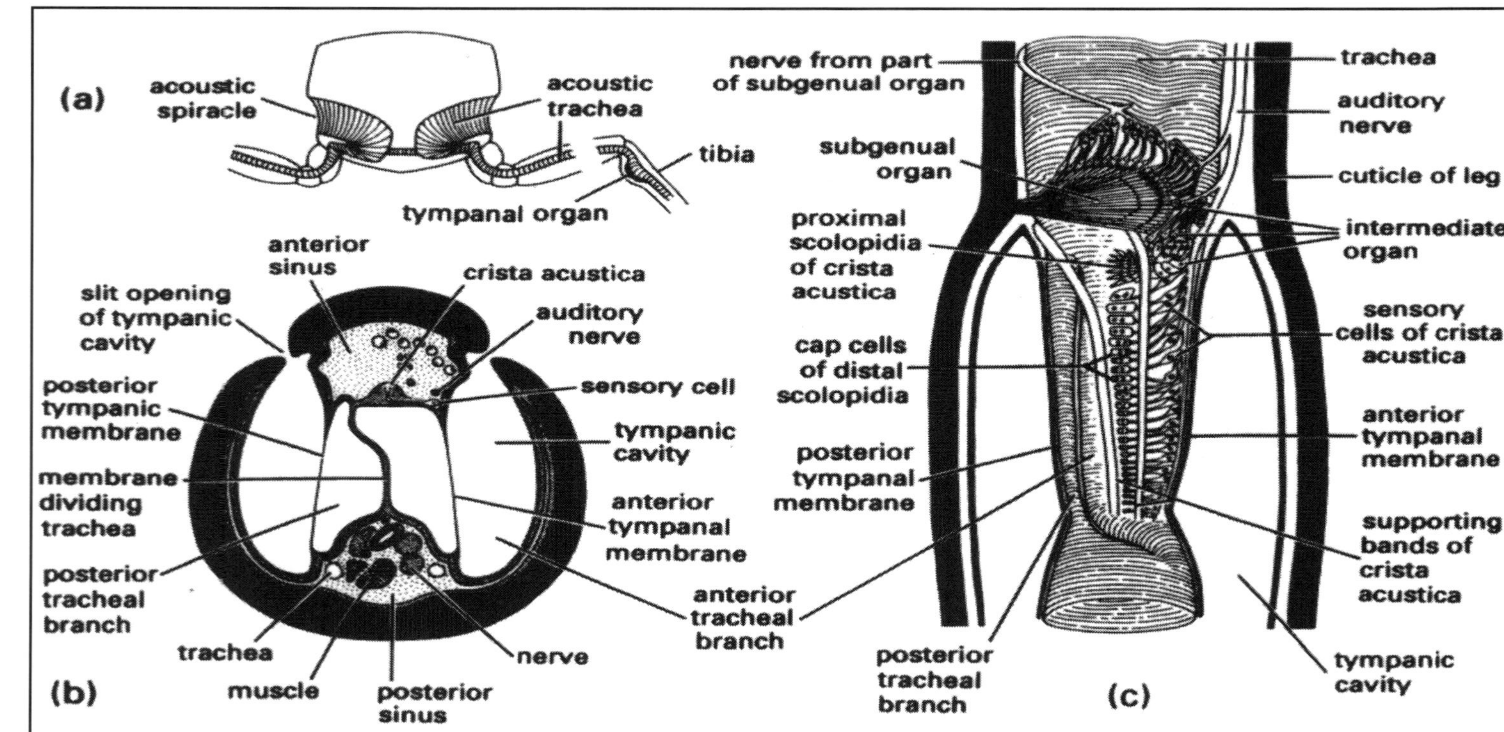

Fig. 77. Tympanal organs if a katydid, *Decticus* (Orthoptera: Tettigoniidae): (a) transverse section through the fore legs and prothorax to show the acoustic spiracles and tracheae; (b) transeverse section through the base of the fore tibia; (c) longitudinal break-away view of the fore tibia. After *Schwabe,* 1906, in Michelsen and *Larsen,* 1985.

portion of the foregut just behind the oesophagus that functions as a site for temporary food storage; an expansible part of the foregut that holds food until it can be passed into the midgut; The food storage area of the digestive system, posterior to the oesophagus (Fig. 41).

Crop hygiene (= Phytosanitation)**:** The removal and destruction of heavily infested or diseased plants from a crop so that they do not form sources of reinfestation.

Crop rotation: The practice of planting botanically unrelated crops on the same land in successive years; the practice of sequentially growing different crops in a field to achieve pest management, soil enhancement, and soil conservation; or alternating the type of crop planted at a location.

Cross fertilization: Union of an egg cell from one individual with a sperm cell from another. Opposite of self-fertilization.

Cross pollination: Transfer of pollen between different plants of the same species.

Cross resistance: Resistance of an insect to one insecticide providing resistance to a different insecticide; the development of resistance to one pesticide that results in resistance to another pesticide that the organism has not been exposed to; the phenomenon of insect resistance to one type of insecticide providing resistance to other insecticides with similar modes of action.

Cross veins: The cross veins are transverse supporting structures in the insect wing that connect the adjacent longitudinal veins.

Crown: (i) The point where the stem and the root meet in a plant. (ii) The upper branches of a tree.

Cruciate: Crossing; shaped like a cross.

Crustacea: A class of arthropods, mostly aquatic in habit, to which crabs and cray fish belong (Fig. 40).

Crustacean: The anthropod class that includes the crayfish, lobsters, pillbugs, crabs, and shrimp (Fig. 40); characterized by having two pairs of antennae, one pair of mandibles, two pairs of maxillae, and gills for respiration.

Crypsis: Camouflage by resemblance to environmental features; combination of form, colour, and pattern that facilitates "hiding" from predators and/or parasites; or close resemblance of an animal to its physical or biotic background (also called protective coloration).

Cryptic species: Sibling species (q.v.).

Cryptic: Hidden, camouflaged, concealed.

Cryptobiosis: The state of a living organism during which there are no signs of life and metabolism virtually ceases; Survival of an organism during a dry period whereby metabolic activity is reversibly ceased and activity stops; some Chironomidae in dried- up pools in Africa.

Cryptonephric system: A condition of the excretory system in which the **Malpighian tubules** form an intricate contact with the rectum, allowing production of dry excreta.

Cryptonephridia: Distal ends of Malpighian tubules encased by rectal tissues (Fig. 78).

Cryptzoic: Fauna concealed under stones, tree bark, in crevies,or; otherwise hidden in darkness.

Crystalline cone: A hard crystalline translucent structure or transparent body just beneath the corneal lens of an ommatidium; the clear conical shaped struicture lying immediately beneath the cornea in each unit (ommatidium) of insect compound eyes.

Crystalline lens: A lens lying beneath the cuticle of the **stemma** of some insects (Fig. 79 (a)).

Ctenidium (pl. ctenidia): Comb-like structures on any part of an insect body; specialized, thickened setae arranged into combs characteristic of many species of fleas; or a row of stout bristles like the teeth of a comb (Psocoptera; Siphonaptera, Fig. 80).

Cubito-anal cross vein: A cross vein between the cubitus and an anal vein e.g. cu-a.

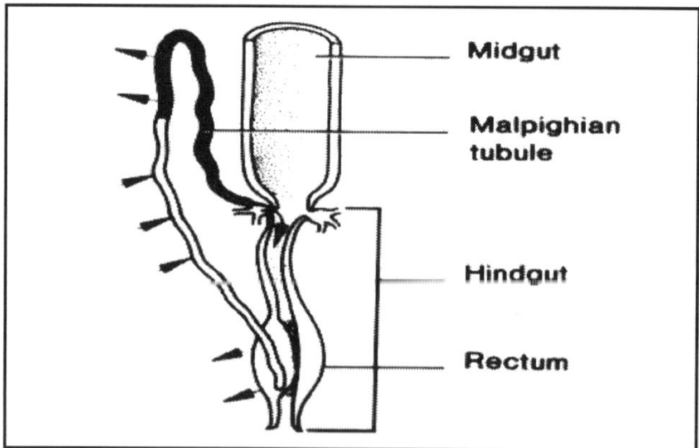

Fig. 78. Lepidopteran type of Malpighian tubule-hindgut systems.

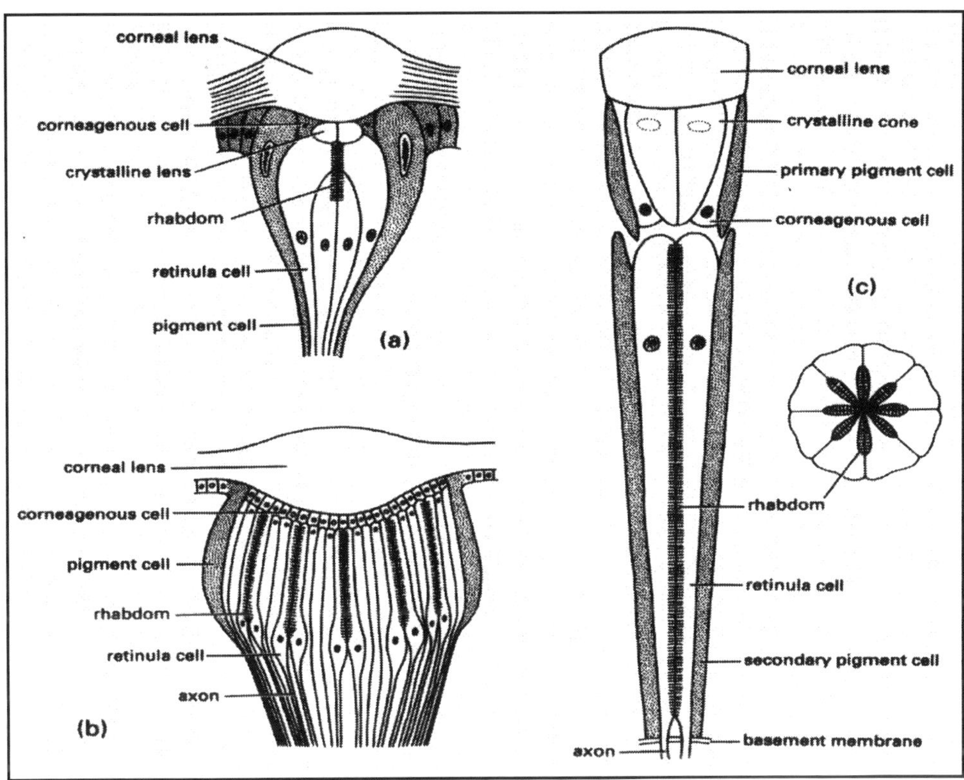

Fig. 79. Longitudinal sections throught the rhabdoms in simple and compound eyes: (a) A simple stemma of a lepidopteran larva; (b) dorsal ocellus of a adult bug; (c) an ommatidium from a compound eye, with enlargement showing a transverse section. (a) After *Snodgrass*, 1935; (b) After *Link*, 1909; (c) After CSIRO, 1970.

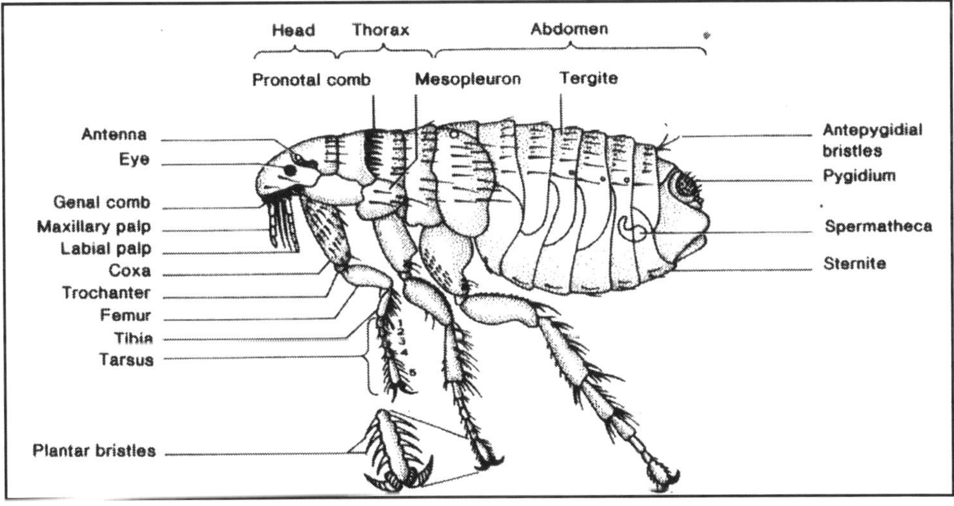

Fig. 80. Siphonaptera, a flea.

Cubito-anal: Region of wing between the cubitus and anal veins.

Cubitus (Cu)**:** The sixth longitudinal vein, lying posterior to the media, often divided into an anterior branched CuA and a posterior unbranched CuP (Fig. 81).

Cultivar (alternative term, variety)**:** Short for cultivated variety; obtained by crossbreeding or selection. See *variety.*

Cultivator: Mechanical device attached to a tractor and dragged across the soil to till and aerate soil and kill weeds.

Cultural control: Manipulation of cultural practices to provide control of a pest; manipulation of cropping system or specific crop production practices to reduce or avoid pest damage; or modification of the environment-for example, by tillage-to make it less attractive to pests.

Cultural management: When factors in the egg cytoplasm prevent fusion of sperm and egg nuclei.

Cultural management phase: Second phase of an IPM program; it involves the application of various cultural control tactics.

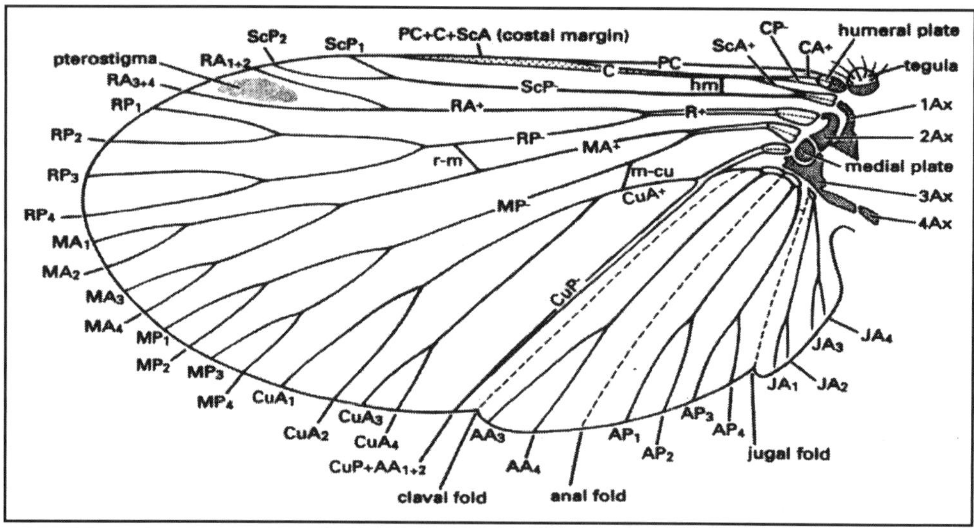

Fig. 81. A generalized wing of neopteran insect (any living winged insect other than Ephemeropytera and Odonata), showing the articulation and the Kukalova-Peck nomenclatural scheme of wing venation. Notation as follows: AA, anal anterior; AP, anal Posterior; Ax, axillary sclerite; C, costa; CA, costa anterior; MA, media anterior; m-cu, cross-vein between medial and cubital areas; MP, media posterior; PC, precosta; R, radius; RA, radius anterior; r-m, cross-vein between radial and median areas; RP, radius posterior; ScA, subcosta anterior; ScP, subcosta posterior. Branches of the anterior and posterior sector of each vein are numbered, e.g. CuA_{1-4}.

Cultural practices: The methods and tactics used in crop production.

Cumulative effect: An effect that increases relative to the summation of a causal stimulus.

Cuneus: Part of the forewing in some Heteroptera; a triangular postero-lateral part of the corium and separated from it by the costal fracture; a more or less triangular apical piece of the corium, set off from the rest of the corium by a suture; the distal section of the corium in the heteropteran wing, separated from the embolium by a costal fracture (Fig. 82).

Cursorial: Legs adapted for running and walking (Fig. 83).

Cutaneous ventilation: Gaseous exchange directly across the cuticle of the exoskeleton.

Cutaneous: Pertaining to the skin. (L. *cutis* = skin).

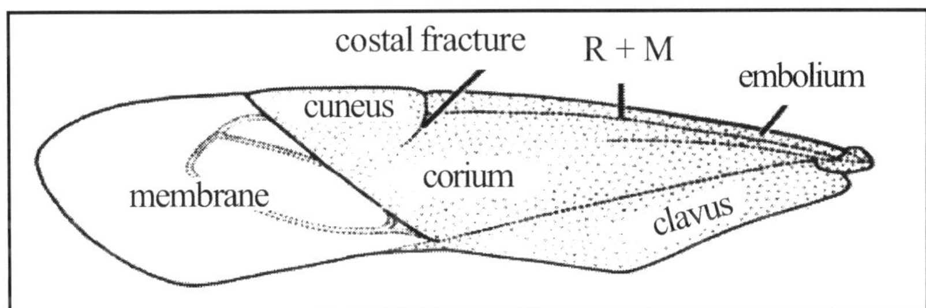

Fig. 82. Fore wing or hemelytron of a mirid bug (Hemiptera: Heteroptera: Miridae) showing three wing areas – the membrane, corium and clavus.

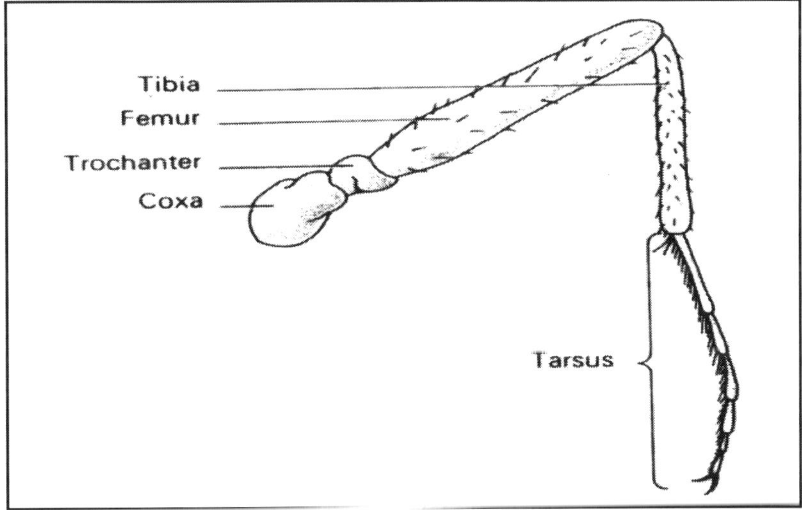

Fig. 83. Cursorial (running) forlegs of a cockroach.

Cuticle (or cuticula): The noncellular outer layer of the body wall of an arthropod; the noncellular outermost layer of the insect integument, secreted by the epidermis (L. *cutis* = skin) (Fig. 84); (i) The waxy cutin-containing covering of aerial plant parts. (ii) The external body covering of insects and nematodes that contains chitin in insects and keratin in nematodes; the external skeletal structure, secreted by the epidermis, composed of **chitin** and protein comprising several differentiated layers (Fig. 84).

Cuticular gills: See spiracular gills.

Cuticular phase: The period of time between the first appearance of new cuticle beneath the old and ecdysis.

Cuticulin layer: A layer of epicuticle, probably lipoprotein that covers the entire integumental surface, including tracheoles and gland ducts.

Cuticulin: The tough, insoluble substance making up the outer surface of the epicuticle, containing cross-linked lipid and protein molecules.

Cutin: A waxy substance in the plant cuticle that protects and prevents moisture loss.

Cutworm: Larva of certain Noctuidae that lives in the soil, emerging at night to eat foliage and stems; serious pests of many crops as seedlings, and root crops.

Cycloalexy: Forming aggregations in defensive circles (Fig. 85).

Cyclodienes: A class of organochlorine insecticides.

Cyclomorphosis: A seasonal (and thus cyclic) nongenetic change of phenotype in species of planktonic fresh-water organisms, particularly cladocerans and rotifers.

Cyclorrhapha: A group of Diptera, regarded as highly evolved; characterized by the puparium inside which the adult develops.

Cyst: The hardened, decay-resistance body of a dead, swollen female nematode that is filled with eggs. The genera *Heterodera* and *Globodera* are cyst nematodes of agricultural importance; or a resistant protective covering formed over a protozoan or other small organisms during unfavourable conditions or reproduction; a small sac or capsule. (Gr. *Kystis* = bladder).

Cysy: A sac, vesicle, or bladderlike structure.

Cyte, cyto-: Cell.

Cytogenetics: The study of cell structures in relation to the phenomena of heredity.

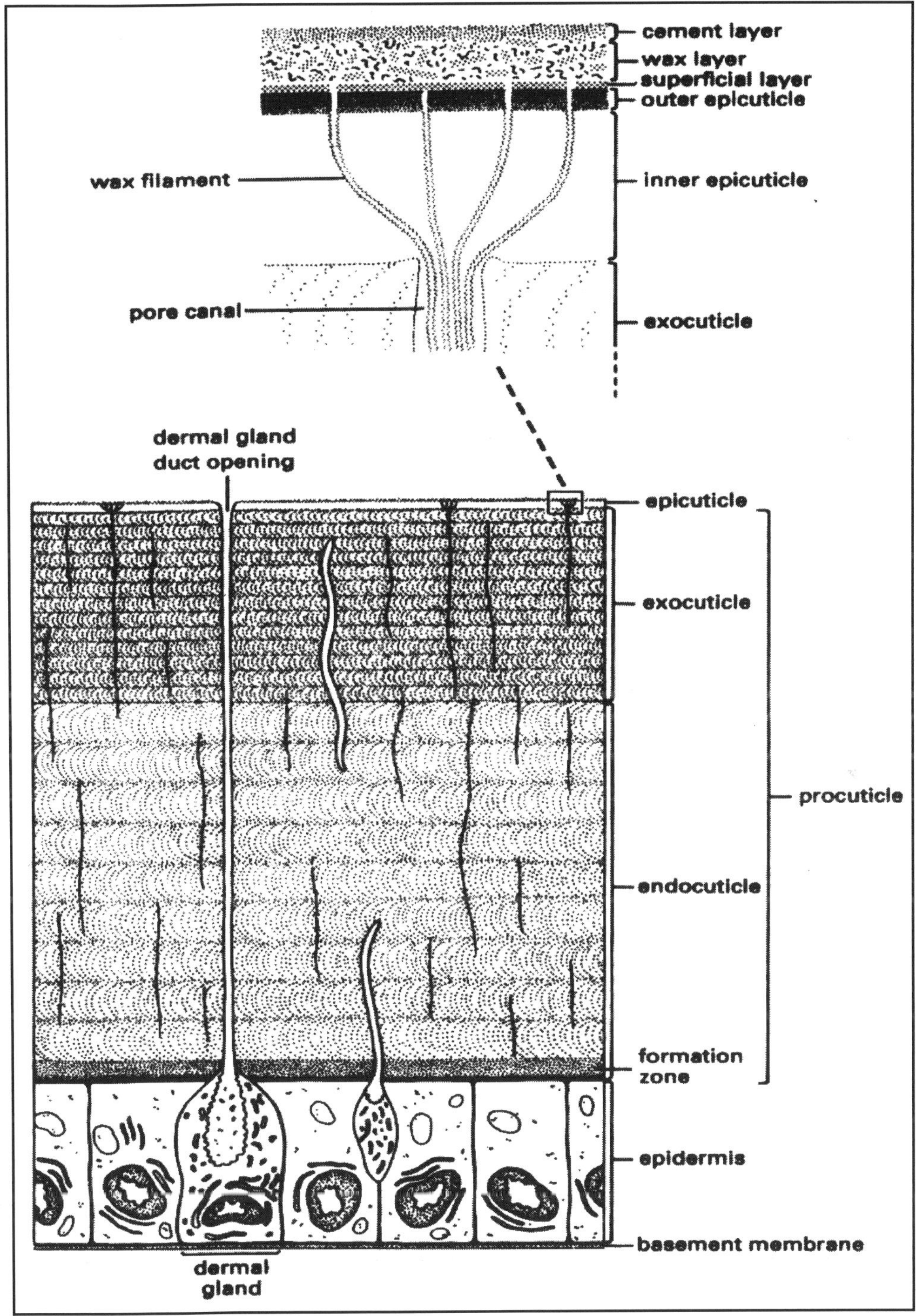

Fig. 84. The general structure of insect cuticleintegument; the enlargement above shows details of the epicuticle. After *Hepburn*, 1982; *Hadley*, 1986; and *Binnington*, 1993.

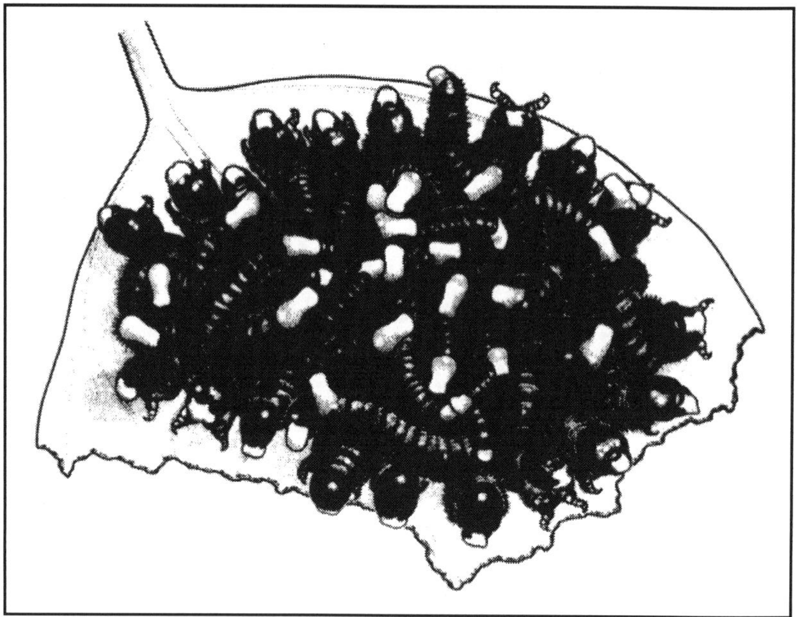

Fig. 85. An aggregation of sawfly larvae (Hymenoptera: Pergidae: *Perga*) on a eucalypt leaf. When disturbed, the larvae bend their abdomens in the air and exude droplets of sequestered eucalypt oil from their mouths.

Cytology: The study of the structure and physiology of the cell and its parts.

Cytoplasm: Part of a cell lying outside the nucleus and within the cell membrane. (Gr. *Kytos* = hollow + *plasma* = form).

Cytoplasmic Polyhedrosis Virus (CPV): A virus that develops in the cytoplasm of host cells, chiefly in the midgut.

Cytoplasmic resistance: Host-plant resistance conferred by mutable substances in cell cytoplasm.

D

Dall code: A code of nomenclature prepared by W.H. Dall at the direction of the American Association for the Advancement of Science (1877).

Damage boundary: The level of injury where damage can be measured.

Damage threshold: Pest population above which crop loss occurs.

Damage: Measurable reduction in crop growth or yield; or a measurable loss of host utility, most often including yield quantity, quality, or aesthetic appeal.

Damping off: Infection of newly formed roots and stems of young seedlings by fungi or bacteria, typically at soil level, resulting in decay or seedling death.

Danger: A signal word used on pesticide labels to inform the user that the pesticide is highly toxic, as defined by FIFRA. The word poison and the skull and crossbones symbol always accompany the signal word danger.

Darwinian fitness: Differential reproduction, in terms of the number of genes an individual passes to the next generation.

Day-degrees: A measure of physiological time, the product of time and temperature above a **threshold.**

Ddensity independent (factor): A factor regulating populations of living organisms, whose influence varies with the population density of the organism concerned.

Ddeposit: The amount of an insecticide on a surface immediately following application.

DDT (dichlorodiphenyltrichloroethane)**:** A synthetic chlorinated hydrocarbon insecticide.

Dead heart: The young caterpillar bores into the stem of the plant either from the top or side, eats it away and makes tunnels inside, as a result of which "dead hearts" appear.

Deciduous: Having a part or parts that may fall off or be shed or detached (e.g. at maturity).

Decline: Gradual reduction over time in the health and vigor of a plant, eventually ending in death.

Decomposer: Organisms which obtain their nutrition by breaking down dead organic matter into simple molecules.

Decticous pupa: An **exarate pupa** in which the mandibles are articulated.

Decticous: Having functional mandibles used for escape from a cocoon or pupal case.

Decumbent: Bent downward.

DEET (N, N-diethyl-m-toluamide): Commonly used biting fly and mosquito repellent.

Defecate: To discharge food residues (faeces) through the anus. (L. *de* = from + *faex* = dung).

Definition: In taxonomy, a formal statement of characters which sets limits to a taxonomic category (cf. Description, Diagnosis, Differential diagnosis, Taxonomic category).

Deflexed: Bent downward.

Deflocculating agent: Material added to a spray suspension to delay sedimentation.

Defoliant: Spray which induces premature leaf-fall.

Defoliation: (i) Reduction in the amount of foloage, possibly due to insects, fungi or other agents as distinguished from natural leaf fall. (ii) Removal of leaves by application of a defoliant.

Degradation: The process of decomposition to simpler component parts. The breakdown of pesticide residue.

Degrade: Decompose; break down into less complex constituent parts.

Degree-day or **day degree (^{0}D):** An accumulation of heat units above some threshold temperature for a 24-hour period; or a measure used to express time relative to the influence of temperature on the growth rate of cold-blooded organisms. Degree-days are based on cumulative time above and below developmental threshold temperatures; or Number of days above a particular threshold temperature; used in prediction of insect development.

Dehisce: The splitting of plant tissues along lines of weakness, such as the opening of a seed pod.

Delayed effect: Of a defensive chemical, an effect that appears after a time lapse from first encounter; *see also* **Immediate effect.**

Delayed parasitism: Parasitism in which hatching of the parasite (or parasitoid) egg is delayed until the host is mature.

Delayed voltinism: Life cycle requiring more than one year for completion.

Delusory parasitosis: A psychological illness in which the inflicted persons imagine themselves to be infected with living parasites; a psychotic illness in which parasitic infection is imagined.

Deme: A population within a species; a subpopulation of interbreeding individuals.

Dendrite: A fine branch of a nerve cell; the process on a nerve cell that conducts impulses to the cell body; often branched. (Gr. *Dendrom* = tree); tiny branching processes associated with or near the cell body of a neuron (Fig. 86 a, b, and c).

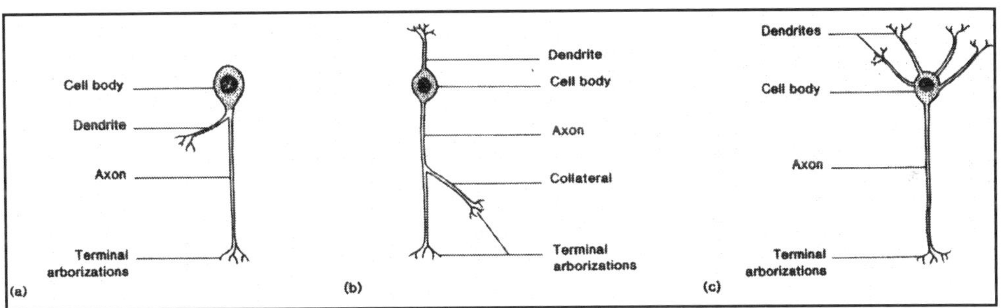

Fig. 86. Structural types of neurons. (a) Unipolar. (b) Bipolar. (c) Multipolar.

Dendrogram: A diagrammatic drawing in the form of a tree designed to indicate degrees of relationship as suggested by degrees of similarity (cf. Phylogenetic tree) See phenogram.

Dengue: A virus disease of man marked by severe pains in head, eyes, muscles and joints and transmitted by certain mosquitoes.

Density: The number of individuals per unit of measure (e.g., larvae per square meter).

Density-dependent factor: A population-regulating factor that changes in intensity with changes in population density (e.g., intraspecific competition); or a factor that causes a level of mortality that varies with the number of individuals in the population.

Density-independent factor: A population-regulating factor that causes a level of mortality and is unrelated to a population's density (e.g., weather factors); An environmental factor that causes a level of mortality independent of the number of individuals in the population.

Dentate: Toothed, or with toothlike projections.

Denticulate: With minute toothlike projections.

Deodorant: A material added to a pesticide formulation for masking unpleasant odors (e.g., cedar oil, pine oil).

Depolarization: The reduction of the electrical potential of a neuron's membrane.

Deposit: (dried) Amount and pattern of active ingredient deposited per unit area of plant surface.

Deposit: (spray) Amount and pattern of spray or dust deposited per unit area of plant surface; or the pesticide remaining on the target immediately following application.

Deposition velocity: Velocity at which the spray impinges on the target.

Depressed: Flattened dorso-ventrally, (L. *de* = down + *permo* = press).

Derm-: Skin; covering.

Derma: Pertaining to the skin, especially the inner connective tissue layers of vertebrate skin. (Gr. *Derma* = skin).

Dermal: Through, by, of, or about the skin. Dermal pesticide absorption is a measure of potential hazard to workers.

Dermal gland (s): A unicellular epidermal gland that may secrete cements, wax, etc. and probably pheromones; modified epidermal cells on the outer surface of the insect body that secrete irritants, poisons, wax, scents, silk, and so on (Fig. 84).

Dermaltoxicity: The toxicity of an insecticide (to higher animals) when applied to the intact skin.

Dermaptera, order: (*derm,* skin; *ptera,* wings) the order of insects comprising the earwigs (Fig. 87).

Dermis: The deeper of "true" portion of the skin beneath the epidermis in a vertebrate, derived from mesoderm. (Gr. *Derma* = skin).

Description: In taxonomy, a more or less complete formal statement of the characters of a taxonomic category without special emphasis on those which set limits to the category or distinguish it from coordinate taxonomic units (cf. Definition, Diagnosis, Differential diagnosis Taxonomic category).

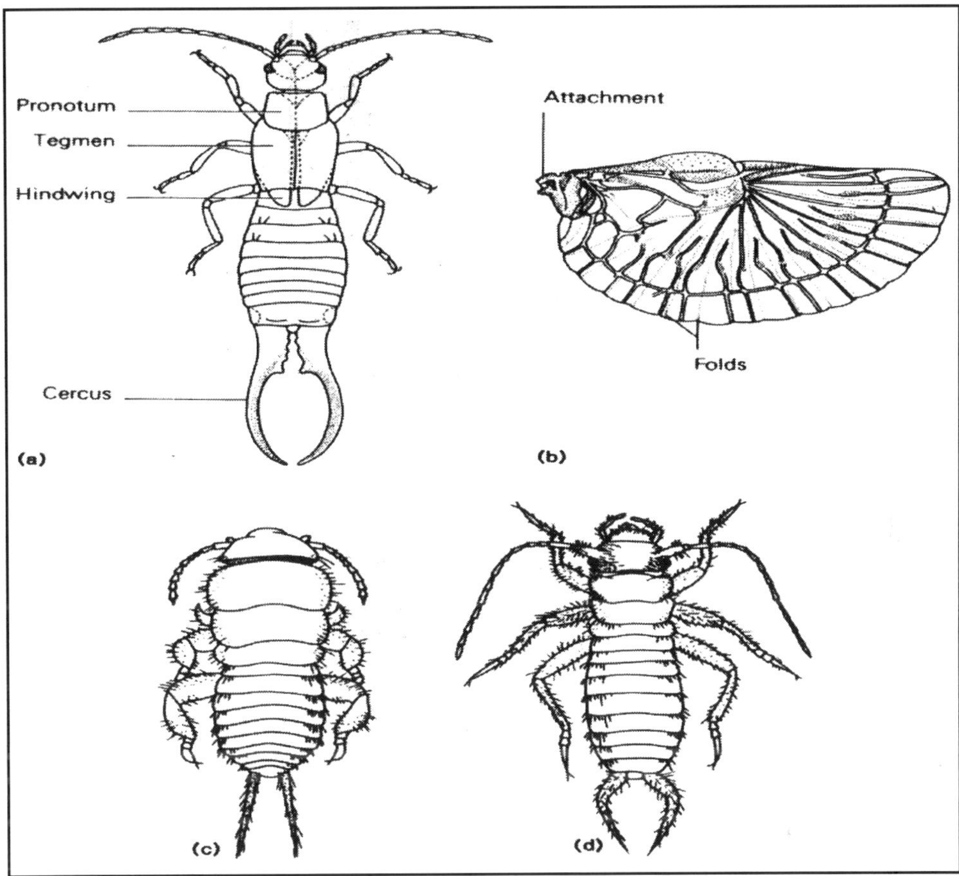

Fig. 87. Dermaptera. (a) Suborder Forficulina; European earwing, *Forficula auricularia*, male (Me.) (b) Same, right hindwing unfolded. (c) Suborder Hemimerina; *Hemimerus talpoides*. (d) Suborder Arixeniina; *Arisenia jacobsoni*.

Desiccant: Chemicals that promote dehydration are called desiccants. Chemical which kills vegetation by including excessive water loss.

Detection (detection survey): The procedures and protocols used to discover the presence of pests to monitor outbreaks or introductions.

Determinate growth: With a distinctive final, adult instar; *see also* **Indeterminate growth.**

Deterrent: A chemical that prevents feeding or oviposition by insects. cf. stimulant.

Detritivore (adj. detritivorous): Animal feeding upon organic detritus or of animal or plant origin; saprophage.

Detritus food chain: The passage of nutrients and energy through an unstructured assemblage of decomposer organisms.

Deuterotoky: see **Amphitoky.**

Deutocerebrum: The middle part of the insect brain; the ganglion of the second segment, comprising antennal and olfactory lobes; the lobe of the insect brain that receives both sensory and motor fibers from the antennae; or the medial pair of lobes of the brain that innervate the antennae (Fig. 88).

Deutonymph: The third instar stage in the development of mites and ticks. cf. protonymph.

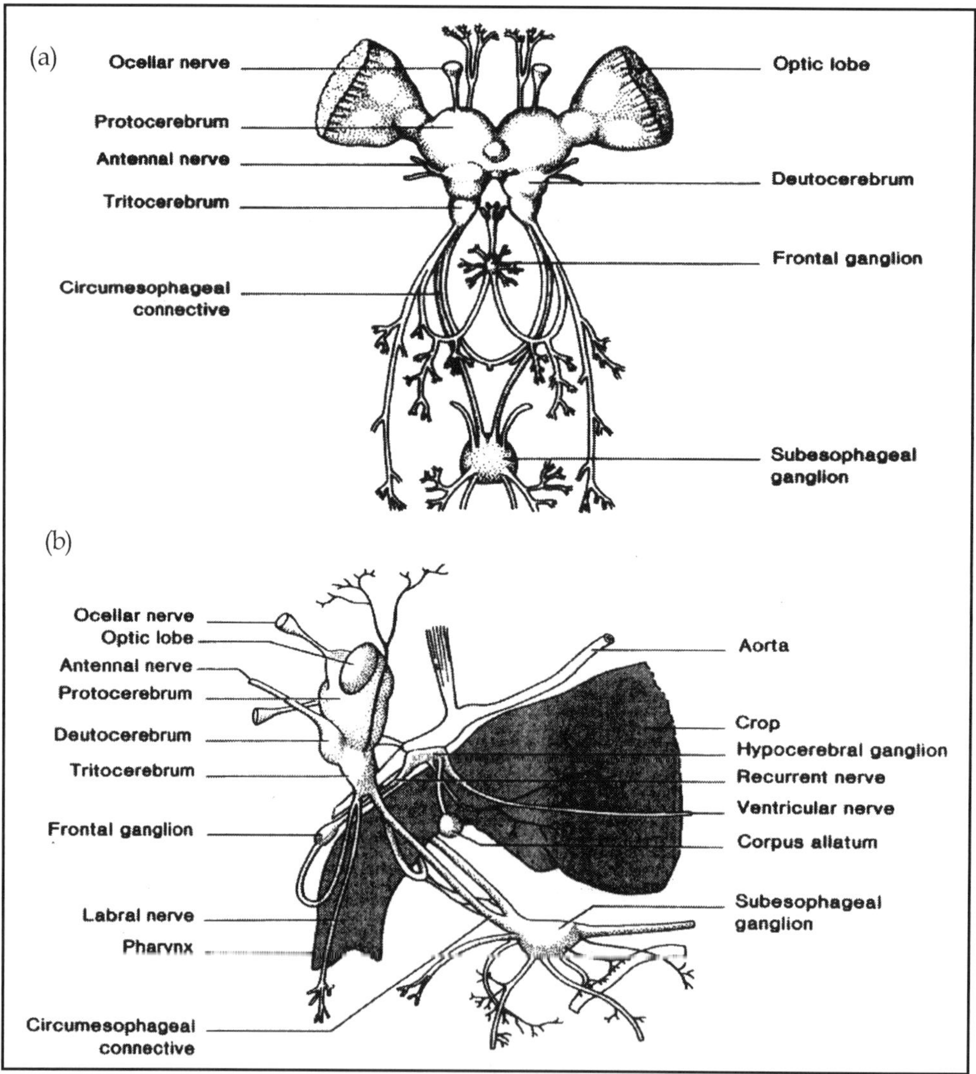

Fig. 88. Brain and stomatogastric nervous system of the grasshopper *Dissosteria Carolina;* (a) Anterior view. (b) Lateral view. After *Snodgrass,* 1935.

Deutoplasm: Substance other than the nucleus and cytoplasm in a cell, especially yolk in an egg cell; (Gr. *Deutoros* = second + *plasma* = form).

Developmental (or growth) threshold: The temperature below which no development takes place.

Developmental threshold (lower or upper): The coldest temperature below which, or the highest temperature above which, growth, development, or activity of a cold-blooded organism stops; the minimum temperature required for development to proceed.

Dew period: Length of time dew forms on plants; important in defining opportunities for infection by plant pathogens that require free water to infect.

di-: Two.

Diagnosis: In taxonomy, a formal statement of the characters (or most important characters) which distinguish a taxonomic category from other similar or closely related coordinate categories (cf. Differential diagnosis, Definition, Description, Taxonomic category).

Diagnosis: The recognition of the presence and extent of a pest infestation and determination of the causal agent responsible.

Diapause hormone: A hormone produced by neurosecretory cells in the suboesophageal ganglion that affects the timing of future development of eggs.

Diapauses-(Dimilin)**:** A chitin synthesis inhihitor; relatively non-toxic to mammals; especially effective against foliar feeding lepidopterous insects.

Diapause. Spontaneous state of dormancy during development in many insects; usually physiologically induced and broken to avoid a long period of inclement weather, usually winter; delayed development controlled by particular environmental conditions; It is mediated by hormones and can occur at any growth stage in the life cycle; or a period of arrested development; or a physiological state of arrested metabolism, growth, and development that occurs at one stage in the life cycle; a state of suspended activity and metabolism which may occur at any stage in the life cycle of an insect as a means of surviving unfavourable conditions (especially reduced daylength) and requires a particular stimulus for its termination.

Diaphragm: A thin membrance that partitions areas of the body cavity; a dividing membrane, as the diaphragm of the ear; the muscular partition between the thoracic and abdominal cavities in mammals. (Gr. *dia* = through + *phragma* = fence).

Diastasis: A period of rest between successive heartbeats.

Diastole: Relaxation phase of the heartbeat cycle (Fig. 89).

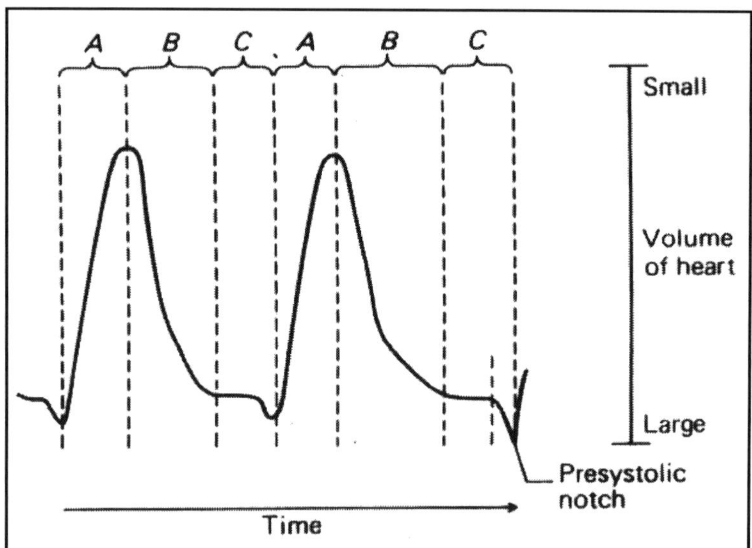

Fig. 89. Mechanical recording of two heartbeats (cardiac cycles) in an isolated cockroach, *Periplaneta Americana,* heart. (a) Systolic phase; (b) diastolic phase; (c) relaxation phase with presystolic notch.

Diatom: Tiny, unicellular, freshwater green alga.

Dichoptic: The eyes separated above (Diptera).

Dichotomous: Double-branching, as in a taxonomic key Sample dichotomous key that differentiates the insects from centipedes and millipedes is given below:

Dicondylar: An articulation (as of a mandible) with two points of articulation (condyles).

Dicondylic joint: A joint consisting of two condyles and restricting movement to a single plane.

Dictyoptera; The order of insects comprising the cockroaches (Fig. 37) and mantids.

Dieback: Progressive death of twigs, branches, or stems from the tip back toward the main stem of the plant.

Diecious: Having the male and female organs in different individuals, any one individual being either male or female.

Differentiae specificae: The descriptive method developed by Linnaeus whereby a series of descriptive words was used to distinguish each species from all others.

Differential diagnosis: A formal statement of the characters which distinguish a given taxonomic unit from other specifically mentioned equivalent units (cf. Diagnosis, Definition, Description, Taxonomic category).

Differentiation center: A location in the insect egg that induces differentiation of embryonic cells.

Diffraction: The splitting of white light into its component spectral colors by a series of fine grooves or ridges.

Diffuse coevolution: see **Guild coevolution.**

Diflubenzuron: A chitin synthesis inhibitor for insect control.

Digestion reducers: See carbon-based defenses.

Digestion: The process of preparing food for absorption and assimilation.

Digestive juice: Secretion of digestive glands which act on different categories of food and thus help in digestion *e.g.* pancreatic juice secreted by the pancreas.

Digit: A finger or toe; one of the terminal divisions of a limb in tetrapods. (L. *Digitus* = finger).

Digitigrade: Walking on the toes. (L. *digitus* = finger + *gradior* = to step).

Dilated: Expanded or widened.

Diluent: Any of several accessory carriers, solvents, emulsifiers, and wetting agents, that are used to formulate pesticides that dilute the active ingredient; or component of spray or dust that reduces the concentration of the active ingredient, and may aid in mechanical application but does not directly affect toxicity; or a substance used as a carrier for concentrated pesticide; can be liquid or solid (e.g., refined oil, organic flour).

DIMBOA (2, 4-dihydroxy-7-methoxyl-1, 4-benzoxazine-3-one)**:** A compound in corn that confers resistance against first-generation European corn borer.

Dimorphism: Having two different life forms in the same species; the most common example being sexual dimorphism wherein males and females differ in appearance; differing sexes, seasonal forms, or geographical forms (Gr. *di* = two + *morphe* = form). (cf. sexually dimorphic, Polymorphism).

Dinitrophenol: A synthetic organic insecticide characterized by nitro groups (NO_2) attached to a phenol (C_6H_5OH) ring (e.g., dinoseb).

Dioecious: Plant species in which the male and female flowers are on separate plants; or having male and female elements on separate individuals of the same species; or male and female organs in separate individuals. (Gr. *di* = two + *oikos* = house).

Dioptric apparatus: The parts of an ommatidium (cornea, crystalline cone, and pigment cells) responsible for the collection of light (Fig. 90).

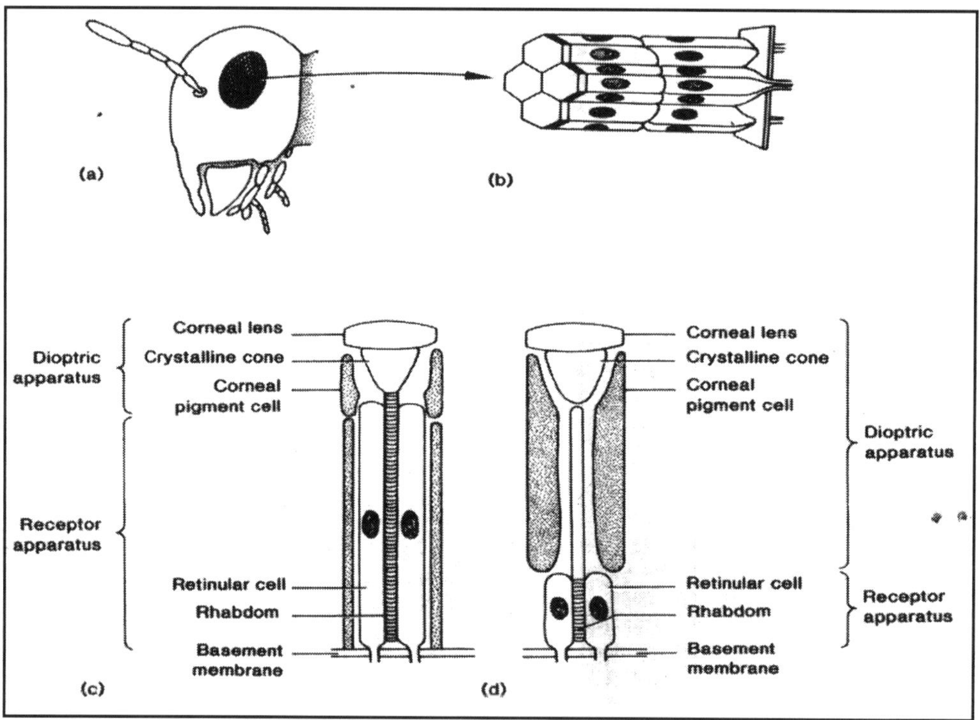

Fig. 90. Compound eye structure. (a) Head with compound eye. (b) Four ommatidia removed and magnified. (c) Apposition ommatidium. (d) Superposition ommatidium.

Diploblastic: Animals whose body is made up of only two cellular layers (sponges and coelenterates). Rest of the animals above the coelenterates in the evolutionary scale are triploblastic.

Diploid: Double; possessing the full complement of maternal and paternal chromosomes; or the dual or somatic number of chromosomes; the normal number in all but the matured germ cells of any particular organism. (Gr. *diplous* = double + *eidos* = form); having a double set of chromosomes (2_n), the normal chromosome number of the cells (except for mature germ cells) of a particular organism derived from a fertilized egg (cf. Haploid, Polyploidy, Chromosome).

Diplopoda: The order of arthropods within the class Myriapoda that comprises the millipedes; characterized by cylindrical bodies and two pairs of appendages on each abdominal segment.

Diplura, order: (*dipl,* two; *ura,* tail) One of the apterygote orders of insects that comprises the two pronged bristle tails i. e. campodeids and japygids (Fig. 91).

Diploid: Having the chromosomes present in homologous pairs, i.e., with a chromosome complement of 2n. cf. aploid.

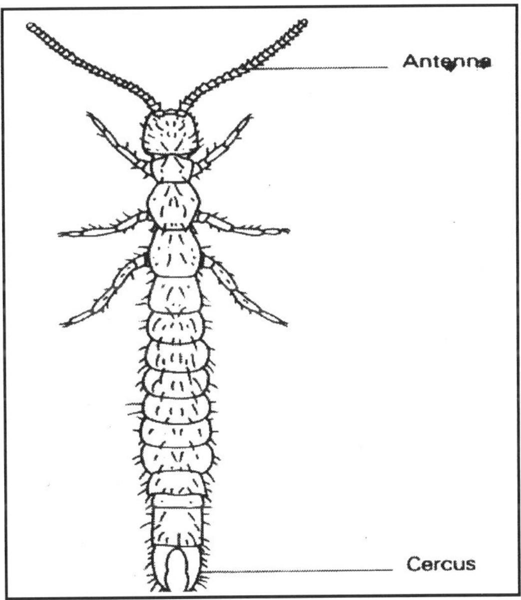

Fig. 91. Diplura, *Japyx diversiungus.* (Mi.).

Diptera, order: (*di,* two; *ptera,* wings)The order of insects that comprises the true flies and mosquitoes which are characterized by possession of only one pair of wings (Fig. 92).

Dipterous: Two-winged; having characters of the Diptera.

Direct flight muscles: Muscles that power flight that are attached directly to the wing (Fig. 93); *see also* **Indirect flight muscles.**

Direct loss (direct injury): Devaluation of a marketable commodity owing to the presence of insects or insect damage; injury to yield-forming organs.

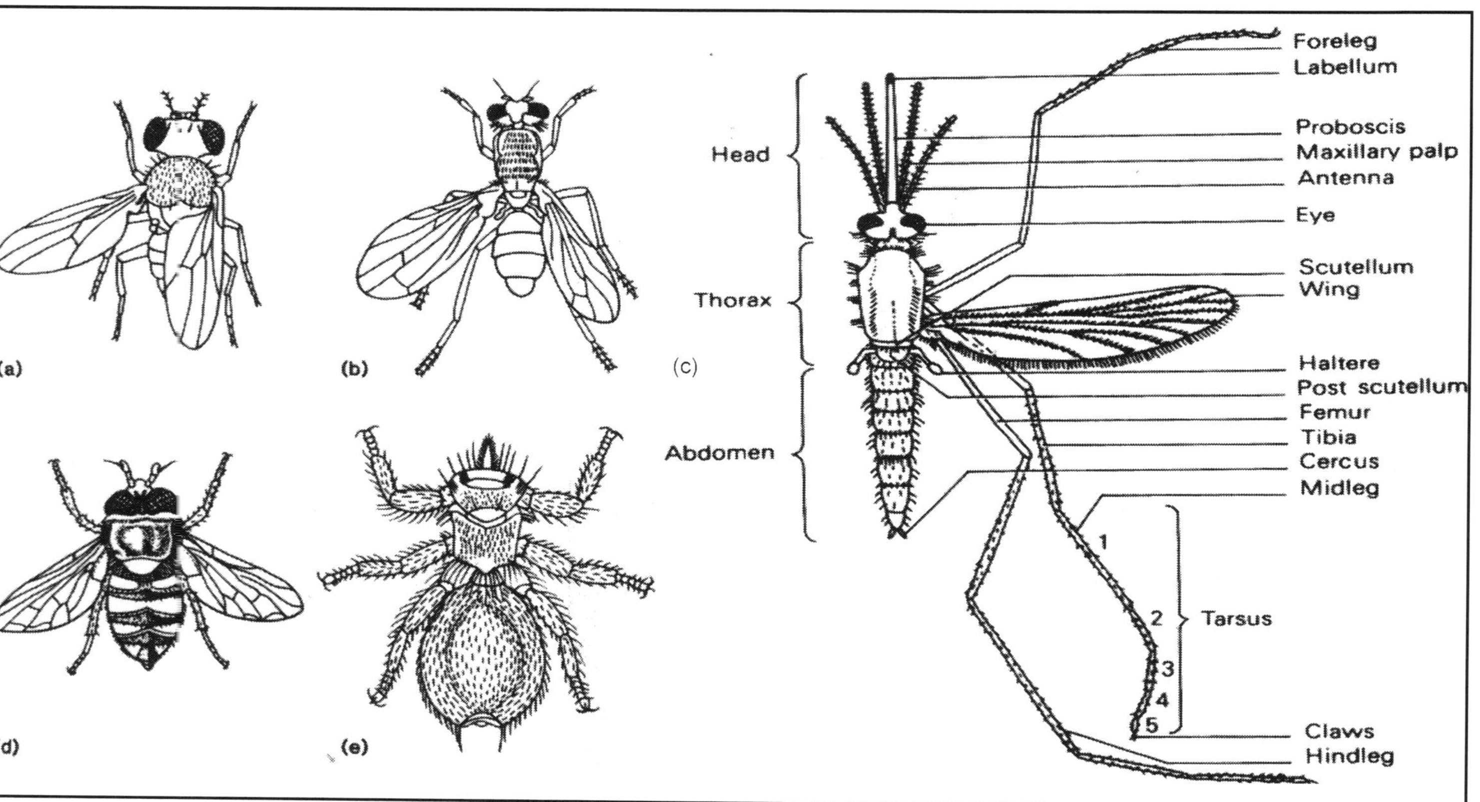

Fig. 92. Diptera. (a) Fruit fly (Drosophilidae). (b) Picture-winged fly (Otitidae). (c) Mosquito, *Anopheles maculipennis* (Culicidae) (female). (d) Hover fly or flower fly (Syrphidae). (e) Sheep ked, *Melophagus ovinus* (Hippoboscidae).

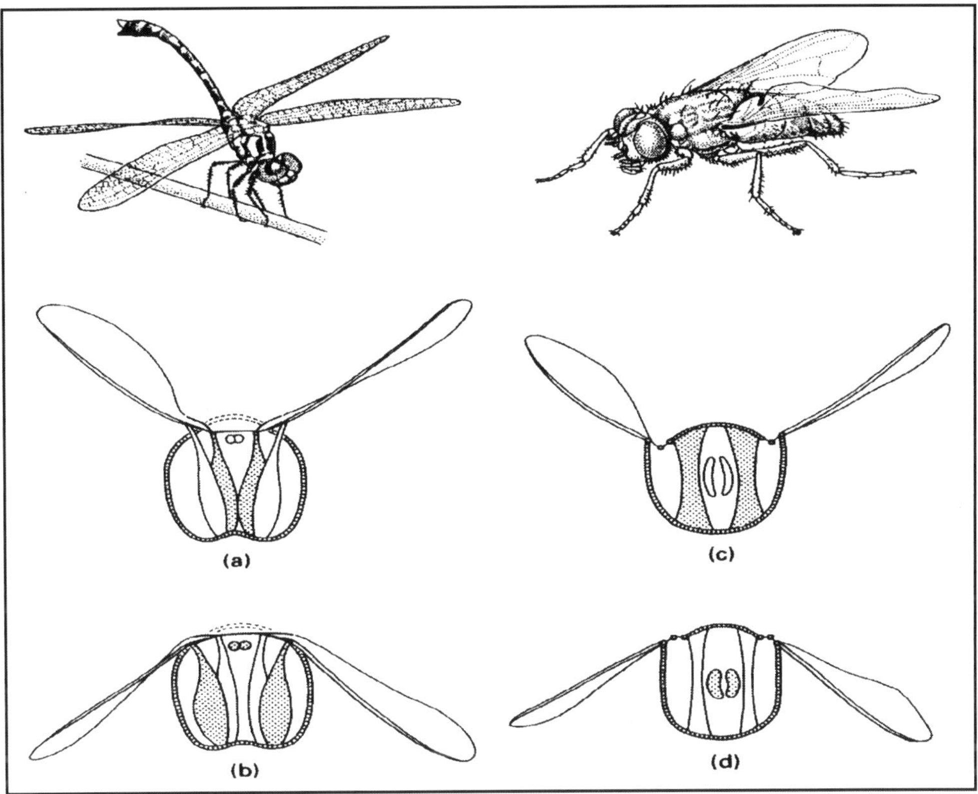

Fig. 93. Direct ((a), (b)) and indirect ((c), (d)) flight mechanisms. Dragonfly thorax during (a) upstroke and (b) downstroke of the wings. House-fly thorax during (c) upstroke and (d) downstroke of wings. Stippled muscles are those contracting in each illustration.

Direct pest: A pest insect that attacks a part of a plant that is harvested, as contrasted to an indirect pest.

Direct wing muscles: Muscles that attach to the wing bases (Fig. 94).

Directed spray: Targeting a specific area or part of a plant for pesticide application in order to reduce the amount of pesticide applied to the crop.

Disc: Soil tillage tool made from multiple groups of concave, circular metal disks. Often used in heavy clay soils because it does not form plough pans.

Discal cell: A more or less enlarged cell in the basal or central part of the wing (Homoptera, the R cell).

Discal cross vein: A cross vein behind the discal cell (Diptera).

Discoidal cell: The cell near the middle of the wing (Hymenoptera).

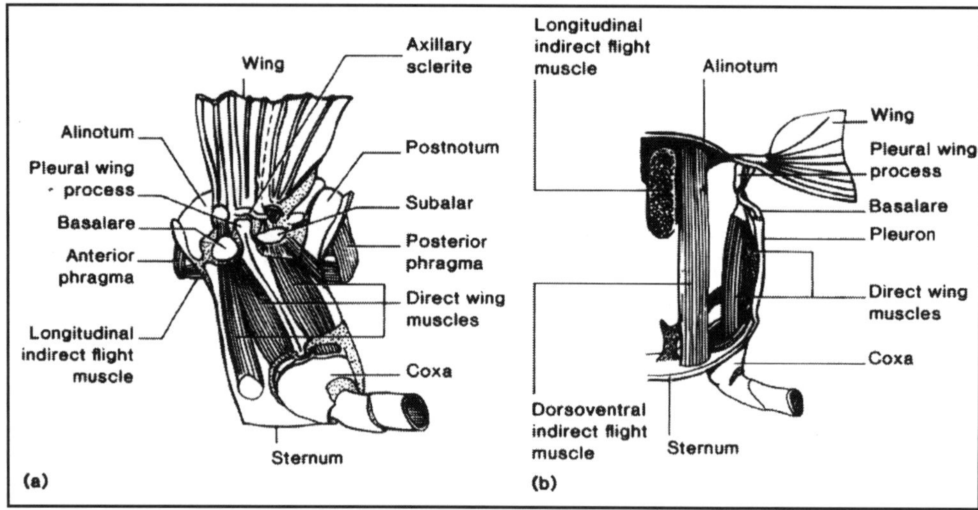

Fig. 94. Generalized wing-bearing segment. (a) Lateral view. (b) Cross-sectional view.

Discoidal vein: The vein forming a continuation of the median vein beyond the end of the transverse median vein, and extending along the posterior margin of the first discoidal cell (Hymenoptera).

Discontinuous variation: Variation in which the individuals of a sample fall into definite classes which do not grade into each other, as variation in qualitative characters (cf. Continous variation).

Disease: Abnormal physiology and metabolism in an organism. Disease may be caused by pathogens, abiotic conditions, or genetic disorders.

Disease-resistant: Plant varieties bred for the characteristic that they do not support the development and reproduction of particular parasites.

Disinfect: 1. To free from infection by destruction of the pest or pathogen established in or on plants or plant parts. 2. To kill or inactivate pests or pathogens present upon the surface of plants or plant parts, or in the immediate vicinity (e.g. in soil).

Dispersal: Movement of an individual or population away from its birth site; movement of individuals out of a population (emigration) or into a population (immigration); or the process of scattering in various directions.

Dispersion: The spatial arrangement of individuals.

Disruptive coloration: A color pattern that breaks up the outline of the body such that it does not stand out against a variable background.

Disruptive: Treatment that interferes with the ability of beneficial species to control pest populations.

Dissemination: The spatial spread of organisms resulting from organism activity, or passive movement by wind, rain, insects, and people.

Distad: Away from the body, toward the end farthest from the body.

Distal: Distant or farthest from the point of attachment or origin; or near or toward the free end of an appendage; that part of a segment or appendage farthest from the body; or away from the point of attachment or place of reference (L. *disto* = stand apart).

Diurnal: Active during the daytime; or recurring over a 24-hour cycle (L. *dien* = day); *see also* **Crepuscular; Nocturnal.**

Divaricate: Extending outward and then curving inward or toward each other distally (divaricate tarsal claws, Fig. A).

Divergent: Becoming more separated distally; going further a part; separating from a common source. (L. *di* = doubly + *vergo* = incline).

Diverticula: Blind sacs that form as evaginations from the alimentary canal, for example, the esophageal diverticula of mosquitoes (Fig. 95).

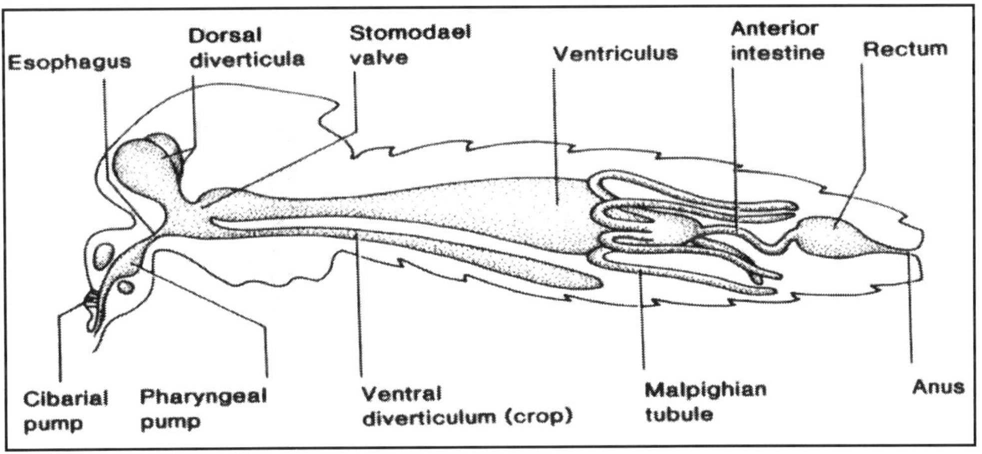

Fig. 95. Alimentary canal of a mosquito.

Division: See Section.

Dockage: Value depreciation of a product owing to the presence of insects or other foreign matter.

Dollo's rule: That structures or functions once gained may be lost, but once lost they can never be regained.

Domatia: Plant chambers produced specifically to house certain arthropods, especially ants.

Dominant character: A character from one parent that manifests itself in offspring to the exclusion of a contrasted (recessive) character from the other parent. (L. *dominor* = rule).

Dominant: In ecology, genetics, and psychology, superior in frequency or rank; a gene that is expressed in the phenotype in like manner regardless of whether the individual is homozygous or heterozygous for this gene (cf. Recessive, Homozygous, Heterozygous).

Dormancy: A seasonally recurring period in the life cycle when growth, development, and reproduction are suppressed; a state of quiescence or inactivity; suppression of activity associated directly with changes in the abiotic environment.

Dormant oil: A Petroleum oil applied to trees only when foliage is not present.

Dormant: Alive but not growing; buds with an unbroken cover of scales; quiescent; inactive; a resting stage; a state in which spores, seeds, or plant organs become inactive and do not germinate or grow; a state of reduced metabolic activity and movement in some insects, mites, and nematodes.

Dorosoventral: From top to bottom, or from the upper to the lower surface.

Dorsal: Toward the back or top.

Dorsal closure: The embryological process in which the dorsal wall of an embryo is formed by growth of the germ band to surround the yolk.

Dorsal diaphragm: Fibromuscular septum that encloses the alary muscles associated with the heart and divides the hemocoel into the pericardial and perivisceral sinuses (Fig. 96); also called *pericardial septum;* a muscular sheet under-laying the heart which assists in the flow of blood.

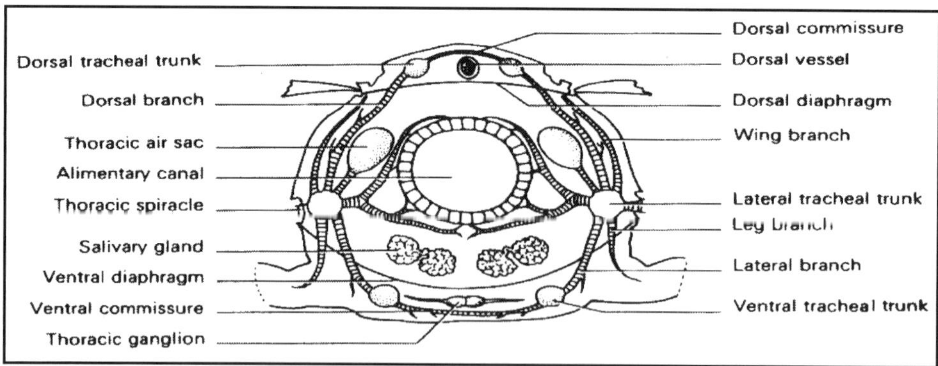

Fig. 96. Cross section of insect thorax showing the major tracheal branches.

Dorsal light reaction: A process in which longitudinal and transverse axes of the body are kept perpendicular to a directed source of light.

Dorsal longitudinal muscles: Muscles running longitudinally, dorsally in insect segments, in the thorax powering the down stroke of the wings of most insects.

Dorsal longitudinal trunks: Major tracheae located on either side of the dorsl midline of the body.

Dorsal ocelli: Simple eyes usually accompanying compound eyes and found dorsally on the insect head.

Dorsal vessel: Primary organ of hemocoel circulation; extends from the abdomen to the head and consists of a posterior heart and anterior aorta; or principle conducting tube of the circulatory system, extending the length of the dorsal midline; the 'aorta' and 'heart', the main pump for haemolymph; a longitudinal tube lying in the dorsal pericardial sinus (Fig. 97).

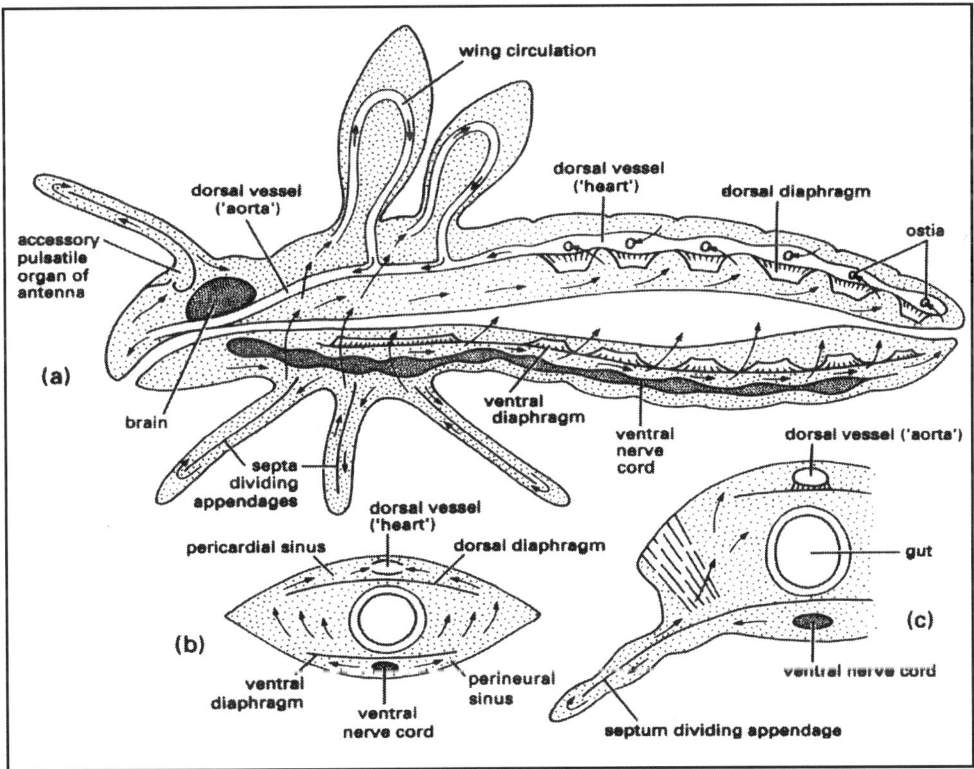

Fig. 97. Schematic diagram of well-developed circulatory system: (a) longitudinal section through body; (b) transverse section of the abdomen; (c) transverse section of the thorax. Arrows indicate directions of haemolymph flow. After *Wigglesworth,* 1972.

Dorsal: Pertaining to the top surface or uppermost back or upper side of the body of an animal. (L. *dorsum* = back). cf. ventral.

Dorsocentral bristles: A pair of bristles on the dorsal portion of the scutellum, one on each side of the midline (Diptera).

Dorsoventral muscles: Muscles inserting on the dorsum of the thorax and originating ventrally, powering the upstroke of the wings of most insects.

Dorsoventral: From top to bottom or the upper to lower surface; dorsoventral muscle.

Dorsum: The back or top (dorsal) side; the upper surface (back) of an animal body.

Dose; dosage: The quantity of a pesticide applied to a target. Also, a constant measure used to define pesticide toxicity; or Quantity of pesticide applied per individual, or per unit area, or per unit volume, or per unit weight.

Double authority citation: See Authority citation.

Douville code: A code of nomenclature prepared by H. Douville (1881) for the International Congress of Geology and designed to set the procedures for the naming of fossils.

Dredging: The process of taking up mud or zoological specimens like sponges and oysters from the bottom of the sea by a machine called dredger.

Drift control: Modifications of application techniques designed to reduce drift. May include additives to spray mixtures or equipment modifications to reduce fine droplets (those smaller than 100 um).

Drift: See Genetic drift.

Drift: The movement of pesticide to non-target locations by wind or atmospheric circulation of fine droplets or dust; or Spray or dust carried by natural air currents beyond the target area; Passive movement due to water or air currents.

Drone: A male bee.

Drone: The male bee, especially of honey bees, bumble bees wasps derived from an unfertilized egg.

Drop spectrum: Distribution, by number or volume of drops, of spray into different droplet sizes.

Droplet: Spherical body of fluid that is produced by a spray nozzle or spinner.

Drop-nozzle application: Lowering nozzles off the spray boom so that they reach into the crop canopy to permit directed spray or basal application.

Duckfeet: A type of cultivation implement called a sweep used to cultivate fields. Sweeps cut weeds below the soil level to provide mechanical control

Duct: A tube by which a liquid or other product of metabolism is conveyed, as of a secretion from a gland; usually opening on a surface or in a larger compartment. (L. *duco* = lead).

Ductule cell: A duct forming cell associated with a simple exocrine gland (Fig. 98).

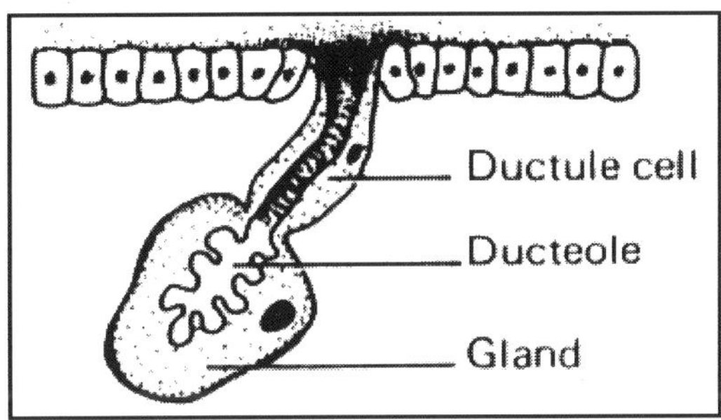

Fig. 98. Exocrine gland of Unicellular gland with ductile cell.

Dufour's gland: In aculeate hymenopterans, a sac opening into the poison duct near the sting (Fig. 99), site of production of pheromones and / or poison components; an exocrine gland on the ventral, posterior part of the abdomen of female Hymenoptera, the source of pheromones serving diverse functions.

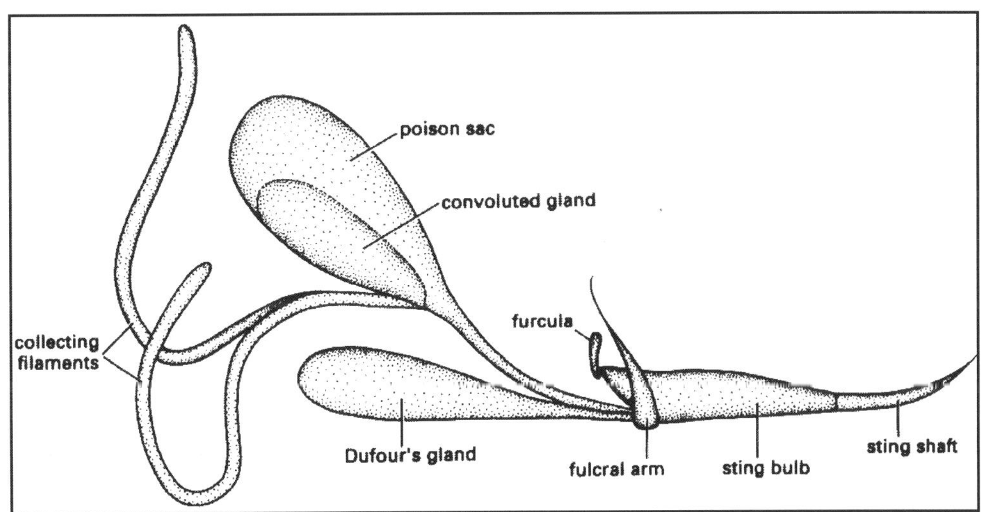

Fig. 99. Diagram of the major components of the venom apparatus of a social aculeate wasp.

Dulosis: A slave-like relationship between parasitic ant species and the captures brood from another species.

Dun: (Alternative term, sub-imago)- Winged but sexually immature mayfly adult which has to undergo another moult.

Dust (D): A formulation of pesticide that contains an active ingredient combined with dry and finely powdered carrier material such as talc, nut hulls, or clay. The possibility of drift is greatest with dust formulation; or a dry pesticide formulation prepared by milling the pesticidal compound into a fine powder that is then diluted with a dry material like organic flour.

Duster: Equipment for applying pesticide dusts to a crop.

D-vac: A portable vacuum device that sucks up insects. It is used to sample insects.

Dyar's rule: An observational 'rule' governing the size increment found between subsequent instars of the same species; the ratio of a given linear dimension in one instar to the same dimension in the next instar is constant throughout the larval instars and for a given species.

E

Early postemergence: The application of herbicide after plant emergence and during the initial growth phases of crop or weed seedlings.

EC: See emulsifiable concertrate.

Ecdysial cleavage lines: Preformed lines of weakness along which the shed cuticle of the cranium splits during ecdysis; cuticle in the region of these lines lacks the exocuticle (Fig. 100).

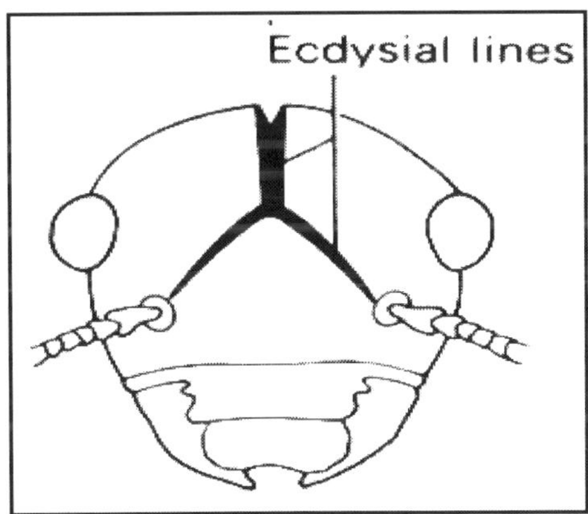

Fig. 100. Splitting of head capsule along preformed ecdysial line.

Ecdysial glands: See prothoracic glands.

Ecdysiotropin: See brain hormone.

Ecdysis (pl., ecdyses): (alternative term, moulting) The process of shedding the skin (exoskeleton) undertaken periodically by insects and other arthropods; the moulting (shedding of the skin) of larval arthropods from one stage of development to another- the final moult leading to the formation of the puparium or chrysalis; the process whereby arthropods shed their exoskeleton, mediated by the hormone ecdysone. (adj. ecdysial): The final stage of moulting, the process of casting off the cuticle.

Ecdysone (molting hormone, prothoracic hormone, MH)**:** A hormone produced in the prothoracic gland; initiates growth and molting activities of epidermal cells; the hormone that induces molting in arthropods; the steroid hormone secreted by the prothoracic gland that is converted to 20-hydroxyecdysone, which stimulates the epidermal moulting.

Ecdysteroid: The general term for steroids that induce moulting (Fig. 101).

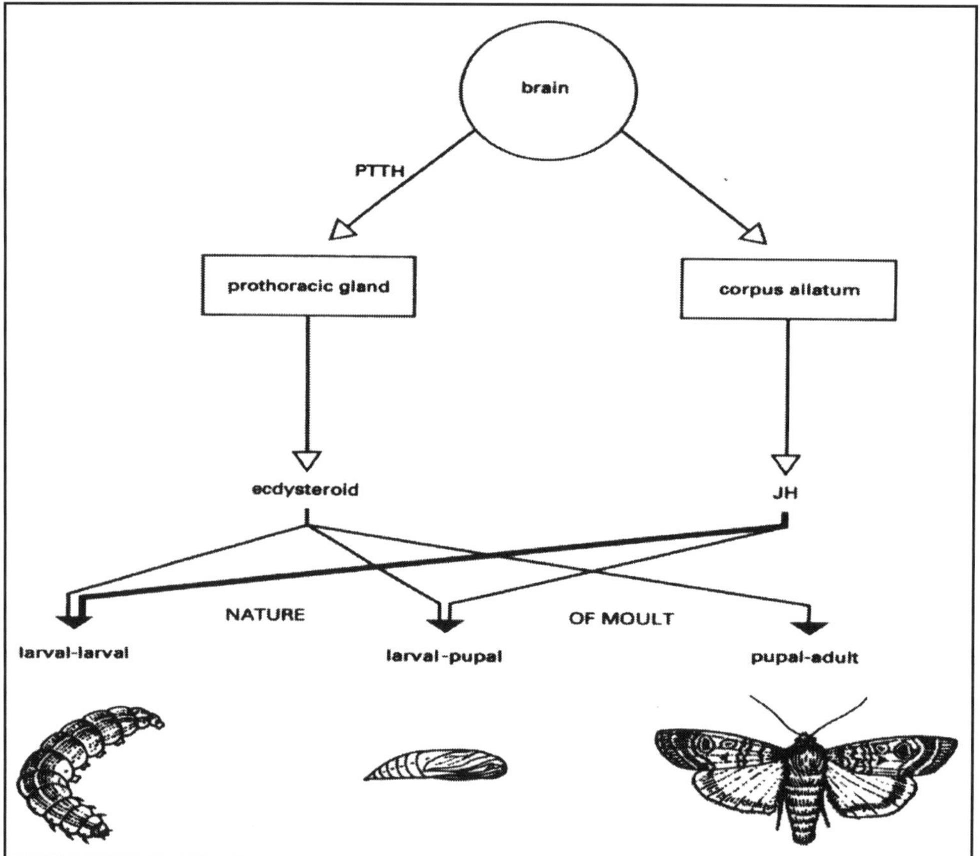

Fig. 101. Schematic diagram of the classical view of endocrine control of the epidermal processes that occur in moulting and metamorphosis in an endopterygote insect. This scheme simplifies the complexity of ecdysteroid and JH secretion and does not indicate the influence of neuropeptides such as eclosion hormone. JH, juvenile hormone; PTTH, Prothoracicotropic hormone.

Ecdysterone: An old term for 20-hydroxyecdysone, the major moult-inducing steroid.

Ecesis (=Oikesis)**:** The establishment of an organism in a new habitat; accidental dispersal and establishment in a new area.

Eclosion hormone: A hormone, synthesized by the median neurosecretory cells, which influences behavior associated with pupal-adult ecdysis; a substance secreted by certain cells in the nervous system that initiates activities associated with ecdysis.

Eclosion hormone: A neuropeptide with several functions associated with adult eclosion including increasing cuticle extensibility.

Eclosion: The emergence of a nematode or insect from its previous life cycle stage, e.g., a larva emerging from an egg, or an adult emerging from a pupa; the act of an insect's leaving the egg or emergence of the adult insect at the terminal molt; the emergence of an immature stage from the egg; sometimes used to refer to pupal-adult ecdysis; hatching of the egg, or emergence of the adult insect at the terminal molt.

Ecoclimate: Climate within the plant (crop) community.

Ecogeography: The study of the role of environment in animal distribution.

Ecological backlash: The counter-responses of pest populations or other biotic factors in the environment that devalue the effectiveness of insect management tactics.

Ecological energetics: An essential aspect of ecological studies is concerned with the energy transformation occurring within ecosystems. The quantity of incident energy from the sun per unit area of the ecosystem and the efficiency of conversion of this energy to other organism is a basic principle. For the effective functioning of the ecosystems, energy transferring mechanism from organism to organism as observed in food chains are of prime importance. Ecological energetics deals with these aspects of ecological studies.

Ecological homolog: One of two or more species having most niche parameters in common.

Ecological homologues: Species that have the same fundamental ecological requirements and therefore cannot occupy the same area indefinitely without one species displacing the other.

Ecological isolation: A condition in which interbreeding between two or more otherwise sympatric populations is prevented by mating in different ecologic niches (cf. Reproductive isolation, Geographic isolation, Genetic isolation).

Ecological management: Purposeful mainipulation of the cropping environment to reduce rates of pest increase and damage.

Ecological niche: The total of needed resources (food, shelter, mates, and so on) of a species.

Ecological race: Subspecies (q.v.).

Ecology: The study of the relationship between organisms and their environment; the study of the relationships among organisms, their environment, and each other; study of the distribution and abundance of organisms; the study of all the living organisms in an area and their physical environment; a fundamental as well as applied study to establish relationship of an organism to its surrounding environment (Gr. *oeikos* = house + *logos* = knowledge).

Economic control: Control of a pest such that expenditure on control measures is more than compensated by increased value of yield.

Economic damage: The amount of pest-induced injury that justifies the cost of applying pest control measures; the injury done to a crop which will justify the cost of artifical control measures.

Economic injury level (EIL): Pest population density sufficient to cause economic losses that are greater than the economic cost of the control action to reduce pest densities; the level of damage to a crop that is equal in value to the cost of suppressive measures; the lowest density of pest population that causes economically significant damage; the lowest numbers of insects that will cause economic damage; the causing a crop loss of about 5-10%, according to definition; pest population level at which expenditure on control measures is just balanced by the value of yield resulting from control of the pest.

Economic poison: The legal definition of a pesticide; any chemical used to control or mitigate pests; legal classification for a substance used for controlling, preventing, destroying, repelling, or mitigating any pest.

Economic threshold (ET, action threshold**):** The pest population density at which a tactic should be initiated in order to stop the pest density from reaching the economic injury level (EIL); the level of damage by a pest that serves to warn the agriculturalist of impending problems; the pest density at which management action should be taken to prevent an increasing pest population from reaching the economic injury level.

Ecophenotypic variation (habitat variation): A nongenetic modification of the phenotype by specific ecological conditions, particularly those of a habitat.

Ecophysiology: Ecological studies having a direct bearing on the survival of a population as a result of functional adjustments of organisms.

Ecospecies: A group of populations so related that they are able to exchange genes freely without loss of fertility or vigor in the offspring (Turesson).

Ecosystem: The ecological system of a given area; composed of living plants and animals and non-living components; the entirety of physical environment and all organisms in a defined region or area; the interacting system of the living organisms in an area and their physical environment; an assemblage of interacting plant and animal communities and nonliving environmental components; a biological community considered in relation to its physical environment.

Ecotype: A descriptive term applied to plant races of varying degrees of distinctness which owe their most conspicuous characters to the selective effects of local environments (cf. Subspecies); a specific strain of a species adapted to a particular set of environmental conditions.

Ectad: Outward relative to the midline of the body.

Ecto-: Outside; external.

Ectoderm: The outermost germ layer; the outermost of the three primitive germ layers of the developing embryo; gives rise to the body wall, foregut, hindgut, nervous system, respiratory system, and many glands (Fig. 64); the outer germ layer or cell layer of an early embryo. (Gr. *ektos* = outside + *derma* = skin).

Ectognathous: Having mouthparts not pulled into the head; having mouthparts that are exposed, not enclosed within a cavity.

Ectoparasite: An organism that lives parasitically on the surface of its host, usually in the fur or feathers of mammals or birds; most insects in this category are blood suckers; a parasite that lives externally on, and at the expense of, another, which it does not kill(Gr. *ektos* = outside + parasite) *e.g.* tick.

Ectoparasitoid: A parasite that lives externally on, and at the expense of, another, which it kills.

Ectoperitrophic space: The space between the **peritrophic membrane** and the midgut wall (Fig. 102).

Ectotherm: Organism that relies on heat from the external environment for thermoregulation.

Ectothermia (adj. ectothermic): The inability to regulate the body temperature relative to the surrounding environment.

Edaphic factor: The influence of soil properties on organisms (especially plants).

Eelworm: See nematode.

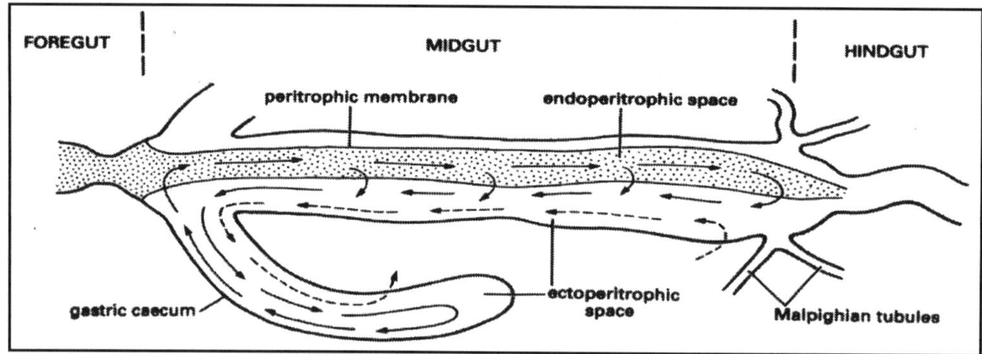

Fig. 102. Generalized scheme of the endo-ectoperitrophic circulation of digestive enzymes in the midgut.

Effective environment: All environmental factors that directly influence the fate of a population; the elements in an ecosystem that have a direct influence on reproduction, survival, and movements of a subject's species.

Efficiency of a pest control measure: This is shown by the number of pests remaining after control treatment.

Effector: A structure leading away from a given point of reference, as an efferent artery. (L. *ex* = out + *fero* = carry).

Efferent neurons: See motor neurons.

Effevtive dose (ED): The ED_{50} is the dose that has an effect with 50% of tested subjects.

Efficacy: A measure or appraisal of the effectiveness of a control tactic with respect to the extent that the tactic controls the target pest.

Egest: To discharge unabsorbed food or residues from the digestive tract. (L. *e* = out + *gero* = bear) *e.g.* in case of amoebas, sponge and *Hydra.*

Egg: A germ cell produced by a functional female organism; an ovum.

Egg-development neurosecretory hormone (EDNH)**:** Secretion of median neurosecretory cells of mosquitoes that induce ovaries to release ecdysone; a product of the brain neurosecretory cells that stimulates vitellogenesis in the female mosquito.

Eimeria: A sporozoan causing coccidiasis in domestic rabbits and fowls, main symptom is desentry.

Ejaculatory duct: The duct that leads from the fused vas deferens to the gonopore, through which semen or sperm is transported (Fig. 4); a duct formed by the joining of vas deferens from each testis and involved in the propulsion of semen; the terminal portion of the male sperm duct; a median duct that carries the sperm from the internal reproductive system to the exterior; the duct in male insects which conveys sperm to the copulatory organ.

Elaiosome: A food body forming an appendage on a plant seed (Fig. 103).

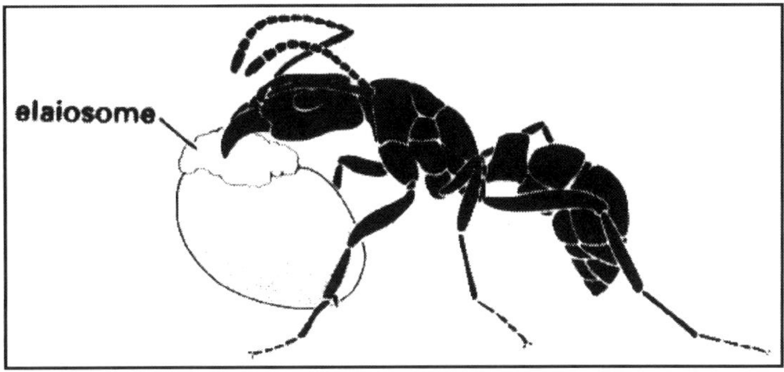

Fig. 103. An ant of *Rhytidoponera tasmaniensis* (Hymenoptera: Formicidae) carrying a seed of *Dillwynia juniperina* (Fabaceae) by its elaiosome (seed appendage).

Elateriform larva: A larva resembling a wireworm, that is, slender, heavily sclerotized, with short thoracic legs, and with few body bristle or hairs.

Elbowed antenna: An antenna with the first segment elongated and the remaining segments coming off the first segment at an angle.

ELISA: Short for enzyme-linked immunosorbent assay. This is an assay that is used for detecting and identifying pest organisms using the affinity of antibodies for antigens that are specific to particular organisms.

Elytron (pl. elytra): A modified thickened, hardened leathery, or horny forewings usually meeting in the mid-dorsal line of the body at rest and that protects membranous hindwings; characteristic of beetles. e.g. (Coleoptera, Dermaptera, some Homoptera; Coleoptera, Fig. 104).

Emarginate: Notched or indented.

Embioptera, order: (*embio,* lively; *ptera,* wing) A small order of tropical insects which live gregariously in silken nests; sometimes referred to as web spinners.

Embolium: A narrow piece of the corium, along the costal margin, separated from the rest of the corium by a suture (Hemiptera).

Embryo, hexacanth: Embryo with six hooks, develops within the egg and on entering the alimentary canal of suitable host comes out of the egg and gets converted into larval form. *e.g.* hexacanath embryo of tape worm.

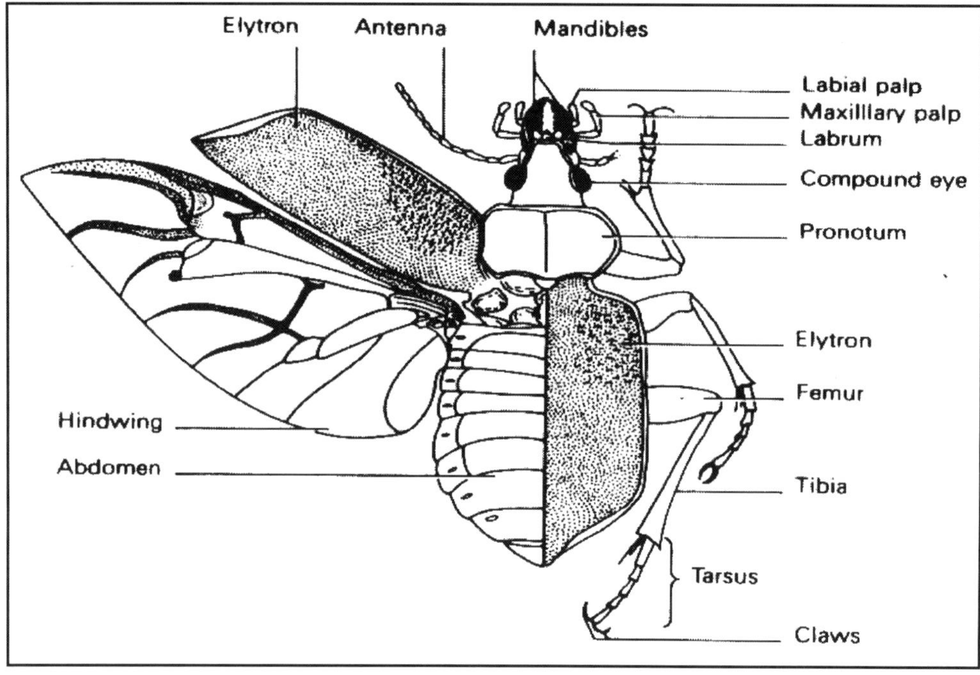

Fig. 104. Coleoptera, ground beetle, *Calosoma semilaeve* (Carabidae), with left elytron and hindwing spread.

Embryo: Animal in the process of development from fertilized or even unfertilized - parthenogenetic ovum, may be contained in egg membrane *e.g.* in hexacanth embryo of tape worm or in maternal body (mammals); the developing organism before it reaches the stage where it leaves the parent body, emerges from the egg, or germinates from a seed.

Embryogenesis: Developmental events between zygote formation and exit from the egg.

Emendation: In nomenclature, an intentional modification of the spelling of a previously published scientific name (Cf. Error, *Lapsus calami*).

Emergence: 1. The act of the adult insect leaving the last nymphal skin, or pupal case. 2. Germination of a seed and the appearance of the shoot; the act of moving out of the protective environment of the body of, or location of a previous life history stage, such as when a young plant breaks through the soil surface or when an insect or nematode comes out of an egg.

Emigration: The movement of individuals out of a population; movement out of an area.

Empodium (pl. empodia): A lobe or padlike or spine or bristlelike structure at the apex of the last tarsal segment, between the claws (Diptera; fig. 407, emp) (Fig. 105).

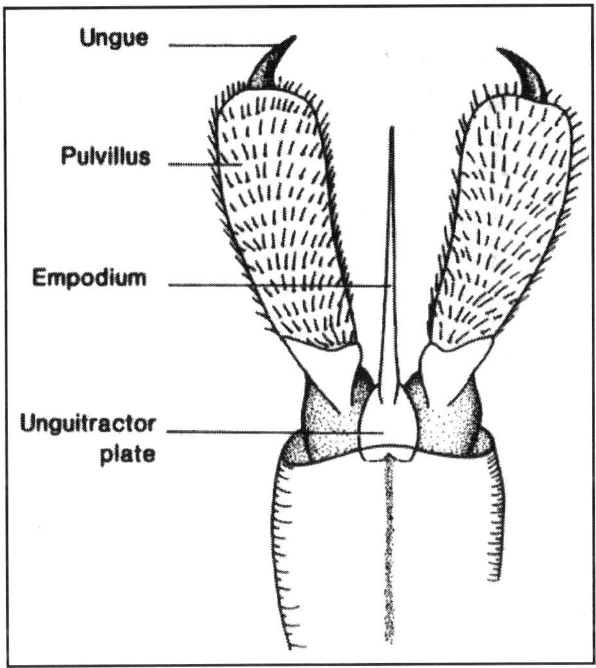

Fig. 105. Robber fly tarsus.

Emulsifiable concentrate (EC): Liquid formulation that when added to water will spontaneously disperse as fine droplets to form an emulsion (= Miscible oil); a liquid formulation of pesticide that forms a milky emulsion when added to water; a concentrated pesticide formulation containing organic solvent and detergentlike emulsifier to facilitate emulsification with water; an insecticide formulation in which the active ingredient is dissolved in a non-aqueous solvent to which emulsifers are added. An emulsion is formed when added to water.

Emulsifier: Spray additive which permits formation of a stable suspension of oil droplets in aqueous solution, or of aqueous solution in oil; surface active chemical that produces stable oil-in-water emulsions; a stable dispersion of oil droplets in aqueous solution, or vice versa; a suspension of microscopic droplets of one liquid in another.

Emulsion, Invert: Suspension of aqueous solution in oil.

Enamel: The dense whitish covering on teeth of vertebrates, the hardest substance produced by animal bodies.

Encapsulated material: A pesticide formulation where the active ingredient is enclosed inside an inert material to allow for decreased hazard in handling and application, and slow release after application.

Encapsulation or microencapsulation**:** (i) A type of pesticide formulation in which the active ingredient is enclosed in a support material (often plastic). (ii) the encapsulation of a pesticide in a non-volatile envelope of gelatin, usually of minute size, for delayed release. (iii) A physiological defense mechanism in insects wherein the eggs or larvae of endoparasites are covered by multiple layers of host blood cells and killed (iv) A protective mechanism in which large numbers of hemocytes become layered around a foreign entity, such as a parasitic worm, that has invaded the hemocoel. (v) The enclosure of a parasitoid larva within the blood of the host by a layer of hemocytes; a reaction of the host to an endoparasitoid in which the invader is surrounded by haemocytes that eventually form a capsule.

Encephalitis (pl. encephalitides): An inflammation of the brain or spinal cord; a group of diseases of the central nervous system caused by mosquito **Aedes** sp -borne viruses.

Encyst: To enclose or become enclosed in a cyst or a vesicle.

Endemic typhus fever: A human disease caused by the oriental rat flea .Clinically similar to epidemic typhus fever except with larva, pupa and adult stages. cf. Exopterygota.

Endemic: Organisms that are naturally occurring in a particular geographic area; a population in equilibrium with its environment (not an epidemic).

Endite: An inwardly directed (mesal) appendage or lobe of a limb segment.

Endite: The basal segment of the spider pedipalp, which is enlarged and functions as a crushing jaw (Fig. 106).

Endo-: Within.

Endocrine glands: A gland that discharges its products (hormones) to the inside (as contrasted to an exocrine gland); Ductless glands whose products are released directly into the insect hemolymph.

Endocrine or entoderm: The layer or group of cells lining the primitive gut, or gastrocoel, in an early embryo, beginning in the gastrula stage. (Gr. *endon* = within + *derma* = skin) *e.g.,* in *Hydra;* a ductless gland; also its internal secretion or hormone, which is diffused into and carried by the blood stream. (Gr. *endon* = within + *krino* = separate).

Endocuticle (or endocuticula): The flexible, unsclerotized inner layer of the procuticle; the inner, unstabilized layer of the chitinous cuticle- which may or may not be pigmented- that is digested during molting; the inner zone of the procuticle, softer and lighter in color than the exocuticle (Fig. 84).

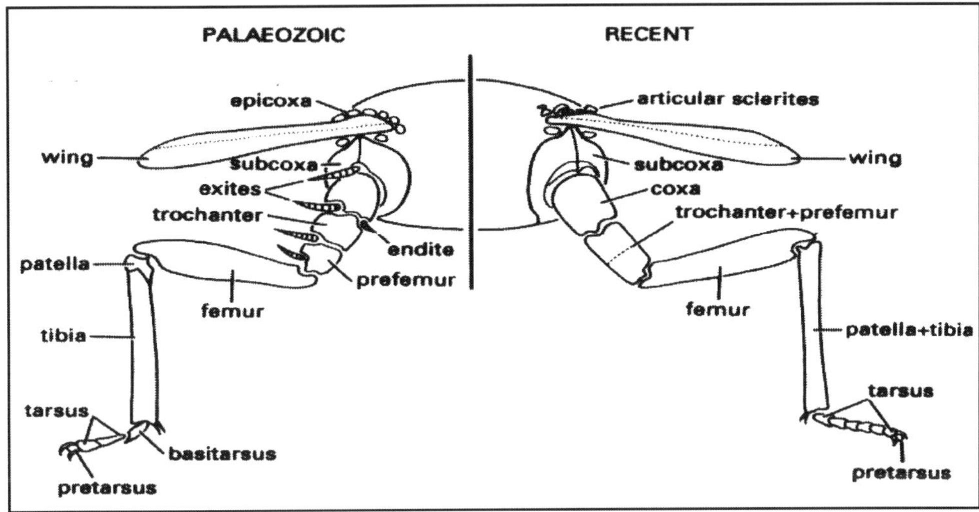

Fig. 106. Appendages of hypothethical primitive Palaeozoic (left of each diagram) and Recent (right of each diagram) pterygotes (winged insects): Thoracic segment of adult showing generalized condition of appendages.

Endoderm: The innermost of the three primitive germ layers of the developing embryo; gives rise to the midgut.

Endogenous activity: Nervous discharges that arise spontaneously, in the absence of stimulation.

Endogenous rhythms: Clock or calendar-like activity patterns, commonly **circadian,** unaffected by external conditions.

Endogenous: Arising from within.

Endoparasite: An organism that lives parasitically inside the body of tne host organism or under the skin surface; a parasite that feeds and develops while inside the body of its host (Gr. *endon* = within + parasite) *e.g.* tapeworms; a parasite that lives internally at the expense of another, which it does not kill.

Endoparasitoid: A parasite that lives internally at the expense of another, which it kills.

Endopeptidase: An enzyme that catalyzes the spilitting of a peptide chain.

Endoperitrophic space: In the gut, the space enclosed within the peritrophic membrane.

Endophallus: The inner, eversible tube of the penis.

Endophilic: Indoor loving, as of an insect that feeds inside a dwelling.

Endopodite: The mesal branch of a biramous appendage.

Endopterygota: The division of Pterygota comprising insects with wings that develop internally; with complete metamorphosis; Refers to insects in which the wings develop internally during the larval stage, i.e., insects that undergo complete metamorphosis; immature instars dissimilar to adults' and usually adapted to different environmental conditions; *syn.* holometabolous; an insect that develops through the immature stages as a larva with wings retained internally as imaginal discs.

Endopterygote: Development in which the wings form within eversion taking place only at the larval- pupal moults (as in the monophyletic grouping Endopterygota).

Endoskeleton: A skeleton or supporting structure on the inside of the body; an internal supporting framework or structure. (Gr. *endon* = within + skeleton) *e.g.* in vertebrates.

Endosperm: The tissue in seeds of angiosperms that contains nutrients and surrounds the embryo.

Endostyle: The ventral ciliated groove in the pharynx of tunicates, amphioxus and larval ampreys, used in food getting; homologus with the thyroid gland of vertebrates. (Gr. *endon* = within + *stylos* = column).

Endothelial cell: Layer of simple squamous cells lining the inner surface of circulatory organs and other closed cavities. (Gr. *endon* = within + *thele* = nipple).

Endotherm: Organism that metabolically generates heat required for thermoregulation.

Endothermy (adj. endothermic)**:** The ability to regulate the body temperature higher than the surrounding environment.

Endysis: Developing a new outer layer.

Energid: In the fertilized egg, the daughter nuclei that have migrated to the egg periphery and become compartmentalized.

Energids: In an embryo, the daughter nuclei cleavage products and their surrounding cytoplasm.

Energy flow: The behaviour of energy in an ecosystem is termed as energy flow. In ecosystem solar energy is trapped by autotrophs and then transferred from one trophic level to other in succession.

Enhanced microbial degradation: Unusually rapid breakdown of soil pesticides by microorganisms.

Entad: Inward relative to the midline of the body.

Enteron: The digestive cavity, especially that part lined by endoderm.

Entire: Without teeth or notches, with a smooth outline.

Ento-: Within.

Entognathous: Having mouthparts inserted in a pocket on the face of the insect; internal mouthparts; having mouthparts somewhat pulled into the head. The opposite is ectognathous.

Entomology. The branch of zoology that refers to the study of the Class Insecta.

Entomopathogen: A pathogen (disease causing organism) that attacks insects particularly.

Entomopathogenic: Organisms that cause disease in insects.

Entomophage (adj. entomophagous): An eater of insects; an animal (or plant) which feeds upon insects.

Entomophilous. Plants that are pollinated by insects; insect-loving.

Entomophily: Pollination by insects.

Entomophobia: An irrational, pronounced fear of insects; irrational and persistent fear of insect infestation or attack.

Entomopox virus: A virus that multiplies in the cell cytoplasm of fat body and blood cells in a variety of insects.

Envenomation: The poisonous effects caused by the bites, stings, or secretions of insects and other arthropods.

Environment: The total of physical, chemical, and biotic conditions surrounding an organism; the surroundings of an area including the water, air, soil, plants, and animals.

Environmental fate: How a substance (usually a pesticide) behaves in the environment, and what becomes of it and its degradation products in the environment.

Environmental manipulation: Alteration of the environment, particularly to enhance natural populations of insect predators and parasitoids.

Environmental Protection Agency (EPA): A federal agency that enforces legislation relating to the protection of the environment in the United States.

Environmental resistance: The collective action of environmental factors that prevents manifestation of the biotic potential of a given species; the physical and biological restraints that prevent a species from realizing its biotic potential.

Enzootic: A disease present in a natural host within its natural range; a disease in an animal population (cf., epizootic)

Enzyme, cytolytic: An enzyme of *Entamoeba histolytica* helping the amoebae to bore into capillary vessels of the host for reaching liver.

Enzyme, digestive: Enzyme which acts upon food taken by the animal and helps in converting into a form assimilable by the animal.

Enzyme: A protein that acts as a catalyst and regulates the rate of chemical reactions of living organisms; an organic/biological catalyst produced by living beings, various kinds, each kind acting only on a very limited range of substances; a biochemical which promotes a specific chemical reaction.

Ephemeroptera, order: (*ephemero,* for a day; *ptera,* wings) The order of insects comprising the mayflies (Fig. 107).

Epicoxa: A basal leg segment, forming the articular sclerites in all extant insects, believed to have borne the exites and endites that fused to form the evolutionary precursors of wings.

Epicranial suture: Coronal plus frontal sutures; a Y-shaped line of weakness on the vertex of the head where the split at moulting occurs (Fig. 108).

Epicranium: The upper part of the head, from the face to the neck.

Epicuticle (or epicuticula): The very thin, nonchitinous, external layer of the cuticle; the outermost layer of the insect body wall (exoskeleton) that protects the insect from drying by a layer of waxes and lipids; unpigmented outer layer of the insect cuticle, 0.05-4.0 micrometers thick, rich in lipid and protein and lacking chitin; The inextensible and unsupportive outermost layer of cuticle, lying outside the **procuticle**.

Epidaphic: Soil fauna living on soil surface *e.g.* Collembola.

Epideictic pheromone (spacing pheromone): An intraspecific chemical messenger that elicits dispersal from potentially crowded food sources

Epidemic typhus fever: A human disease caused by *Rickettsia prowazekii* and transmitted by the body louse, *Pediculus humanus humanus*.

Epidemic: Severe outbreak of a disease over a large geographic area; the spread of a disease from its **endemic** area and / or from its normal hosts.

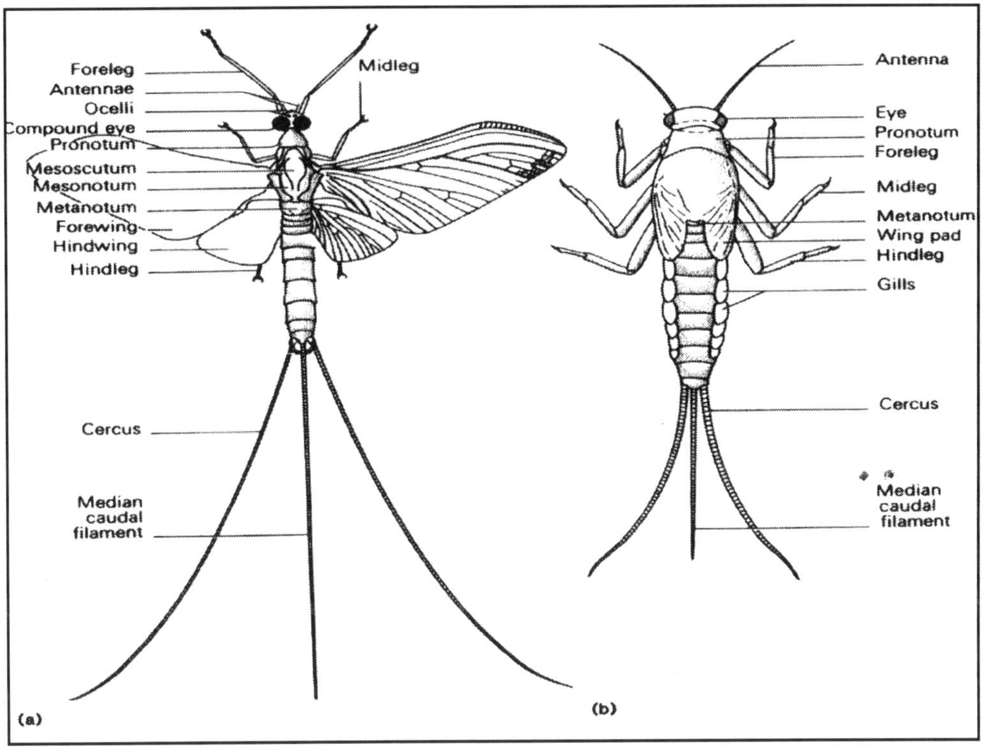

Fig. 107. Ephemeroptera, mayflies. (a) Adult. (b) Nymph.

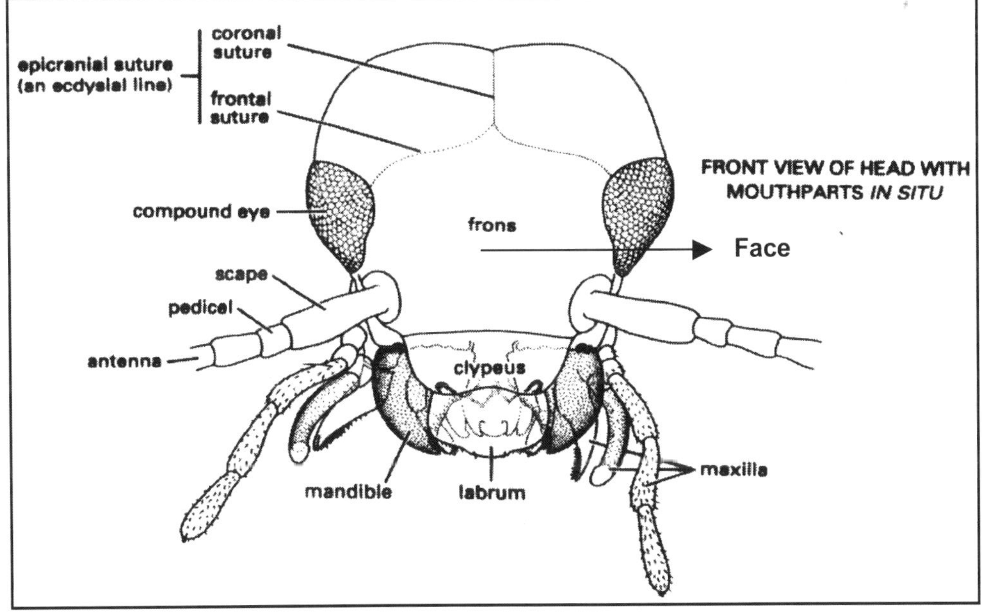

Fig. 108. Anterior view of generalized insect head.

Epidermis: The unicellular layer of ectodermally derived integument that secretes the cuticle; a single layer of cells (sometimes stratified) covering an external surface; the ectodermal portion of the skin of most animals; secretes cuticle in some animals. (Gr. *epi* = upon + *derma* = skin); the cellular layer of the body wall or insect integument which secretes the cuticle (Fig. 84).

Epididymis: Coiled vas deferens in which sperm are stored.

Epigastric furrow: A transverse ventral suture near the anterior end of the abdomen, along which lie the openings of the book lungs and the reproductive organs (spiders).

Epigynum: The external female genitalia of spiders.

Epilimnion: The superficial layer of freshwater media constantly stirred by wind.

Epimeron (pl. epimera)**:** The posterior division of the pleuron of a thorax, separated from the **episternum** by the **pleural suture**.

Epimorphic development: Development in which the segment number is fixed in the embryo before hatching.

Epimorphosis: A type of development in which the insect emerges from the egg with its full complement of body segments (opposite of anamorphosis).

Epinasty: When more rapid growth on one side of a plant part, particularly leaves, causes it to bend or curl in one direction.

Epipharynx (labrum-epipharynx): A fleshy mouthparts structure on the inner surface of the labrum or clypeus; a lobe on the epipharyngeal wall; in chewing insects a median lobe on the posterior (ventral) surface of the labrum or clypeus (Fig. 109).

Epiphysis (pl., epiphyses): A movable pad or lobelike process on the inner surface of the front tibia (Lepidoptera).

Epiphyte: An air plant, one growing nonparasitically upon another plant or upon a nonliving object.

Epiphytotic: A disease epidemic in a plant population.

Epipleura (pl., epipleurae): The bent-down lateral edge of the elytron (Coleoptera).

Epipleurite: A small sclerite in the membranous area between the thoracic pleura and the wing bases (Fig. 16, epp); the more dorsal of the sclerites formed when the pleuron is divided longitudinally; *or* one of two small sclerites of a wing-bearing segment - the anterior **basalre** and posterior **subalare**.

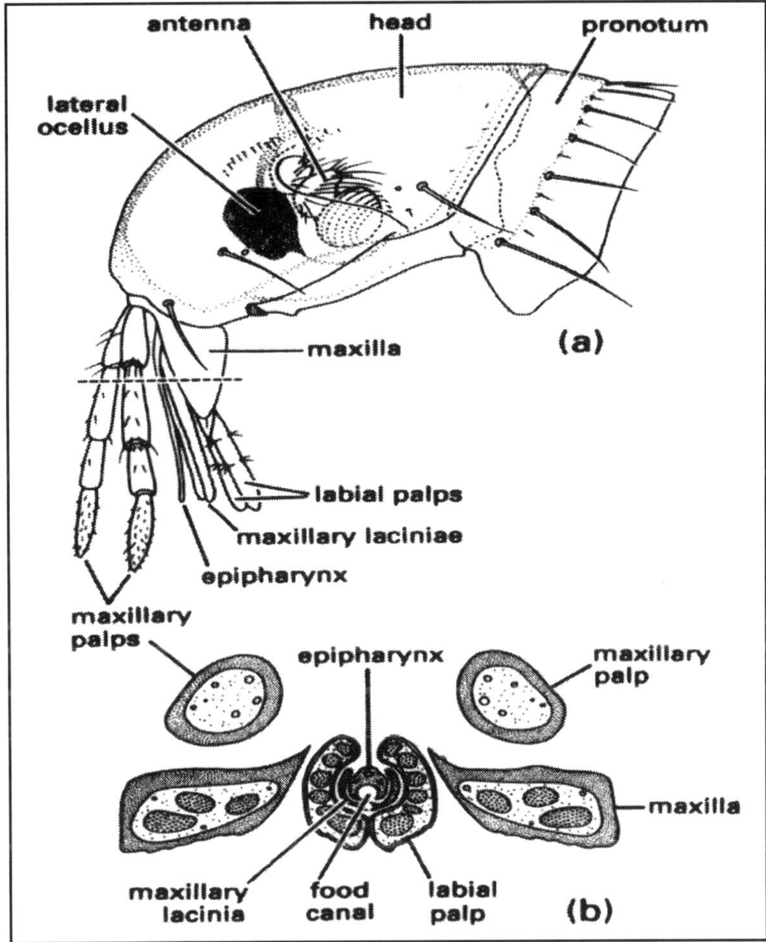

Fig. 109. Head and mouthparts of a human flea, *Pulex irritans* (Siphonaptera: Pulicidae): (a) lateral view of head; (b) transverse section through mouthparts. The plane of the transverse section is indicated by the dashed line in (a).

Epiproct: A process or appendage situated above the anus and appearing to arise from the tenth abdominal segment; actually the dorsal part of the eleventh abdominal segment; the dorsal part of the terminal abdominal segment that often appears as a process situated above the anus; dorsal relic of segment 11 (Fig. 1).

Episternum (pl. episterna)**:** The anterior division of the pleuron, separated from the **epimeron** by the **pleural suture**.

Epistomal suture: Suture above the clypeus and connecting the subgenal sutures; the suture between the frons and the clypeus (Fig. es); See **frontoclypeal suture.**

Epistome: The part of the face just above the mouth; the oral margin (Diptera).

Epithelial cell: A cell that covers a body structure or lines a body cavity.

Epithelium: A layer of cells (sometimes stratified) covering an external surface; the ectodermal portion of the skin of most animals; secretes cuticle in some animals. (Gr. *epi* = upon + *thele* = nipple).

Epizootic: Outbreak of disease in an animal (**insect**) population at the same time; a group phenomenon.

Equilibrium position (EP): In insect bioeconomics, the average density of a potential pest on a specific crop; the population density at equilibrium.

Eradication: The complete elimination of all individuals of a species from a particular area.

Ergatoid (also **apterous neotenic**): In termites, a supplementary reproductive derived from a worker, held in a state of arrested development and lacking wings, able to replace reproductives if they die; *see also* **Neotenic.**

Erinea: Out growths of abdominal hairs and leaves which is produced by certain gall-mites (Eriophyidae), accompanied by some leaf distortion.

Erinium: A growth of hairs in dense patches on plant leaves resulting from the attack of certain Eriophyidae (Acarina).

Error: In nomenclature, an unintentional misspelling of a scientific name, as a typographical error or an error of transcription (cf. Emendation, *Lapsus calami*).

Eruciform larva: A caterpillar; a larva with a more or less cylindrical body, a well-developed head, and with both thoracic legs and abdominal prolegs.

Eruciform: Larva resembling a caterpillar (polypod) in appearance; applies to larval Lepidoptera and some sawfly larvae.

Erythrocyte schizogonous cycle: The stage in the asexual development of a malaria parasite (*Plasmodium* spp.) im which **trophozoties** within the vertebrate host red blood cells divide to form **merozotics** (Fig. 110).

Erythrocyte: A red blood cell.

Esophagus: The narrow portion of the alimentary canal immediately posterior to the pharynx before the crop (oesophagous); a tubular portion of the foregut, behind the pharynx. See oesophagus.

Ester: An acid derivative in which the hydrogen of the acid has been replaced by a hydrocarbon group.

Estuarine habitat: Transitional habitat between the freshwater and marine habitats.

Etoxication: The process by which a poison is rendered harmless.

Euatachian tube: The passage between the pharynx and middle ear in land vertebrates. (Eustachio, an Italian Anatomist).

Eucephalous: An apodous larva with a well- developed head capsule (for example, Diptera, Nematocera).

Eudaphic: Soil fauna living in deeper layers of soil below the surface. *e.g.* earthworm.

Euplantula (pl. euplantulae) (*also* **pulvillus**)**:** Pad-like structure on the ventral surface of some tarsomeres of the leg.

Euryhaline: Usually aquatic organisms adaptable to a wide range of salinity. The opposite is stenohaline.

Euryhygric: Organisms adaptable to a wide range of atmospheric humidity. The opposite is stenohygric.

Eurythermal: Organisms adaptable to a wide range of temperatures. The opposite is stenothermal.

Eurytopic: Having a wide geographical distibution, and tolerant of a wide range of ecological habitats. The opposite is stenotopic.

Eusocial: Exhibiting co-operation in reproduction and division of labour, with overlap of generations.

Eusocial: The highest level of social development. Truly social.

Eusociality: A type of social behavior involving overlap of generations, cooperative brood care, and a caste system in which many colony members are sterile.

Eusternum (pl. eusterna)**:** The dominant ventral plate of the thorax that frequently extends into the pleural region; the ventral plate of a thoracic segment, exclusive of the spinasternum.

Eutrophication: Nutrient enrichment, especially of water bodies.

Evagination: An outpocketing, or saclike structure of the outside; an outpocketing from a hollow structure. (L. *e* = out, from + *vogina* = a sheath).

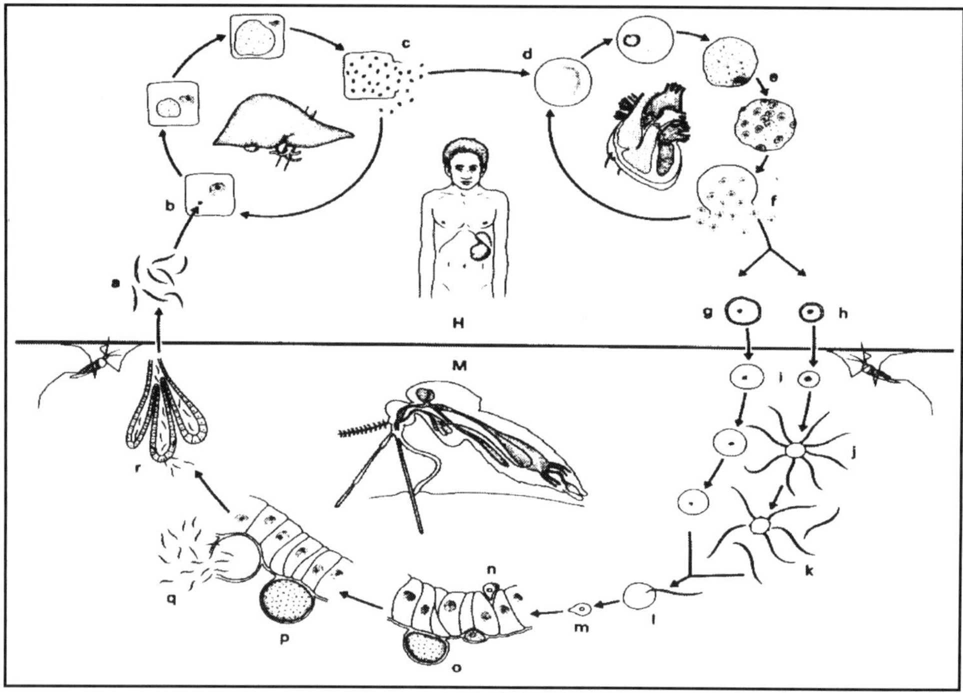

Fig. 110. The life cycle of a *Plasmodium,* the malaria pathogen, in human (H) and mosquito (M): (a) injection of sporozoites into human host by feeding mosquito; (b) sporozoite enters liver parenchyma cell; (c) rupture of schizont to release merozoites; (d) merozoites enter erythrocyte; (e) trophozoite becoming schizzont; (f) schizont releasing merozoites; (g) female macrogamete formation; (h) male gametocyte formation; (i) gametes lacking erythrocyte coat; (j) flagellate microgamete; (k) exflagellation; (l) zygote formation; (m) ookinete; (n) oocyst forming on mosquito inner midgut wall; (o), (p) oocyst maturation and sporogeny; (q) release of sporozoites; (r) sporozoites entering salivary gland. After Keetle, 1984; Katz *et al.,* 1989.

Evaluation survey: A systematic survey to evaluate the current and potential significance of a pest outbreak.

Eversible: Capable of being everted or turned outward.

Evolution: The change in the characteristics of organisms from generation to generation; the process by which living organisms have come to be what they are, structurally and functionally complex forms being derived from simpler forms; hence, descent with modification. (L. *evolve* = unroll, unfold).

Evolutionary systematics: A classification system in which **clades** and **grades** are recognized.

Exarate: Sculptured; an excavated surface; an insect pupa in which legs and antennae are free from the body and the abdomen is movable insect pupae having legs and other appendages separately encased (as in the pupae of beetles, bees, and wasps).

Exarate pupa: A pupa in which the appendages are free and not glued to the body; pupa with free appendages, usually not covered by a cocoon (Fig. 111); as opposed to being cemented; *see also* **Obtect pupa.**

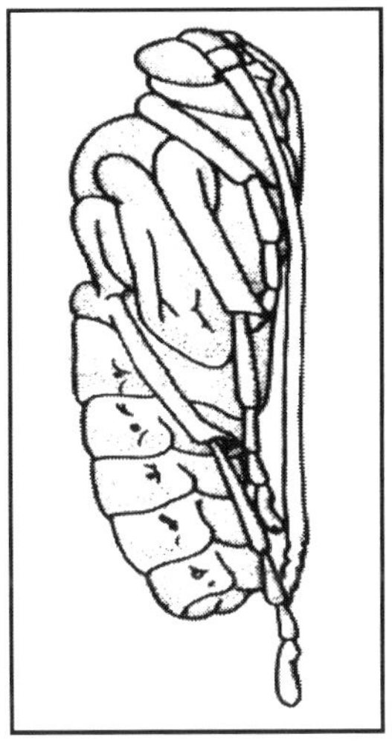

Fig. 111. Exarate, ichneumon wasp.

Excavated: Hollowed out.

Exclusion: Prevention of pests entering or being moved into an area in which they do not already occur.

Excretion: Waste material resulting from metabolism and discharged from the body as useless; also the process of its elimination. (L. *excerne* = separate, secrete).

Exite: An outer appendage or lobe of a limb segment.

Exocrine gland: A gland that discharges its products to the outside (as contrasted to an endocrine gland); glands that discharge their products through apertures or ducts into the external world or into lumens of various viscera (Fig. 112).

Exocuticle (or exocuticula): The layer of the cuticle just outside the endocuticle, between the endocuticle and the epicuticle (Fig. 112); stabilized layer of cuticle between the endo and epicuticular layers; resistant to molting fluid and often pigmented; the middle layer of the cuticle; the rigid, sclerotized outer portion of the procuticle, generally harder and darker than the inner portion (endocuticle); *See also* **Endocuticle.**

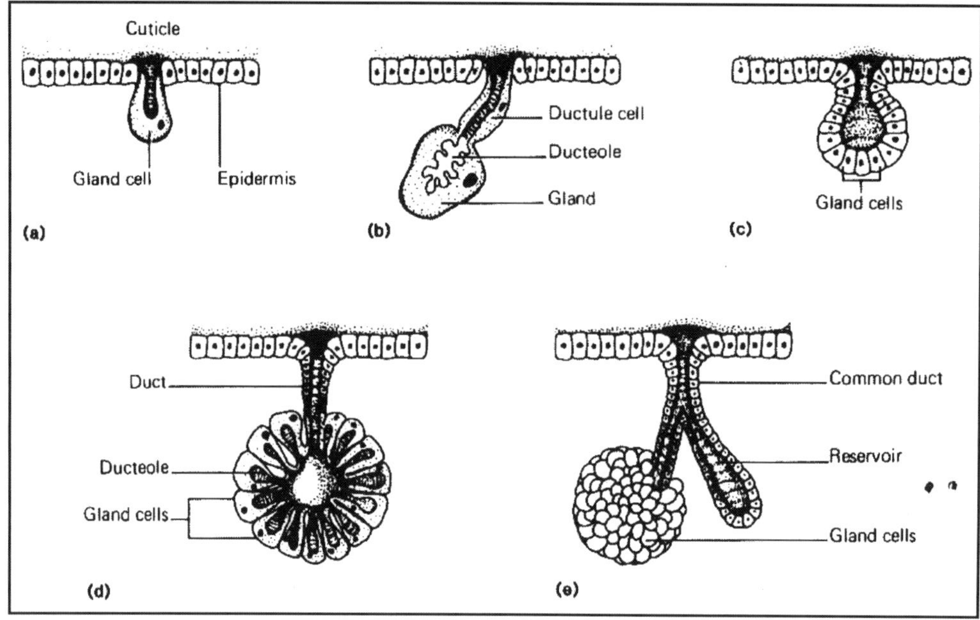

Fig. 112. Example of exocrine glands. (a) Simple unicellular gland. (b) Unicellular gland with ductile cell. (c) Simple aggregation of gland cells. (d) Aggregation of gland cells with common duct. (e) Complex aggregation of gland cells with common duct and reservoir.

Exo-erythrocytic: see **Pre-erythrocytic schizogonous cycle.**

Exogenous: Arising from without.

Exogenous rhythms: Activity patterns governed by variations in the external environment (e.g. light, temperature, etc.).

Exopeptidase: AN enzyme that catalyzes the splitting of terminal peptides from a peptide chain.

Exophily (adj. exophilic): Out-of-doors loving, used of biting insects that do not enter buildings; *see also* **Endophilic.**

Exopodite: The outer branch of a biramous appendage.

Exopterygota: (alternative term, Hemimetabola); The division of Pterygota comprising insects with wings that develop externally and The that undergo incomplete metamorphosis, cf. Endoprerygota.

Exopterygote development: Development in which the wings form progressively in sheaths that lie externally on the dorsal or dorsolateral surface of the body (as in the paraphyletic grouping Exopterygota).

Exopterygote: With the wings developing on the outside of the body, as in insects with simple metamorphosis; an insect that retains its wing pads externally through its immature stages; describes insects having wings that become externally apparent in the immature instars (immatures resemble adults), i.e., undergo simple metamorphosis; syn. Hemimetabolous.

Exoskeleton: The chitinous body wall or external skeleton of insects and other arthropods; a skeleton the muscles attaching to its inner surface; an external hardened cuticular supporting structure or covering to which muscles are attached internally. (Gr. *exo* = outside + skeleton) *e.g.* in insects.

Exotic: Nonnative organism introduced into an area in which it does not naturally occur; Introduced from another country or continent.

Experimental use permit (EUP): Allow commercial-scale application of experimental pesticides, although the amount that can be applied is restricted.

Exposure period: That amount of time workers are exposed to pesticides while mixing, loading, applying (including flagging), maintaining, and cleaning equipment. Also, substantial and prolonged contact with pesticides or their residues.

Exposure: The amount of contact an organism has with a toxin. Physical external contact is not necessarily the same as exposure.

Exserted: Protruding or projecting from the body.

Extant: Presently existing form.

External: The outside; that part away from the center (midline) of the body.

Extinct: Having no living descendant or no longer in existence.

Extra-oral digestion: Digestion that takes place outside the organism, by secretion of salivary enzymes on or into the food, with soluble digestive products being sucked up.

Extrinsic muscles: Muscles that originate on the inner surface of the body wall of the head, thorax, or abdomen and insert into an appendage (Fig. 113).

Extrinsic: Having its origin outside the limits of an organ with which it is associated.

Exudate: Matter that leaks from, is extruded from, or secreted by an organism.

Exuvia: The undigested exocuticle and epicuticle shed during ecdysis.

Exuviae (always used in the plural)**:** The cast skin or cuticles of an arthropod; the cast-off nymphal or larval skin that results from arthropod molts (ecdysis).

Exuvial fluid: See molting fluid.

Exuvial membrane: A thin, homogeneous membrane that may appear in the exuvial space during molting.

Exuvial phase: Period of time between apolysis and the beginning of new cuticle secretion.

Exuvial space: The space between the epidermis (and later the new cuticle) and the old cuticle that forms as the old cuticle is digested during molting.

Exuvium. Cast- off skin (cuticle) of an arthropod after molting

Exuvium: The cast skin of arthropods after moulting.

Eye, compound: (see *compound eye*).

Eye, simple: (see *ocellus*).

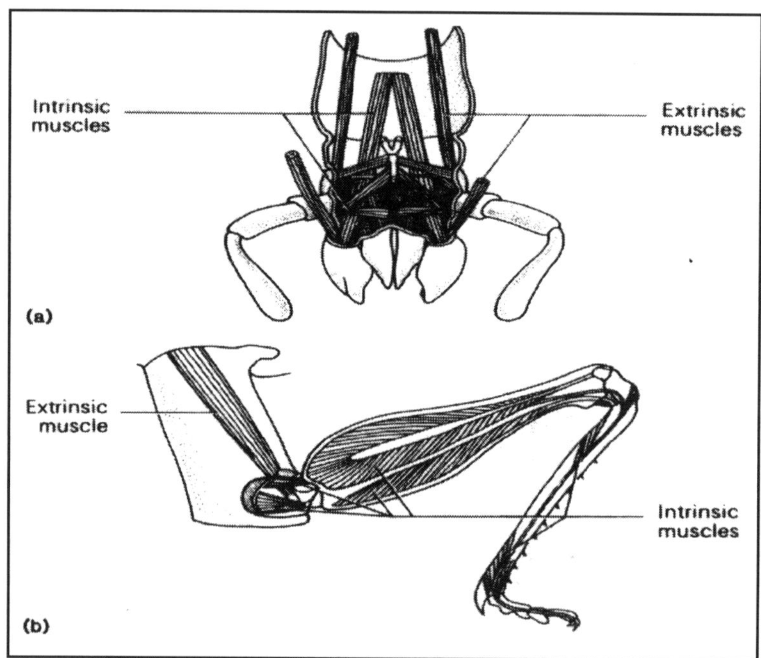

Fig. 113. Extrinsic and intrinsic skeletal muscles. (a) Labium of a cricket. (b) Metathoracic leg of a grasshopper.

Eye-cap: A structure overhanging or capping the compound eye (Lepidoptera).

Eyepiece micrometer: A linear scale in the field of vision of the eyepiece (or one of a pair of eyepieces) of a microscope for use as a measuring device.

F

F_1 generation: The first-generation offspring of a particular mating.

F_2 generation: Derived from a mating within the F_1 generation.

Face: The front of the head, below the frontal suture (Fig. 108).

Facet: The hexagonal external surface of an individual compound eye unit of ommatidium.

Facies: In taxonomy, the general aspect, appearance, or habits of a species or group.

Facultative diapause: More than one generation between the termination of one diapause and the initiation of the next diapause.

Facultative mutualism: Capable of, but not obliged to, form a mutualistic relationship.

Facultative parasite: An organism that is usually saprophytic, but under some circumstances lives as a parasite; a parasite that can complete its life cycle without invading a host but can also take advantage of a host if one is available.

Facultative: Not compulsory, optional behavior, such as facultative parasitism, in which a free-living organism may adopt a parasitic mode of life.

Faeces: Excrement; unabsorbed or indigestible food residues discharged from the digestive tract as waste.

Fallow: Land that is usually used for growing crops that is not planted during the growing season; land left unseeded and usually clean cultivated during a growing season.

Family: A classification category that includes a number of genera sharing one or a number of characteristics or tribes of common phylogenetic origin, which is separated from related similar units (families) by a decided gap, the size of the gap being in inverse ratio to the size of the unit (family); a subdivision of an order, suborder, or superfamily, and containing a group of related genera, tribes, or subfamilies. Family names end in *–idea* (e.g., pyralidae).

Farnesene: An alarm pheromone of aphids, secreted from the cornicles.

Farnesol: An alcohol, C_{15} H_{25} OH, with juvenile hormone activity in insects.

Fastidious: Term applied to parasites that are difficult to isolate or culture on media.

Fastigium: The anterior dorsal surface of the vertex (grasshoppers).

Fat body: A meshwork of loosely organized cells in the body that serve as storage depots and function in intermediate metabolism; accumulation of large cells in the hemocoel that store metabolites and are centers of intermediary metabolism.

Fauna: The animal life in a given region or period (cf. Flora, Biota).

Faunal work: A method of presenting taxonomic materials defined primarily by geographic area rather than by phylogenetic units (cf. Local list, Monograph).

FDA (Food and Drug Administration): Federal agency that oversees the safety of food, drugs and cosmetic that are available to the public.

Feces: Indigestible wastes discharged from the digestive tract.

Fecundity: Capacity to produce offspring (reproduce); power of a species to multiply rapidly; the rate at which females produce ova.

Feedback: The return of apart of the output of a system as input.

Feeding deterrent: A chemical that prevents feeding of pests on treated material.

Felt chamber: Atrium of a spiracle lined with fine hairs.

Femur (pl. femora): The third segment of an insect leg, following the coxa and trochanter; often the stoutest leg segment (Fig. 114).

Fenestrae: Openings in the dorsal diaphragm through which hemolymph can pass.

Feral: Existing in the wild.

Fermentation: Breakdown of complex molecules by microbes, as carbohydrates by yeast.

Fertility: The rate at which fertilized eggs (zygotes) are produced.

Fertilization: Union of two gametes (egg and sperm) to form a zygote and initiate the development of an embryo. (L. *fertile* = fruitful, from *fero* = to bear).

Fetus, foetus: The later stages of development of an embryo within an egg or uterus.

Fiber tracts: Groups of axons within a ganglion that run parallel to one another.

Fibril: The contractile unit of a muscle cell (fiber).

Fidelity: The accuracy with which a population estimates follows actual numbers in a population.

FIFRA: Federal Insecticide, Fungicide, and Rodenticide Act.

Filament: A slender threadlike structure.

Filaria. A type of small, thin, nematode parasite; usually blood parasites in vertebrate hosts.

Filariasis: A disease state caused by infection with filarial roundworms (nematodes) and transmitted by *Culex* sp. of mosquitoes.

File: A toothed or ridged structure used in sound production by stridulation through contact with a **scraper;** a filelike ridge on the ventral side of the tegmen, near the base, a part of the stridulating mechanism in crickets and long-horned grasshoppers.

Filiariasis: Infection with parasitic nematodes, often transmitted by mosquitoes.

Filiform: Hairlike or threadlike; filiform antenna.

Filler: Inert component of pesticide dust or granule formulation.

Filter chamber: A modification of the gut of many Homoptera (such as aphids), permitting much water and some carbohydrates to bypass the midgut; close association between anterior and posterior regions of the alimentary canal, found in insects like spittlebugs, which ingest food with high water content; facilitate passage of water in such a way that the midgut is bypassed (Fig. 115).

Filter feeder: An insect that seives particles from water by means of brushes or webs.

Filth: Food contamination by rodents, insects, birds, or parts thereof, or any other objectionable matter.

Fin: An extension of the body of an aquatic animal; used in locomotion or steering.

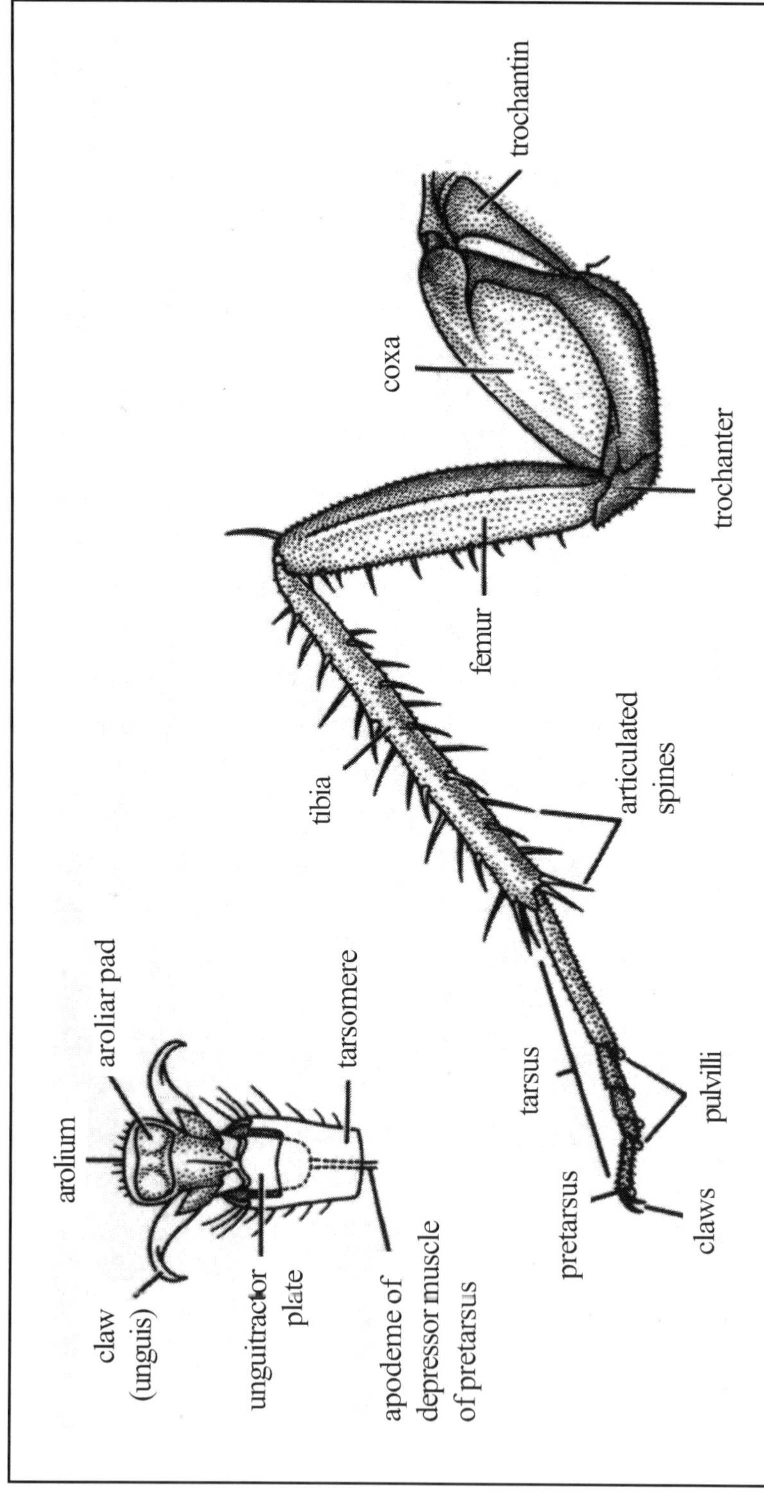

Fig. 114. The hind leg of cockroach, *Periplaneta Americana* (Blattodea: Blattidae), with enlargement of ventral surface of pretarsus and last tarsomere.

Finite rate of increase: Number of individuals added to the population per head per unit of time.

Fire fly: Lampyrid and elaterid beetles have the capacity to emit light. Their winged-adults are erroneously called firefly where as the non-flying adults and larvae of fire fly are called glow worms.

First Reviser: The first author to publish a definite choice of one among two or more conflicting nomenclatural or zoological interpretations which are equally available under the Rules; in order to qualify as first reviser an author must give evidence of a choice between available alternatives.

Fission: A sexual reproduction by division into 2 or more parts usually equivalent. (L. *findere* = split).

Fitness (Darwinian; inclusive, extended): The contribution of an individual to the gene pool through its offspring (Darwinian) or by enhanced success of its kin (inclusive, extended).

Fixed action pattern: A segment of behavior performed in a stereotyped, species-specific manner.

Flabellate: With fanlike processes or projections; flabellate antenna.

Flabellum (pl., flabella): A fanlike or leaflike process (Hymenoptera); in bees, the lobe at tip of the glossae ('tongue') (Fig. 116); the small spoon-shaped tip of the "tongue" of bees.

Flagellate, Flagellata: A class of Protozoa characterized by the possession of flagellum *e.g., Trypanosoma.*

Flagellomere: One of the subdivisions of a 'multisegmented' (actually multiannulate) antennal flagellum.

Flagellum (pl. flagella): The third part of an antenna, distal to the scape and pedicel; more generally, any whip or whip-like structure; a whiplike structure; that part of the antenna beyond the second segment; long, apical part of the antenna, commonly composed of several subsegments (Fig. 2.14); the outermost part of the antenna, beyond the scape and pedicel, usually divided into many subsegments (flagellomeres).

Flagellum: Fine but long thread-like with sinuous movement projecting from a cell *e.g.*, in *Trypanosoma.*

Flagging: (i) A branch with yellowing or dead foliage on an otherwise healthy tree as a result of rapid killing by adverse abiotic conditions, disease agents, or insects. (ii) Identification of an area to be treated (or to be left untreated) by aerial application.

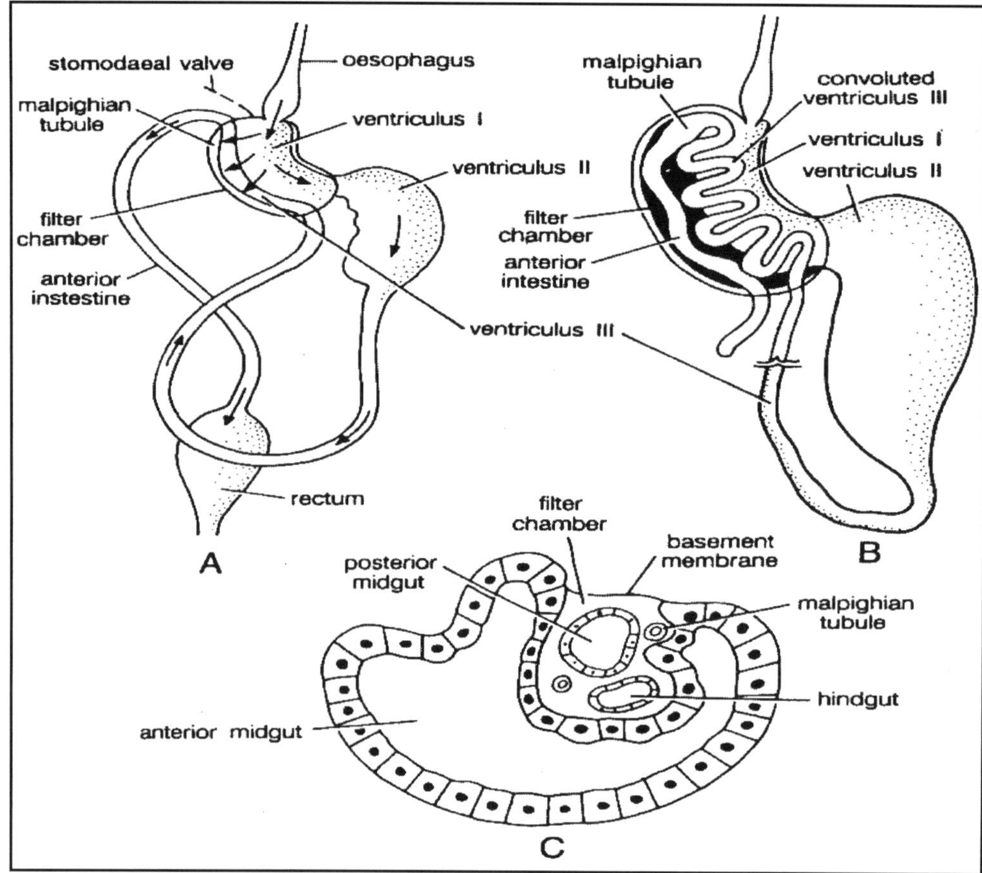

Fig. 115. Filter chambers of bugs. (a) Filter chamber with straight ventriculus, (b) Filter chamber with convoluted ventriculus. (c) T. S. filter chamber.

Flame cell: A type of hollow terminal excretory cell in certain invertebrates that contains a beating (flame-like) group of cilia.

Flaming: Using a tool that emits fire to destroy pests.

Flavines: Greenish yellow pigments found in insects.

Flaxseed stage: The puparium of the Hessian fly.

Flexion-line: A line along which a wing flexes (bends) when in flight (Fig. 14).

Flight peak: The time of maximum flying during a flight period of adult insects.

Float: A partial envelope, more or less inflated, around the anopheline eggs of mosquitoes.

Floating row cover: Layer of lightweight translucent material put over plants as protection from insect attack and environmental stress.

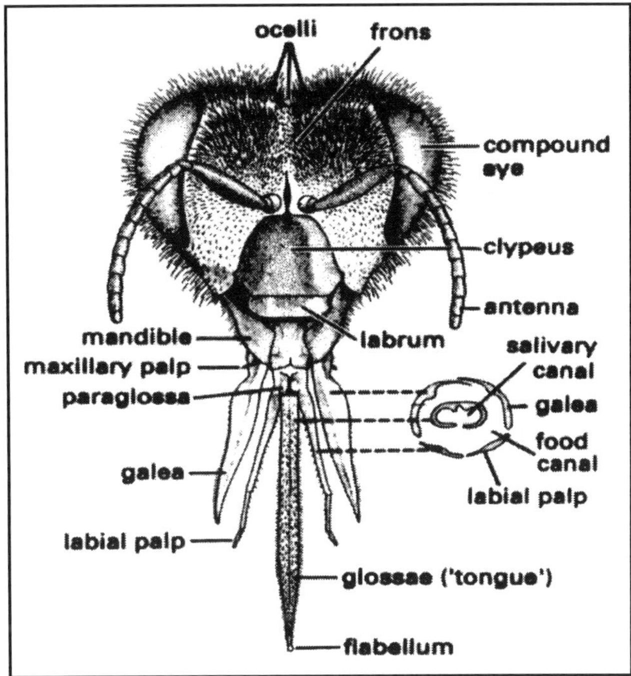

Fig. 116. Frontal view of the head of a worker honey bee, *Apis mellifera* (Hymenoptera: Apidae), with transverse section of proboscis showing how the 'tongue' (fused labial glossae) is enclosed within the sucking tube formed from the maxillary galae and labial palps.

Flocculent: Woolly; like tufts of wool.

Flora: The plant life of a region (cf. Fauna, Biota); the total of plants growing without cultivation in an area. The plants particular to a given geological formation.

[**Flowability:** Property of flowing possessed by dusts, colloids, liquids, and some pastes.

Flowable (FL): A pesticide formulation in which the active ingredient is wet-milled with a clay diluent and water; has a puddinglike consistency and can be mixed with water for spraying; pesticide formulated as a suspension of finely ground active ingredient mixed with liquids and emulsifiers that can be mixed with water.

Fluctuating asymmetry: The level of deviation from absolute symmetry in a bilaterally symmetrical organism, argued to be due to variable stress during development.

Fluorescent tracer: Fluorescent material added to a spray to aid the assessment of spray deposits on plants.

Fogger: Pesticide application equipment that breaks the pesticide into very fine droplets (aerosols or smokes) and blows or drifts the drops onto the target area.

Fold-lines: A line along which a wing is folded when at rest.

Foliar application: Spraying a pesticide or fertilizer onto the leaves of target plants.

Follicle: A minute cavity, sac, or tube; structure composed of oocytes, nurse cell, and follicular epithelium; a tubule of the testis in which sperm are produced.

Follicular epithelium: Layer of cells that invests a developing oocytes and that secretes the chorion (Fig. 117).

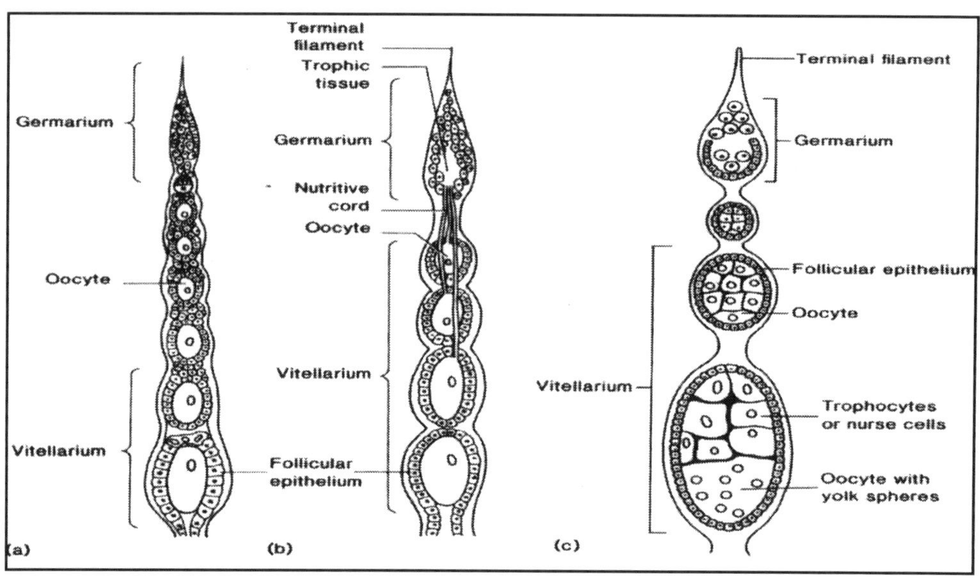

Fig. 117. Ovariole types. (a) Panoistic. (b) Telotrophic. (c) polytrophic.

Follicular relic: Morphological evidence left in the ovary showing that an egg has been laid (or resorbed), which may include dilation of the leumen and / or pigmentation.

Fontanelle: A small, depressed, pale spot on the front of the head between the eyes (Isoptera); depression on dorsum of head in many species of termites.

Food canal: Canal anterior to the cibarium through which fluid food is ingested (Fig. 118).

Food chain: A simplified linear depiction of trophic relationships that illustrates the flow of energy from producers to herbivores to carnivores;

an assemblage of organisms, including producers, consumers, and decomposers, through which energy and materials may move in a community.

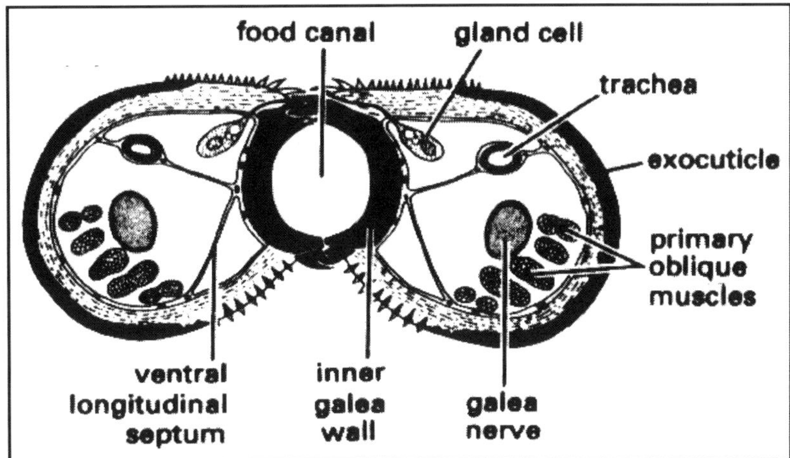

Fig. 118. Mouthparts of the cabbage white or cabbage butterfly, *Pieris rapae* (Lepidoptera: Pieridae): transverse section of the proboscis in the proximal region.

Food Quality Protection Act (FQPA): Act instituted in 1996 to improve food safety in the USA.

Food web: An ordered depiction of the interconnections that occur among the organisms in food chains in a given environment.

Foodspray: A nutrient-rich liquid food that is sprayed onto foliage to provide food for beneficial insects and so to increase their egg production (e.g., lacewings).

Forage: To seek and gather food.

Foramen magnum: The posterior opening of the head capsule through which internal structures (alimentary canal, dorsal nerve cord, and so on) pass to the thorax (Fig. 119).

Fore: Anterior, towards the head.

Fore wings: The anterior pair of wings, usually on the mesothorax (Fig. 1).

Forceps: The claspers or pincers at the apex of the abdomen in some insects; most evident as the male cerci in earwigs (Dermaptera); part of the male genitalia in many different insects used either for copulation or for defense purposes.

Fore-coxae: Coxae of the forelegs (most anterior pair).

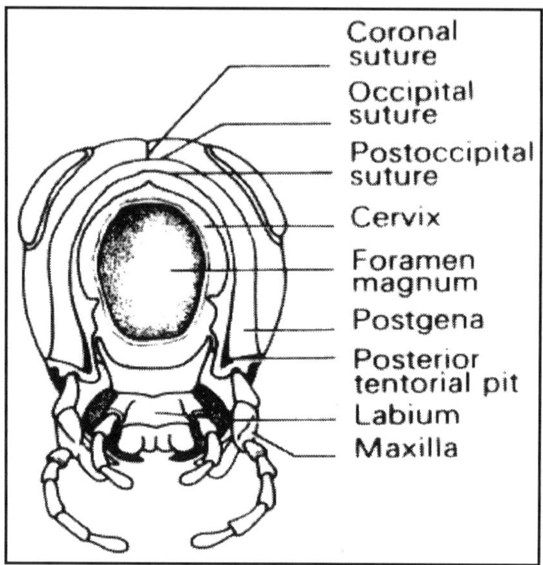

Fig. 119. Posterior view of generalized insect head.

Foregut: Anterior region of the insect alimentary canal, also called the stomodaeum; the anterior portion of the alimentary tract, from the mouth to the midgut derived from the ectoderm (Fig. 120).

Forensic entomologist: A scientist who studies the role of insects in legal matters.

Forensic entomology: The employment of entomological information in criminal investigations such as in assessing time, location, or cause of a human death or a criminal act.

Form: A neutral term for a single individual or taxonomic unit (cf. Group, Neutral term).

Formamidine: A synthetic organic amine insecticide; a recent type of insecticide that appears to mimic some insect neuro transmitters (e.g., chlordimeform).

Formenkresis: A collective category of allopatric subspecies or species (Kleinschmidt); in paleontology, a group of related species or variants.

Formulation (pesticide formulation, insecticide formulation)**:** The form in which an insecticide is prepared for practical use; a mixture of active and inert ingredients for killing pests; the combination of active pesticide ingredients, inert ingredients, and other additives as sold commercially. Specific formulations are designed to permit accurate and safe pesticide application; statement of nature and amount of all constituents of a

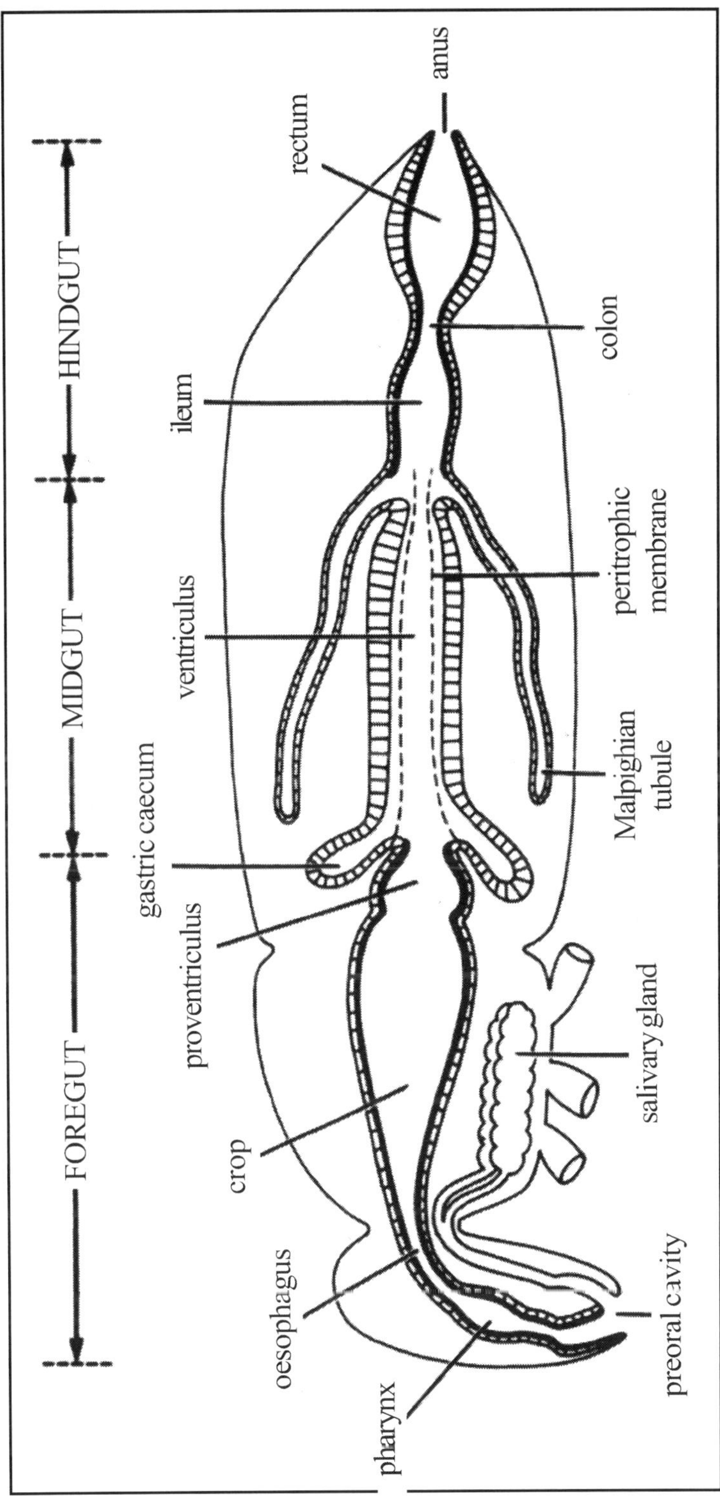

Fig. 120. Generalized insect alimentary canal showing division into three regions. The cuticular linning of the foregut and hindgut are indicated by thicker black lines.

pesticide concentrate; method of preparation of a pesticide concentrate; the components and proportions of additional substances that accompany an insecticide when prepared for application.

Fossil: The remains of plants or animals, preserved in geological past, other than the recent epoch.

Fossorial: Modified for digging; in the habit of digging or burrowing; legs bearing sclerotized claws for digging (Fig. 121).

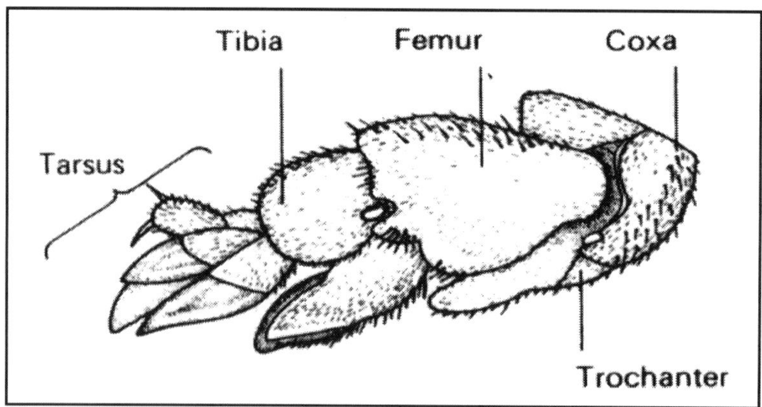

Fig. 121. Fossorial (digging) foreleg of a mole cricket.

Foulbrood: A bacterial disease of honey bee larvae and pupae.

Founder effect: Speciation resulting from the establishment of a small population in an entirely new area and the subsequent divergence of the resulting population from the parent stock.

Frass: The solid faeces of insects; plant fragments made by a wood - boring insect, usually mixed with excrement; insect excreta; usually a mixture of plant fragments and insect excrement produced by plant-boring insects, most commonly wood-boring insects.

Free living: Not attached or parasitic; capable of independent movement and existence.

Freezing tolerance: Survival despite freezing of body fluids.

Frenate coupling: A form of wing-coupling in which one or more hind-wing structures (**frenulum**) attach to a retaining structure (**retinaculum**) on the fore wing.

Frenulum: A bristle or group of bristles arising at the humeral angle of the hind wing (Lepidoptera); spinelike wing-coupling mechanism characteristic of several Lepidoptera (Fig. 122); the spines or bristles on the base of the costal vein in the hindwing of Lepidioptera which serve to lock with the fore-wing retinaculum in flight.

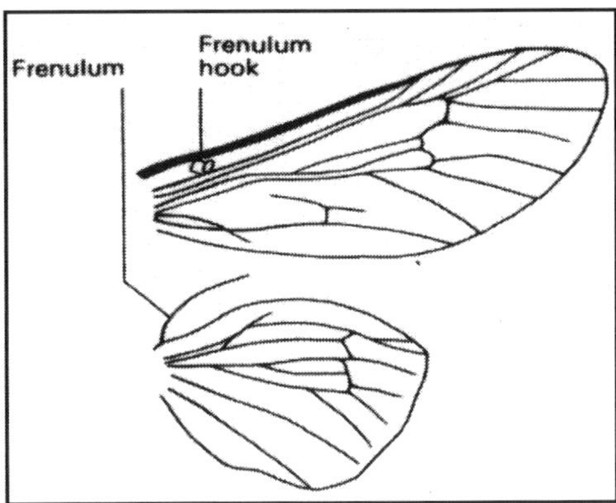

Fig. 122. Wings of a frenate moth (Psychidae).

Frons: Frontal region delimited by the frontal sutures; the head sclerite bounded by the frontal and epistomal sutures and including the median ocellus (Fig. 123); the single medio-anterior sclerite of the insect head, usually lying between the epicranium and the clypeus.

Frontal bristles: Bristles above the antennae, away from the edge of the compound eye (Diptera).

Frontal ganglion: Ganglion of the stomatogastric system, located on the dorsal midline of the foregut just anterior to the brain; the median ganglion of the stomodaeal nervous system; lies above the oesophagus (Fig. 124).

Frontal lunule: A small crescent-shaped sclerite located just above the base of the antennae (Diptera; Fig. 124).

Frontal suture: One of two sutures arising at the anterior end of the coronal suture and extending ventral toward the epistomal suture (Fig. 124); a suture shaped like an inverted **U** with the base of the **U** crossing the face above the bases of the antennae, and the arms of the **U** extending downward on each side of the face (Diptera); two sutures that diverge ventrally across the anterior portion of the head.

Frontal vitta: An area on the head between the antennae and the ocelli (Diptera).

Frontalin: A sex attractant pheromone of male bark beetles.

Frontoclypeal (or **epistomal) suture:** A groove that runs across the insect's face, often separating frons clypeus.

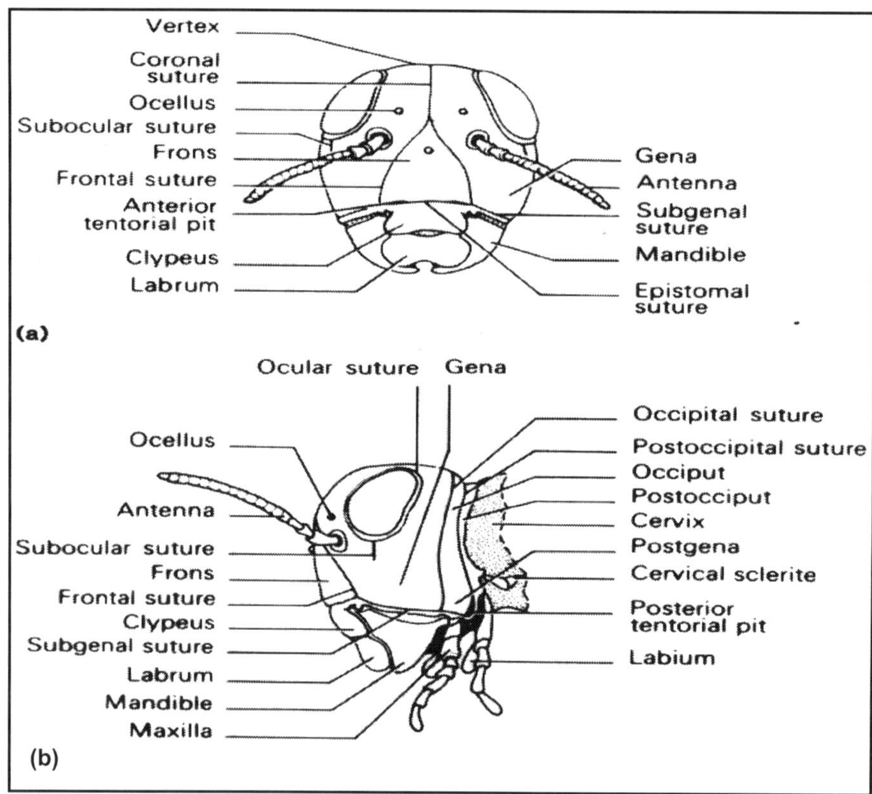

Fig. 123. Generalized insect head (a) Anterior view. (b) Lateral view.

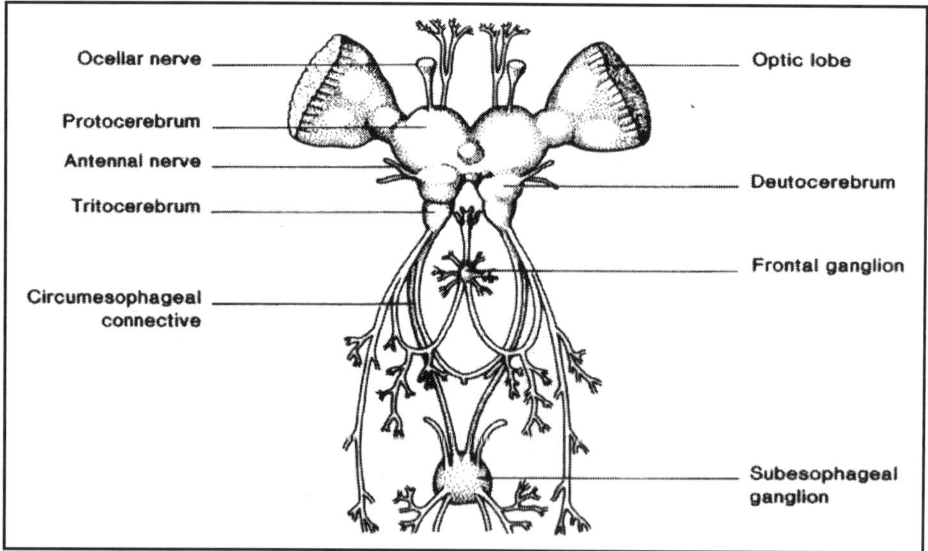

Fig. 124. Brain and stomatogastric nervous system of the grasshopper *Dissosteira Carolina;* Lateral view.

Frontoclypeus: The combined **frons** and **clypeus.**

Fronto-orbital bristles: Bristles on the front next to the compound eyes (Diptera).

Fructivore (adj. fructivorous): A fruit-eater.

Fruiting bodies: The reproductive structures of fungi that contain spores.

Full bibliographical reference: See Bibliographical reference.

Full bibliographical synonymy: A reasonably complete list of references to given taxonomic category arranged so as simultaneously to serve the needs of nomenclature (chronology of names) and zoology (pertinent taxonomic and biological sources) (cf. Synonymy).

Fumigant: Pesticides that are a gas under normal temperature and pressure. They are injected into the soil or into a confined area, such as a grain storage structure, to kill pests; a volatile insecticide that enters an insect via the tracheal system exhibiting toxicity in the vapour phase; insecticides that enter through the spiracles and tracheal system in gaseous form.

Fundatrigenia: Daughter of female aphid living on the primary host plant.

Fundatrix: Apterous, parthenogenetic, viviparous female aphid emerging from the overwintering egg in the spring, on the primary host plant.

Fungicide: A substance or pesticide that kills fungi.

Fungivore (adj. fungivorous): A fungus-eater.

Fungus (pl. fungi): Single to multi-celled eukaryotic organisms that lack chlorophyll and grow on living or dead organisms, and therefore may cause plant disease. They grow from spores and commonly produce tiny thread-like growths (mycelium) that ramify through a substrate to absorb nutrients.

Funiculus (or funicle): The central and major part of geniculate antenna between the pedicle and the club (Fig. 15).

Furca: A fork or forked structure; a forked apodeme arising from a thoracic sternum (Fig. 23,fu); sternal apophysis in the form of a V (Fig. 125); the forked skeletal support in the upper part of the mouthparts of flies.

Furcal pits: External manifestations of the sternal apophyses (Fig. 126).

Furcula: A forked abdominal organ used in springing or jumping of Collembola (Fig. 127); fork-shaped structure on the abdomen of the springtail that engages the tenaculum and is used to propel the insect through the air.

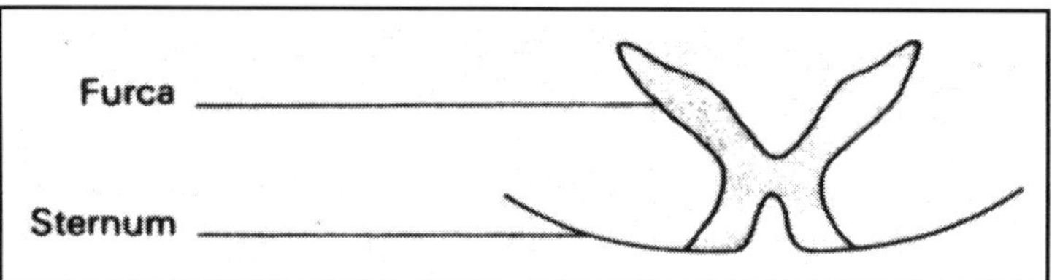

Fig. 125. Sternal apophyses in the form of a furca.

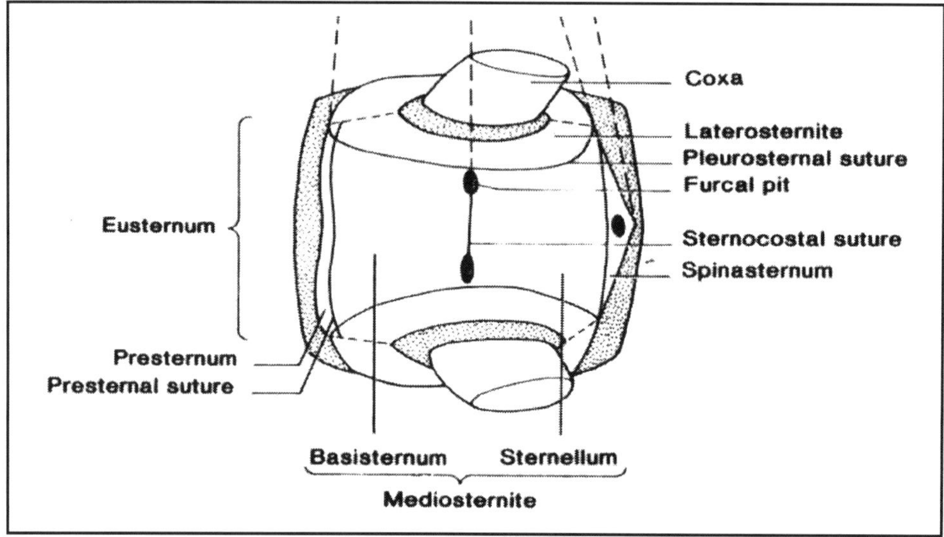

Fig. 126. Sternum of generalized wing-bearing thoracic segment.

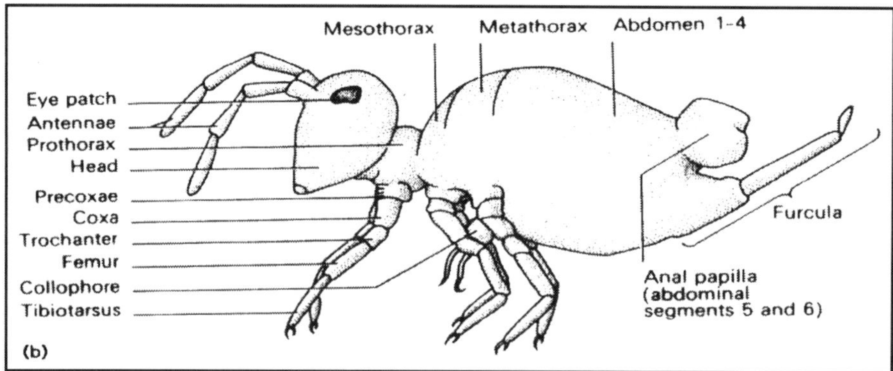

Fig. 127. Collembola, springtails; Symphypleonid.

Furrow application: Placement of pesticides with seed in the furrow at the time of sowing.

G

Gain threshold: The beginning point of economic damage, expressed in amount of harvestable produce; when cost of suppressing insect injury equals money to be gained from avoiding the damage.

Galea: The lateral lobe of the maxillary stipes; tonguelike, distal lobe attached to the stipes of the maxilla; an apical lobe of the maxilla of an insect; the outer two-jointed lobe of the "pincers" present on the maxilla of insect mouthparts; in adult Lepidoptera the two are joined and enlarged to form the proboscis. (Fig. 63)

Gall: Swellings, deformities, or growths of plant tissue that result from the activities various types of pests, including bacteria, nematodes, mites and insects; an abnormal growth of plant tissues, caused by the stimulus of an animal or another plant (Fig. 128).

Gallicolous: Living in a plant gall either as the producer or inquiline.

Gallinaceous: A game bird; member of the Order Galliformes.

Gamete: Reproductive cells that are usually formed by meiosis, such as egg and sperm; reproductive cell whose nucleus fuses with that of another gamete amounting to fertilization; often cytoplasm also of the gametes fuses with that of the other.

Gametocyte, macro: Female gamete or acting as female gamete destined to fuse with the male gamete (microgamete); size bigger than that of micro (male) gametocyte.

Gametocyte, micro: Male gamete or acting as male gamete destined to fuse with the female (macro) gamete; size smaller than that of macrogametocyte.

Gametocyte: Cell which forms gametes by meiosis division.

Gamma diversity: The inclusive organismal diversity of all habitats within an area.

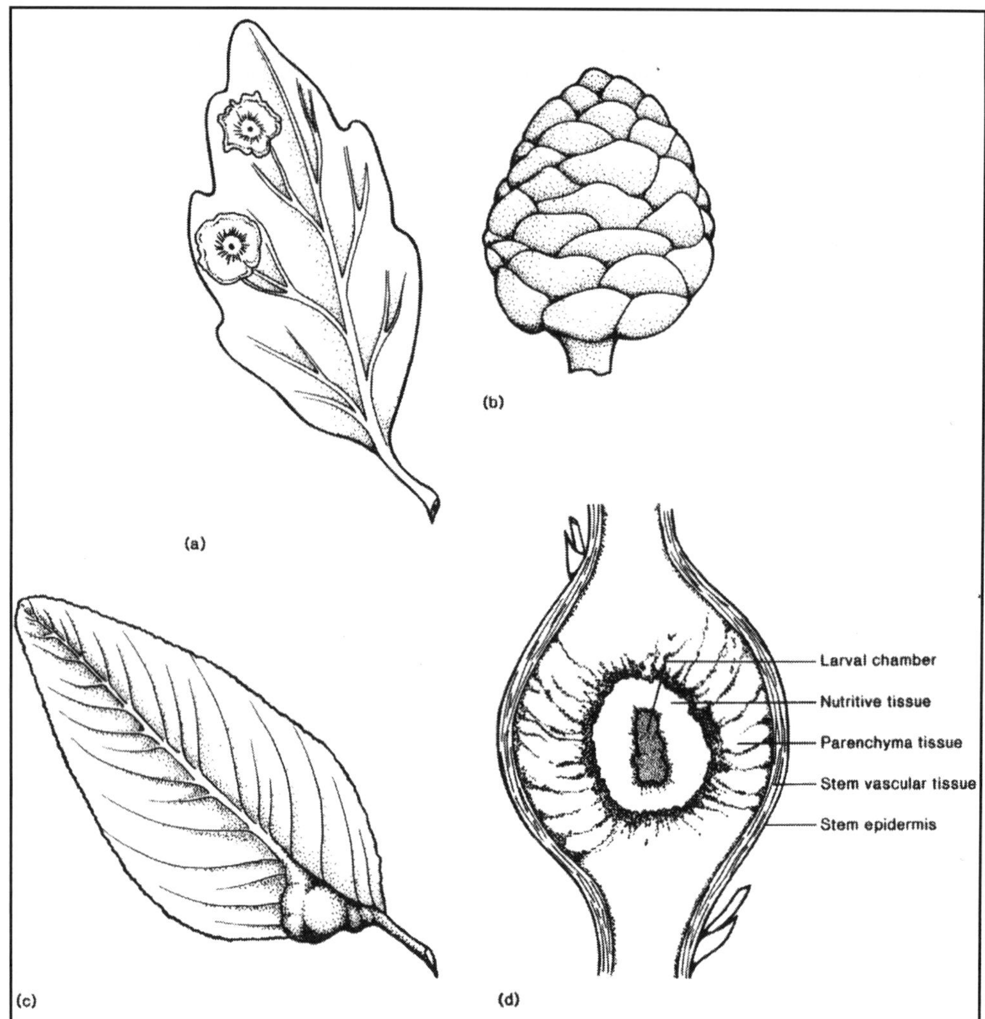

Fig. 128. Insects galls. (a) *Andricus pattersonae* (Hymenoptera; Cynipidae) on leaf of blue oak. (b) Pine-cone willow gall caused by *Rhabdophaga strobiloides* (Diptera; Cecidomyiidae). (c) *Pemphigus betae* (Homoptera; Aphididae) on *Populus angustifolia.* (d) Longitudinal section through the goldenrod ball gall caused by *Eurosta solidaginins* (Diptera; Tephritidae). (a) After Essig; (b) After Boreal Labs key card; (c) from Whitham 1978; (d) from Abrachamson, 1989.

Gamma taxonomy: The level of taxonomy dealing with various biological aspects of taxa, ranging from the study of intraspecific populations to studies of speciation and of evolutionary rates and trends (cf. Alpha taxonomy, Beta taxonomy).

Ganglion (pl. ganglia): A bundle of nervous tissue (Fig. 38); a knotlike enlargement of a nerve, containing a coordinating mass of nerve cells; a mass of nervous tissue, the basic anatomical unit of the central nervous system; a group of neurons; a group of concentration of nerve cell bodies, set apart, and acting as a center of nervous influence; a nerve centre; in insects, forming fused pairs of white, ovoid bodies lying in a row ventrally in the body cavity, linked by a double nerve cord.

Gas exchange: The system of oxygen uptake and carbon dioxide elimination.

Gas gills: Specific respiratory (gas exchange) surfaces on aquatic insects, often as abdominal lamellae, but may be present almost anywhere on the body.

Gaster: Collective term for the hymenopteran abdominal segments posterior to the petiole; the rounded part of the abdomen posterior to the nodelike segment or segments (ants).

Gaster: The swollen part of the abdomen of aculeate Hymenoptera, lying posterior to the **petiole**; and always excluding the propodeum (first abdominal segment) (Fig. 129).

Gastric caeca: An outgrowth from the forepart of the mid-gut of insects.

Gastric caecum (pl. gastric caeca): A group of diverticula commonly found in the midgut; a saclike outpocketing of the anterior portion of the midgut; a fingerlike, anterior extension of the midgut that serves a function in food absorption.

Gastric mill: A grinding structure within the foregut of some orthopteroid insects.

Gastric nerves: Nerves connecting the hypocerebral ganglion and ventricular ganglion.

Gastrocoel: The primitive digestive cavity of metazoan embryo, formed by gastrulation. (Gr. *gaster* = stomach + *koilos* = hollow).

Gastrodermis: The inner of the two cellular layers of coelenterates, equivalent to endoderm.

Gastrovascular: Serving for both digestion and circulation.

Gastrula: An embryo in the process of gastrulation, when a single-layered blastula turns into a three-layered embryo consisting of ectoderm, mesoderm, and endoderm, referring to reproductive organs (L. *gingo* = beget).

Gastrulation: Formation of the mesoderm and endoderm.

Gause's rule: The theory that no two species with identical ecological requirements can coexist in the same place.

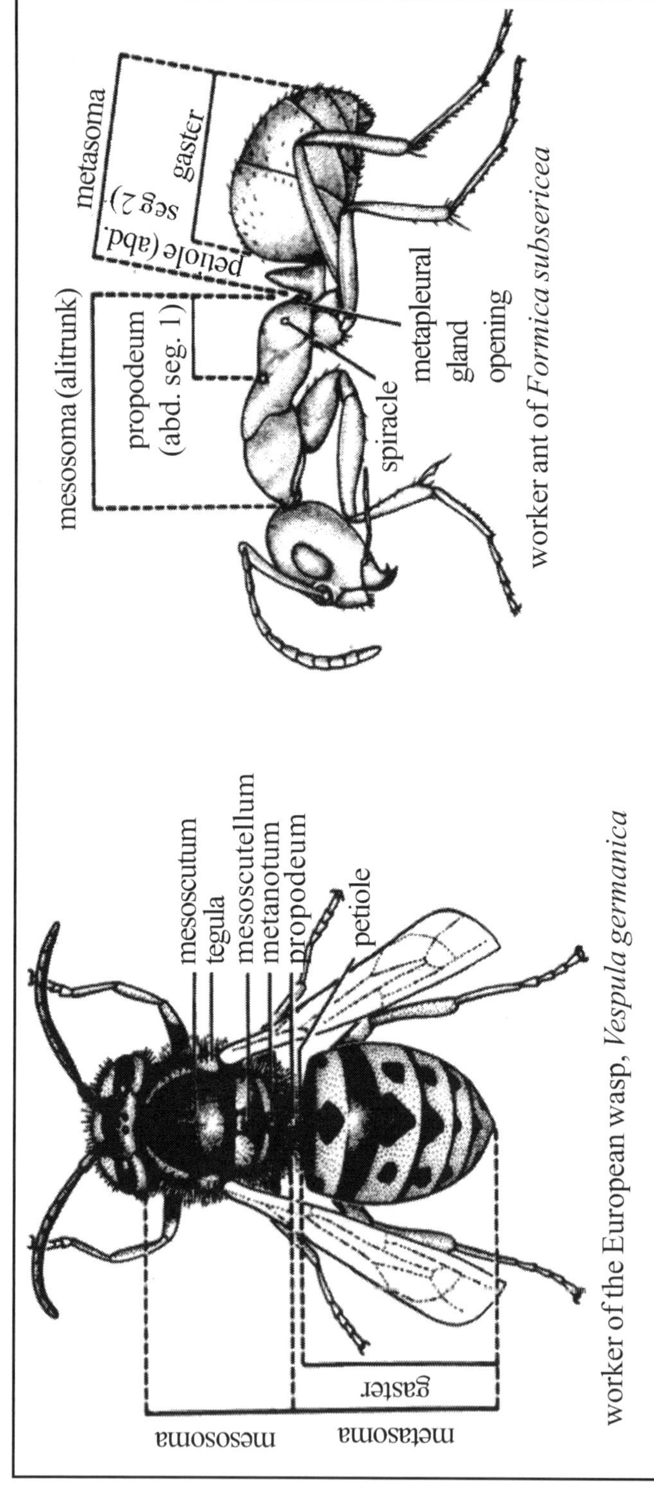

Fig. 129. Hymnoptera (wasps, bees, ants, sawflies and woodwasps) showing gaster.

-gen, -geny: Producing ; generation.

Gena (pl., genae): The part of the head on each side below and behind the compound eyes, between the frontal and occipital sutures; the cheek; lateral "cheek" region of the head delimited by the compound eye, subgenal suture, and occipital suture (Fig. 123).

Genal comb: A row of strong spines brone on the anteroventral border of the head (Siphonaptera).

Gene flow: The exchange of genetic factors between populations owing to dispersal of zygotes or gametes, e.g., pollen.

Gene frequency: The percentage of a given gene in a population (cf. Gene, Population).

Gene pool: The sum total of all the genes of all the individuals in a population.

Gene(s): A hereditary determiner; the unit of inheritance that consists of a sequence of nucleotides in a DNA molecule, contains the hereditary information for a single characteristic, and occurs at a particular locus on a chromosome transmitted from generation to generation by the gametes, and controlling the development of the individual (cf. Chromosome).

Gene-for-gene relationship: The concept of a resistant (or susceptible) allele at agene locus in a particular plant cultivar corresponding to a susceptible (or virulent) allele at the same locus in the pest; a one-to-one relationship between host and pest genotypes.

General equilibrium position (GEP): A population's long-term average density.

Generation: The time required to complete the life cycle of an organism, from birth (hatch, or germination) to sexual maturity and reproduction; the period from any given stage in the life cycle (usually adult) to the same stage in the offspring; a cohort of offspring from a parent population moving through a life cycle together.

Generotype: Genotype in the nomenclatural sense (cf. Genotype).

Genetic control: An approach that emphasizes decreased fecundity instead of increased mortality. See *sterile insect release method (SIRM)*; altering the genetic makeup of organisms to inhibit their reproduction and survival.

Genetic drift: Genetic changes in populations due to random fixation rather than to selection; the so-called "Sewall Wright effect" (cf. Local population).

Genetic engineering: Altering the genetic constitution of organisms by removing, modifying, or adding to the DNA in their chromosomes, or modifying the expression of extant DNA in organisms.

Genetic isolation: A condition in which interbreeding between two or more populations is prevented by sterility barriers (cf. Reproductive isolation, Geographic isolation, Ecological isolation).

Genetic resistance: Genetically controlled plant resistance to insect attack.

Genetically modified organism (GMO): Used to inaccurately describe genetically engineered organism including transgenic crops. Technically, any domesticated organism derived from a wild species by artificial selection is a genetically-modified organism.

Geniculate: Elbowed; abruptly bent; refers to the antennae of weevils (Coleoptera) and Chalcidoidea (Hymenoptera).

Genital chamber: A cavity of the female body wall that contains the gonopore, also known as the **bursa copulatrix** if functioning as a copulatory pouch.

Genital ridge: Embryonic tissue of mesodermal origin that, together with the germ cells, are the precursors of the gonads.

Genitalia: All ectodermally derived structures of both sexes associated with reproduction (copulation, fertilization and oviposition); the external sexual organs and associated structures of insects (male or female); structures associated with the release of sperm or eggs.

Genotype: In nomenclature (not recommended, International Commission, 1948), the type species of a genus (cf. Type species); in genetics, the class in which an individual falls on the basis of its genetic constitution, without regard to visible characters (cf. Phenotype); the particular combination of genes present in the cells of an individual; the internal genetic or hereditary constitution of an organism without regard to its external appearance. Compare phenotype. (Gr. *genos* = race + *tyoos* = impression, form).

Genovertical plate: An area on the head above the antenna, next to the compound eye (Dipteral).

Genus (pl. genera, adj. generic): The name of the taxonomic category ranked between species and family; an assemblage of one or more species united by one or more derived features and therefore believed to be of a single evolutionary origin (i.e. **monophyletic**); a group of closely related species which is separated from related similar units (genera) by a decide gap, the size of the gap being in inverse ratio to the size of the unit (genus); the first name in a binomial or trinomial scientific name. Names of genera are Latinized, capitalized, and when printed are italicized.

Geographic information system (GIS): Computer-based system used for the collection, storage, and analysis of spatial data, including the ability to generate maps.

Geographic isolation: A condition in which interbreeding two or more allopatric populations is prevented by extrinsic barriers or geographic discontinuity (cf. Reproductive isolation, Ecological isolation, genetic isolation).

Geographical race: Subspecies (q.v.).

Geomagnetic receptor: A sense organ involved in the perception of magnetic fields.

Germ: The embryo within a seed.

Germ anlage: In embryology, the germ disc that denotes the first indication of a developing embryo **(Fig. 130).**

Germ band: In embryology, the postgastrulation band of thickened cells on the ventral gastroderm, destined to form the ventral part of the developing embryo (Fig. 130); thickened area of columnar cells of the blastoderm that develops into the embryo; a thickening of the blastoderm that produces the embryo.

Germ cell (s): A reproductive cell in a multicellular organism; cells that differentiate into gametes- eggs and sperm.

Germ layer: One of the two (in diploblastic) or three (in triploblastic) cell layers (ectoderm, endoderm, mesoderm) in an early embryo of a multicellular animal from which various tissues of the adult are formed.

Germ plasm: The material basis of inheritance; the gametes and the cells and tissues from which they form.

Germarium: The apical tissue of a testicular follicle or ovariole from which the germ cells originate.

Germarium: The structure within an **ovariole** in which the **oogonia** give rise to **oocytes** (Fig. 131); an area at the tip of the sperm follicles or ovarioles where sperm or egg formation is initiated.

Germination: The initiation of growth from a seed or spore.

Gestation: Contact chemoreception; "taste."

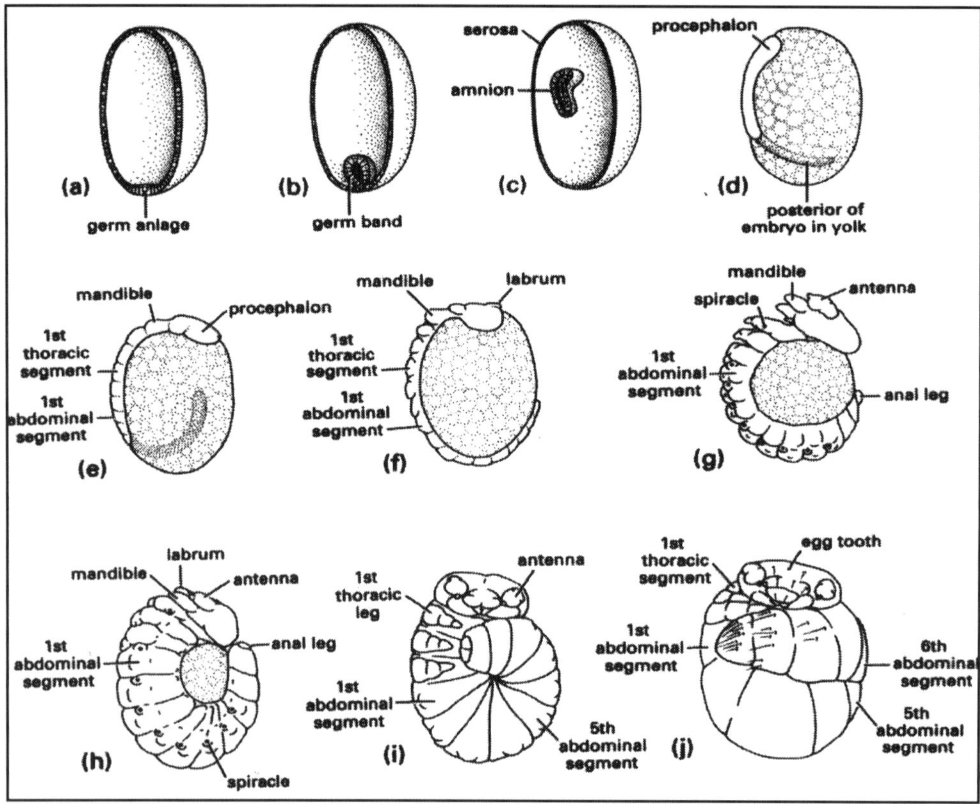

Fig. 130. Embryonic development of the scorpionfly *Panorpodes paradoxa* (Mecoptera: Panorpodidae): showing drawings of egg halves and gross morphology of developing embryos at various ages. After Suzuki, 1985.

Giant axon: A nerve fibre that conducts impules rapidally from sense organ to muscles; a large-diameter axon of an interneuron that traverses several body segments and conducts messages quickly.

Giant chromosomes: Chromosomes that are much larger than ordinary chromosomes, consisting of a very large number of identical DNA molecules; polytene chromosomes.

Gibberellins: Plant hormones that are involved in regulating cell elongation and other processes in plant tissues.

Gill: A gas exchange organ, found in various forms in aquatic insects; a respiratory organ in the aquatic immature stage of many insects that obtains dissolved oxygen from water an organ for aquatic respiration; *e.g.* in fishes; evaginations of the body wall or hindgut, functioning in gaseous exchanges in an aquatic animal.

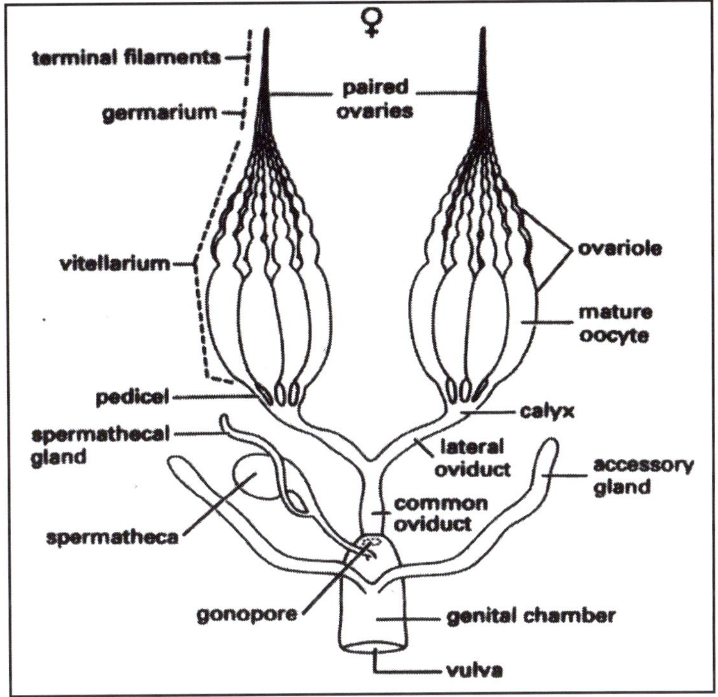

Fig. 131. Female reproductive system.

Girdling: Damage to, or removal of, a complete ring of outer and inner bark from a shoot, trunk, or root. Girdling prevents the normal downward translocation of photosynthate, resulting in death of tissue either above or below the girdle.

Gizzard: A heavily muscled portion of the digestive tract *e.g.* in earthworm and birds; the portion of the gut of insects immediately behind the crop; often thick -walled and muscular. See **Proventriculus**

Glabrous: Smooth, without hairs.

Gland, pl. glands: An organ with the function to build up one or more specific secretions which may be poured out of the glands by ducts (exocrine) or without ducts (endocrine).

Glands, accessory: Glands connected with the genital organs *e.g.* in insects.

Glands, ductless (endocrine): A gland that elaborates and secretes a hormone, or "internal secretion", directly into the blood stream; an endocrine gland *e.g.*, pineal/pituitary bodies of frog.

Glands, mammary: Glands secreting milk for the off-spring and important characteristic feature of mammals- the most highly evolved group in the animal world.

Glands, salivary: Glands secreting saliva which is passed on to mouth *e.g.* salivary glands of cockroach/man, variously modified also in some; *e.g.* in silkworm larvae, to secrete silk.

Glands, scent: Give secretion with characteristic odour which helps to track species/brings together opposite sexes of the species/acts as a means of defence.

Glands, sweat: Another skin gland of mammals secreting out a dilute solution of salts *etc.*- excretory products – on skin surface, due to evaporation of which on skin surface, body has an overall cooling effect *e.g.* sweat glands of man.

Glial cell: A cell surrounding the axon of a neuron.

Glial cells: Cells that invest neurons and aid in the functioning of these neurons.

Global positioning system (GPS): A system that uses triangulation to multiple satellites to allow a user to determine their geographic location with great accuracy.

Globose, Globular: Spherical or nearly so.

Glomeruli: Spheroidal collections of neuropile in each antennal lobe of the brain.

Glossa (pl. glossae)**:** The 'tongue', one of a pair of distal lobes on the inner apex of the prementum of the labium (Fig. 132).

Glottis: The opening from the pharynx into the trachea. (Gr. *glotta* = tongue).

Glycogen: A carbohydrate (polysaccharide) stored in the muscles and liver; "animal starch". (Gr. *glykys* = sweat + *gen* = come into being).

Gnathal segments: The posterior three segments of the developing insect head that give rise to the mouthparts.

Gnathobasic mandibles: Appendage with basal process that functions as a jaw.

Gnathochilarium: A platelike mouth-part structure in the Diplopoda, representing the fused maxillae and labium.

Gonad: A reproductive organ (ovary, testis) in which gametes (ova or sperm) are produced. (Gr. *gonos* = generation, seed).

Gonapophysis (pl. gonapophyses): A valve (part of the shaft) of the **ovipositor**; also in the genitalia of many male insects (Fig. 133); a mesal posterior process of a gonopod, in the female forming the ovipositor.

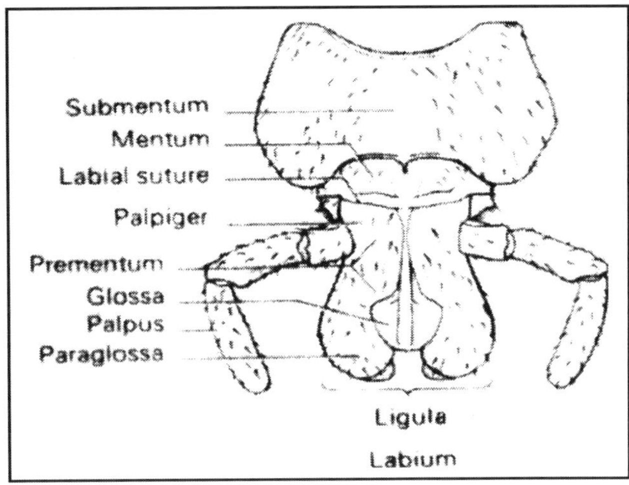

Fig. 132. Generalized mandibulate mouthparts as illustrated by a cockroach as showing Glossae.

Gonochorism: Sexual reproduction in which males and females are separate individuals.

Gonocoxite: The base of an appendage formed of **coxa** plus **trochanter,** of a genital segment (8 or 9) (Figs. 133); *see also* **Valvifer.**

Gonoduct: The duct leading from the genital organs, opening to the outside by the opening called gonopore.

Gonopod (s): A modified leg that forms a part of the external genitalia; appendages composing the ovipositor.

Gonopore: The external opening of the reproductive organs, opening of genital duct, in the unmodified female the opening of the common oviduct, in the male the opening of the ejaculatory duct.

Gonostyle: The style (rudimentary appendage) of the ninth segment, often functioning as a male clasper (Fig. 173).

Grade: A paraphyletic group, one which does not include all descendants of a common ancestor, united by shared primitive features.

Gradual metamorphosis: Alterative term for incomplete or partial metamorphosis.

Graft transmissible: A virus or phytoplasma that can be spread plant-to-plant through grafting.

Graft union: Where the rootstock joins the scion or shoot of a grafted tree or vine.

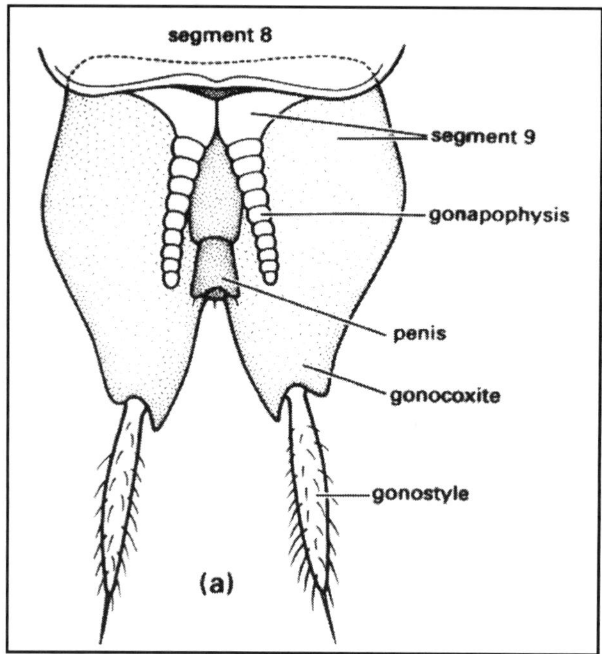

Fig. 133. Male external genitalia: gonapophysis.

Granaries: Storage structures for the grain.

Granular (G): A type of pesticide formulation prepared by applying liquid insecticide to coarse particles of porous material like clay, corncobs, or walnut shells.

Granule applicator: Machine designed to apply measured quantities of granules.

Granules (G): Dry, particulate pesticide insecticide formulation which consists of relatively large (1-3 mm) particles of an inert material (carrier – pumice, Fuller's earth, clay, corn cob, or walnut shell rice husks) which are impregnated with insecticide or on which; the insecticide is coated used mainly for soil application, but sometimes for foliar application (pumice formulation).

Granulosis virus (GV): A virus that multiplies in both the nucleus and cytoplasm of host cells, usually in the fat body.

Gravid: An organism that contains fertile eggs.

Grease band: Adhesive material (e.g. resin in castor oil, or 'Sticktite') applied as a band around a tree to trap or repel ascending wingless female moths, or ants.

Green manure crop: Plants grown and then incorporated into the soil to provide nutrients, organic matter, and to improve soil structure. Often legumes or grasses.

Gregarious: Forming aggregations.

Gregarious: Living in groups; animals which have the habit of moving together and living together *e.g.* Desert locust, parrots *etc.* (L. *grex* = head).

Gressorial: Walking, or adapted for walking.

Ground cover: The degree to which chemical applied covers the ground area being treated

Group: A neutral term for a number species within a genus (cf. Complex, Neutral term, Section).

Growth form (population growth form): The shape of a population density curve (Figure 5.4).

Growth regulator: A chemical that alters the normal growth or reproduction of an organism.

Growth threshold: see **Development threshold.**

Grub (White): A scarabaeiform larva; thick-bodied, with a well-developed head and thoracic legs, without abdominal prolegs, usually sluggish in behaviour; general term for larvae of Coleoptera.

Grylloblattodea, order: (*gryll,* cricket; *blatta,* cockroach) rock crawlers or ice bugs (Fig. 134); A small order of primitive insects related to the cockroaches and crickets for which there is no common name

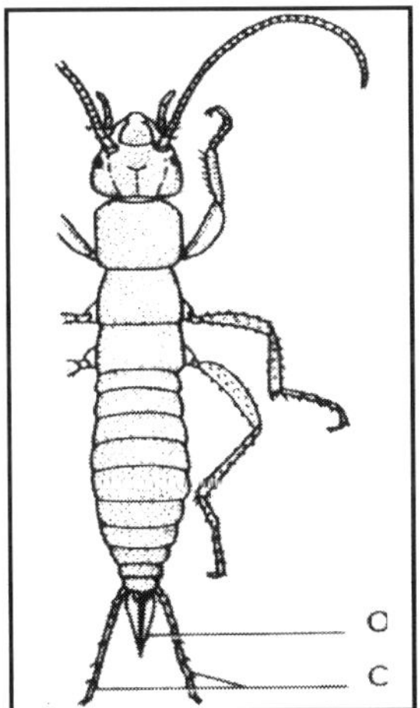

Fig. 134. Grylloblattodea. A rock crawler, *Grylloblatta campodeiformis.*

Guild (or diffuse) coevolution: Concerted evolutionary change that takes place between groups of organisms, as opposed to between two species; *see also* **Specific coevolution.**

Guild: Organisms that have similar resource requirements and trophic habits, consequently filling similar roles in the community.

Gula: A sclerite on the ventral side of the head between the labium and the foramen magnum (Fig. 197, gu); structure formed from the fusion of the postgenae, characteristic of many beetles and neuropterous insects (Fig. 135); a ventromedian sclerotized plate on the head of prognathous insects.

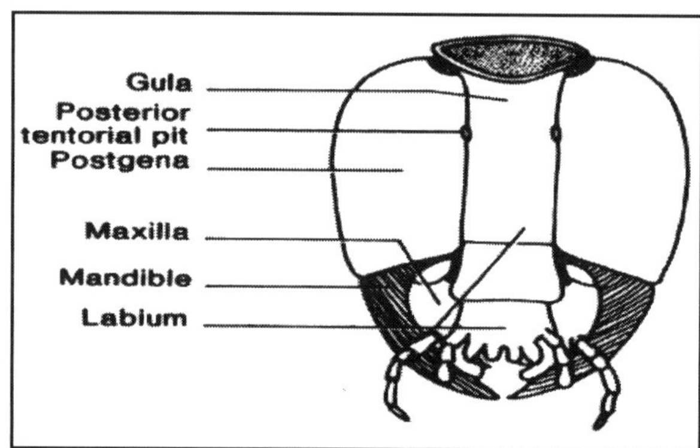

Fig. 135. Frontal view of the head and dissected mouthparts of an adult of the European earwing, *Forficula auricularia* (Dermaptera: Forficulidae). Note that the head is prognathous and thus a gular plate, or gula, occurs in the ventral neck region.

Gulamentum: Fusion of the gula and mentum.

Gular sutures: Longitudinal sutures, one on each side of the gula.

Gummosis: The abnormal exudation of gum, latex, or sap from a plant that results from parasite infection.

Gustation: The act of sensation of tasting.

Gynandromorph: An abnormal individual containing structural characteristics of both sexes; an individual in which one part of the body is masculine, the other feminine; most frequent are bilateral gynandromorphs, in which the left and right halves are of different sex; insects displaying some typically female body parts and some typically male body parts.

Gyne: A reproductive female hymenopteran, a **queen.**

Gynoecious: Plants that produce only female flowers.

H

Hard pesticide: One which persists for a long time in the environment; particularly applies to organochlorine insecticides.

Habit, Habitus: The general appearance of an organism; facies.

Habitat modification: Alteration of the environment where an organism occurs, e.g., cultivating land for agriculture.

Habitat: The environmental conditions and associated organisms that occur in a particular site; the place where plants and animals live, usually with a distinctive boundary (e.g. field, pond, sand-dune, rocky crevice); the natural or usual dwelling place of an individual or group of organisms. (L. *habitus* = condition).

Habituation: A form of associative learning in which a given stimulus no longer elicits a response; learning not to respond to a stimulus that provides no reward or punishment.

Habituation: Reduction in the response to a stimulus with repeated exposure, through modification of the central nervous system.

Haematophagous: Blood-sucking; blood –feeding.

Haemocoel (also **hemocoel):** The main body cavity of many invertebrates, including insects, formed from an expanded 'blood' system that is filled with blood (haemolymph) within which the internal organs lie.

Haemocoelous viviparity: Viviparity (producing living offspring) in which the immature stages develop within haemocoel of the parent female, as for example in Strepsiptera.

Haemocyte (also **hemocyte):** Insect blood cells.

Haemolymph (also **hemolymph):** The fluid filling the **haemocoel;** the unpigmented blood of insects.

Hair: A slender filamentous growth on the skin of mammals. A cuticular extension, also called a **macrotrichium** or **seta.**

Hair pencil: Found in some male Lepidoptera; a scent brush for dispersing pheromones, protrusible and usually located at the end of the abdomen.

Hair pencils: Tufts of fine setae serving to dust pheromone-coated particles onto a member of the opposite sex; eversible structures covered with fine hairs, providing extensive surface area that facilitates dispersal of pheromones.

Hair plate: A group of tiny trichoid sensory hairs located in appressed or overlapping areas of the body, that act as a **proprioceptor** for movement of articulating parts of the body (Fig. 136).

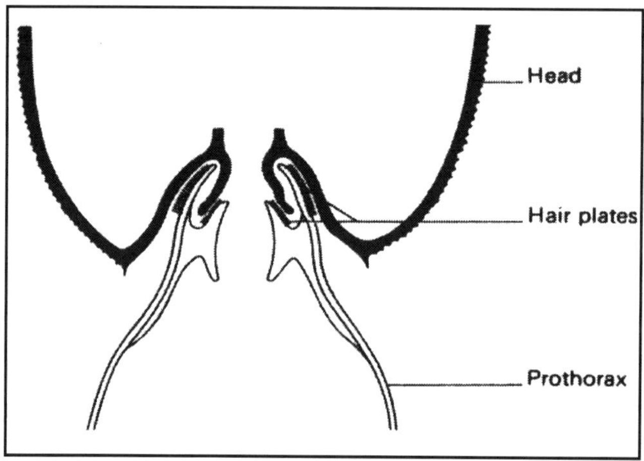

Fig. 136. Transverse section of the head of the ant *Formica polyctena,* showing hair plates on the prothorax.

Half-life: The time required for 50% of a pesticide to degrade.

Hallucination of arthropod infestation: A condition in which a person imagines he or she is being molested by small, difficult-to-locate forms, which localize on the body despite extraordinary preventive measures.

Halter (pl., halteres): A small knobbed structure on each side of the metathorax representing the hind wings which act as balancing organs (Diptera); a modified hind wing of a fly (Diptera), acting to maintain flight stability (Fig. 137).

Hamuli (sing., hamulus, adj. hamulate): Tiny hooks on the costal margin of the hind wings that attach to the anal margin of the forewings, affecting wing coupling; characteristic of the Hymenoptera (Fig. 138); minute hooks; a series of minute hooks on the anterior margin of the hind wing, with which the front and hind wings are attached together (Hymenoptera).

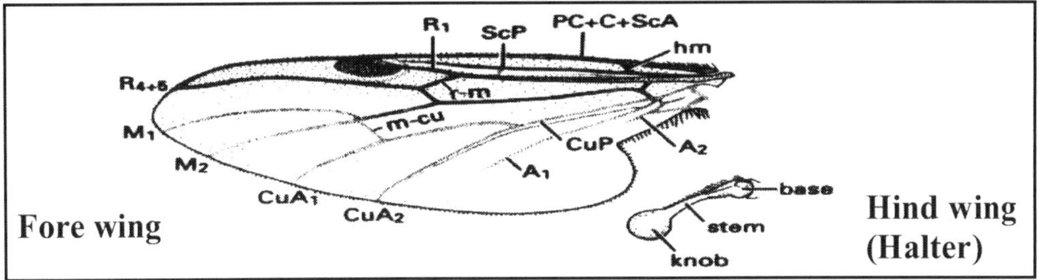

Fig. 137. Fore wing and haltere of a fly of *Bibio* (Diptera: Bibionidae).

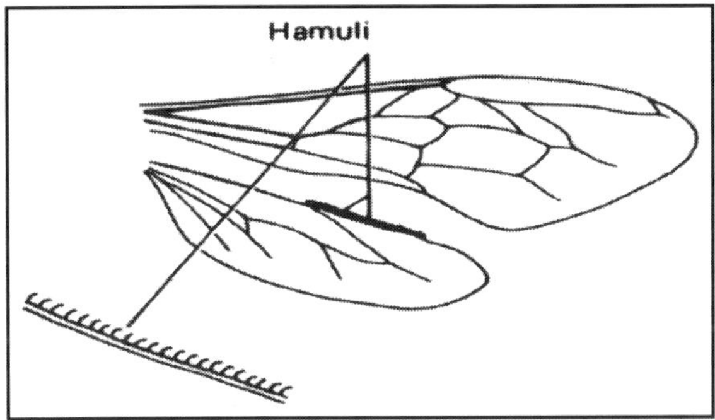

Fig. 138. Wings of the honey bee: Hamuli.

Handbook: In taxonomy, a publication designed primarily as an aid to field and laboratory identification rather than for the presentation of new taxonomic conclusions (cf. Manual, Monograph).

Haplodiploidy (adj. haplodiploid) (facultative parthenogenesis)**:** Reproduction via a combination of sexual and asexual modes resulting in both haploid offspring (from unfertilized eggs) and diploid offspring (from fertilized eggs); a reproductive process in which females develop from fertilized (diploid) eggs, whereas males develop from unfertilized (haploid) eggs, as in the social Hymenoptera.

Haploid: Having only a single (material) complement of chromosomes; single or halved number of chromosomes as found in matured germ cells. (Gr. *haploos* = single + *eidos* = form) (cf. Diploid, Polyploidy, Chromosome).

Harmonic growth: A growth pattern in which all parts of the insect body, and the body as a whole, increase by the same ratio during each molt; also called *isogonic growth.*

Harrow: A type of cultivation equipment that is used to level ground, stir soil, and cover seed.

Haustellate: The sucking mouthparts of some insects composed of styles and a proboscis; the opposite is mandibulate.

Haustellum: General term for a proboscis. In adult Diptera it is the distal part of the proboscis representing the prementum (Fig. 139).

Haustorium: The specialized absorbing organ of some pathogens and parasitic higher plants; inserted into host cells to obtain nutrition.

Hazard: The danger that injury will occur with use of a particular pesticide; depends on both toxicity and exposure; the potential for damage as a function of exposure, for example, to a toxic material.

Head capsule: The hardened external skeleton of the head of insects.

Head: The anterior of the three major divisions (**tegmata**) of an insect body which bears the eyes, antennae, and mouth parts and houses the brain (Fig. 1).

Headland: The area at the end of a cultivated field where machinery turns around, or the area that is not treated by aerial application of pesticides to provide a safety zone for flaggers or so aircraft can avoid hazards such as utility lines.

Heart: Posterior division of the dorsal blood vessel; associated with alary muscles and contains ostia (Fig. 20); a muscular tube extending dorsally and longitudinally through the insect abdomen, continuous with the aorta, serving in circulation of blood and acts as a circulatory pump.

Heat unit accumulation: The sums of the daily average day degrees above the lower and below the upper developmental threshold temperatures.

Hectare: Metric unit of land area equaling 10,000 square meters or 2.47 acres.

Hellgrammite: A larva of a dobsonfly. Megaloptera.

Hematophagous: Blood-feeding.

Hematozoa: Certain blood-inhabiting parasitic Protozoa.

Hemelytron (pl. hemelytra): Anterior wings of Heteroptera with thickened basal part and distal part membranous (literally, half an elytron; applied chiefly to Hemiptera) (Fig. 140).

Hemicephalous: Larvae of certain Diptera (Tipulidae and Brachycera) with a reduced head capsule which can be withdrawn into the thorax.

Hemidaphic: Soil fauna living in upper layers of soil below the surface.

Hemimetabola: (altenative term, Exopterygota) The subdivision of insects that undergo incomplete metamorphosis. cf. Holometabola.

Hemimetabolous: Having a simple metamorphosis; lacking a pupal stage includes insects that go through three distinct growth stages during their

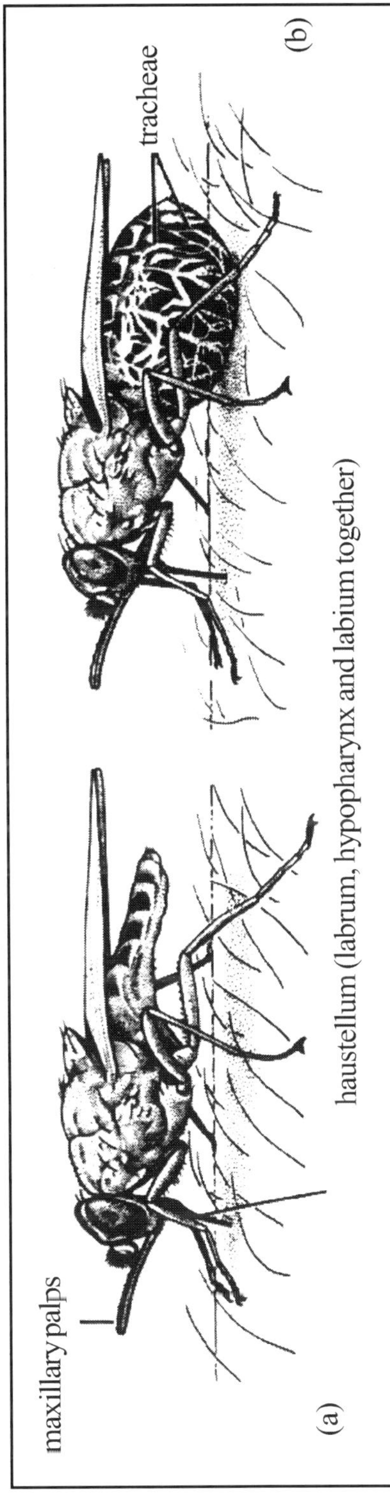

Fig. 139. A tsetse fly, *Glossina morsitans* (Diptera: Glossinidae): (a) at the commencement of feeding; (b) fully engorged with blood. Note that the tracheae are visible through the abdominal cuticle in (b).

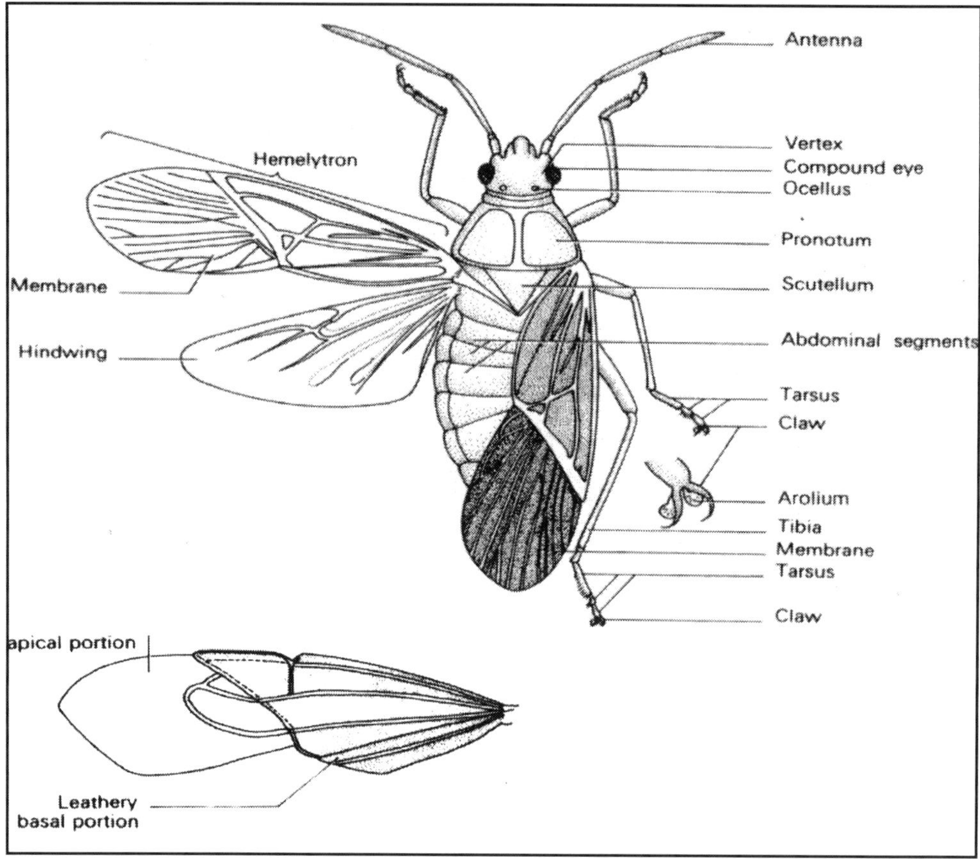

Fig. 140. Hemiptera, Heteroptera, true bugs showing major external features of the boxelder bug, *Leptocoris trivittatus* (Rhopalidae), and Hemelytron (forewing) of a true bug.

life cycle: egg, nymph, and adult. Nymphs look different than adults and often live in aquatic environments. Also termed incomplete metamorphosis that is, showing gradual change from molt to molt, with externally developing wing pads like that in the Odonata, Ephemeroptera, and Plecoptera, Orthoptera, Heteroptera, and Homoptera. Examples are mayflies, stoneflies, and dragonflies.

Hemimetaboly: With development in which the body form gradually changes at each moult, with wing buds growing larger at each moult; incomplete metamorphosis; *see also* **Holometaboly.**

Hemipneustic: Number of spiracles reduced in comparison with most generalized pattern.

Hemiptera, order: (*hemi,* one-half; *ptera,* wing) the order of insects comprising the true bugs cicadas, leafhoppers, scale insects, aphids, and many other groups; characterized by the possession of piercing and sucking mouthparts (Fig. 140).

Hemocoel: Body cavity of the insect in which the viscera are located and the hemolymph circulates, the body cavity in which the blood or hemolymph is normally filled and flows.

Hemocoelic insemination: See traumatic insemination.

Hemocyte: A cell of the hemolymph.

Hemocytes: Insect blood cells.

Hemocytopoietic organ: Organs that produce blood cells.

Hemogram: A count of the total number of hemocytes and an estimate of the numbers of different types of hemocytes in an insect at a given time.

Hemolymph: Insect blood that, as implied by its name, carries out the functions of both blood and lymph, which are different fluids in vertebrates; the "blood" of insects, combining functions of the lymph and blood of vertebrates (other than respiration).

Hemorrhagic fever: Fever resulting from the infection of blood cells by a parasite.

Hemostasis: Inhibition of hemolymph loss; coagulation.

Herbaceous: Plants that have fleshy tissues rather than persistent woody tissues.

Herbicide: Pesticide that kills plants. Used to kill weeds.

Herbiphagy: Herbivorous; phytophagous; feeding on plants.

Herbivor: Any animal (including insects) which feeds on living plants.

Herbivorous: Feeding on plants (phytophagous).

Herding: Assemblage of individuals for feeding, shelter, pairing, hibernation, etc.

Hermaphrodite: An individual having both male female reproductive organs and capabilities in the same individual (cf. Intersex).

Hermaphroditism: Having individuals that possess both testes and ovaries.

Hetero-: Other; different.

Heterocyclic: Containing more than one type of atom joined in a ring.

Heterodynamic life cycle: A life cycle in which there is a period of dormancy.

Heteroecious: An organism that must feed on, or parasitize two unrelated host plants in order to complete its life cycle, rust fungus, lettuce root aphid.

Heterogametic: Possessing the XO or XY chromosome configuration.

Heterogamy: Alternation of bisexual reproduction with parthenogenetic reproduction.

Heterogonic growth: Growth patterns in which some parts of the insect body develop at different rates than others.

Heterogony: Reproduction with alternation of generations, often involving sexual and asexual forms.

Heteromerous: The three pairs of tarsi differing in the number of segments (Coleoptera; for example, with a tarsal formula of 5-5-4).

Heteromorphosis (also **hypermetamorphosis):** Undergowing a major change in morphology between larval instars, as from **triungulin** to grub.

Heteromorphous: Describes occurrence of a regenerated appendage different in form from the appendage it has replaced.

Heteroptera: A sub - order of the insect order Hemiptera. The Heteroptera includes the true bugs. cf. Homoptera.

Heterotrophs: Organisms which feed on other organisms.

Heterozygote: Having two different alleles of a given gene.

Heterozygous: Having different alleles at one locus (cf. Allele, Locus, Homozygous).

Hibernaculum (pl. hibernacula): Silken shelter in which larvae of some Lepidoptera hibernate or aestivate.

Hibernate: To enter a state of dormancy over the winter period. cf. aestivate.

Hibernation: A period of dormancy that usually occurs in winter, or cold season (cf., aestivation); passing winter season in an inactive condition during which period the rate of metabolism is very slow. (L. *Hiberno* = pass the winter).

Hierarchy: In classification, the system of ranks which indicates the taxonomic level of various taxonomic categories (i.e., kingdom to species) (cf. Taxonomic category).

High volume spraying: The application of pesticides in dilutes form to thoroughly wet the object being sprayed; usually involves the application of 1120 or more liters per hectare (100 gals per acre).

Higher category: A taxonomic category of rank higher than the species (i.e., from subgenus to kindom) (cf. Supraspecific).

Hind wings: The wings on the metathoracic segment.

Hind: At or towards the posterior.

Hindgut (also **proctodeum**): The posterior section of the gut, extending from the end of the midgut to the anus; the posterior portion of the alimentary tract, between the mid-gut and anus (Fig. 41 and 120).

Hindwings: The posterior pair of wings attached to the insect metathorax.

Hist: Tissue.

Histogenesis: The formation of new body tissues.

Histolysis: A breakdown of body tissues.

Hollow-cone: Spray jet with a core of air breaking to give drops in an annular pattern.

Holoblastic: In embryology, a complete cleavage of the egg.

Holoblastic cleavage: Method of blastoderm formation characteristic of Collembola and certain parasitic wasps and associated with eggs that contain little yolk; the entire zygote divides (compare with meroblastic cleavage).

Holocrine secretion: Cell products released by complete breakdown of a cell.

Holometabola: The group of insects that undergo complete metamorphosis.

Holometabola: an alternative term for Endopterygota the subdivision of insects that undergo complete metamorphosis. cf. Hemimetabola.

Holometabolous: Insect life cycle that includes four growth stages: egg, larva, pupa, and adult. Also known as complete metamorphosis as in the Diptera, Hymenoptera, Coleoptera, and Lepidoptera (e.g., beetles, flies, wasps, and moths).

Holometaboly (adj. holometabolous): With development in which the body form abruptly changes at the pupal moult; complete metamorphosis, as in the group Holometabola; *see also* **Hemimetaboly.**

Holophytic: Obtaining nutrition like green plants *i.e.* manufacturing its own food by sunlight in the presence of chlorophyll (the green pigment of plants) *e.g.* nutrition of *Euglena.*

Holopneustic: Pertaining to open ventilatory system that consists of two lateral rows of 10 spiracles each.

Holoptic: Compound eyes extended dorsally so that they are contiguous along the midline (dragonflies and many male Diptera).

Holotype: The single specimen designated or indicated as 'the type' by the original author at the time of the publication of the original description [*see Bull. Zool. Nomencl.,* **4:** 186 (1950)].

Holozoic: Nutrition obtained in a normal animal-like fashion.

Homeo-, homo: Like; similar.

Homeostasis: The feedback mechanism by which an organism maintains physiological processes within a particular range; maintenance of a functionally steady state in the body, in the colony of social insects, or in an ecosystem.

Homodynamic life cycle: A life cycle in which there is continuous development, without a period of dormancy.

Homogametic sex: Possessing two X chromosomes.

Homogeneous: Uniform in structure or composition.

Homoiothermous: Please see warm-blooded.

Homologous: Organs or structures that are similar in their position and developmental origin, but not necessarily similar in appearance or function; similarity of organs, parts, or functions with comparable features in another species or group as a result of a structural pattern derived from a common ancestor (cf. Analogous); characters in two or more species that are fundamentally similar and attributable to common ancestry.

Homology (adj. homologous**):** Morphological identity or similarity of a structure in two (or more) organisms due to common evolutionary origin.

Homonym: In nomenclature, one of two or more identical but independently proposed names for the same or different taxa (cf. Senior homonym, Junior homonym, Primary homonym, Secondary homonym).

Homoplasy: Characters acquired convergently or in parallel, rather than by direct inheritance from an ancestor.

Homoptera, order: (*Homo,* uniform; *ptera,* wing) the order of insects comprising the cicadas, leafhoppers, aphids, scale insects, and others (Fig. 141).

Homoptera: A sub order of the insect order Hemiptera. The Homopetera includes leafhoppers, cicadas, aphids and scale insects (Fig. 141). cf. Heteroptera.

Homozygote: Having two similar alleles of a given gene.

Homozygous: Having identical alleles at one locus (cf. Allele, Locus, Heterozygous).

Honeydew: Exudate produced by sucking insects derived from plant sap, produced by aphids and scale insects; the liquid excretions of sucking insects (Homoptera), consisting largely of water and sugars discharged from the anus of some Homoptera(Aphidoidea, Coccoidea, some Fulgoroidea).

Hook, hooks: organ/organs of attachment developed especially in certain internal parasites like tapeworm, to keep their attachment to wall of the alimentary canal of the host.

Hooked trichomes: Minute, sharp, hooked hairs found on certain plants that provide defense against insect attack.

Hopperburn: A leaf necrosis resulting from the feeding of leafhoppers.

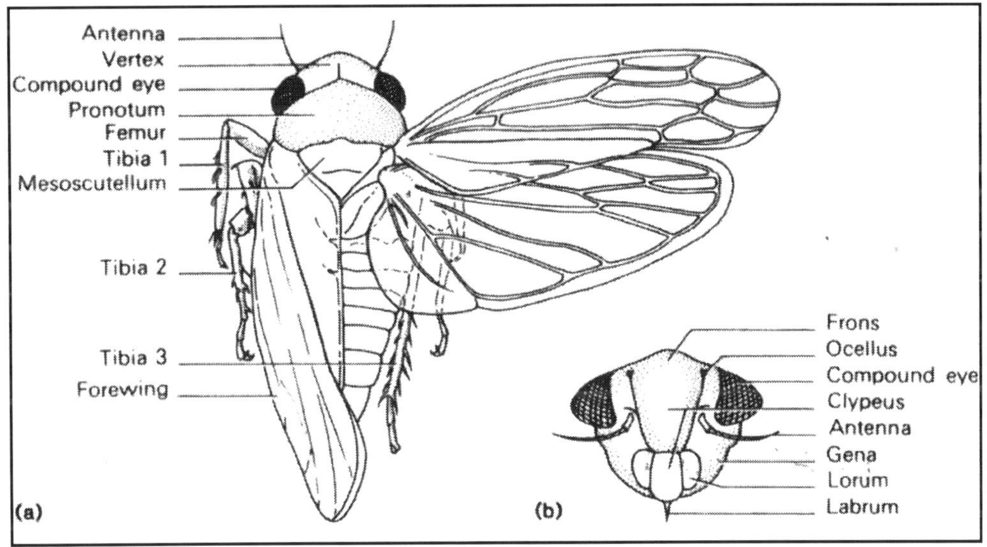

Fig. 141. Homoptera. Major external structures of a leathopper, *Paraphlepsius irroratus* (Cicadellidae). (a) Dorsal view. (b) Anterior view. (S.)

Horizontal classification: Classification based upon organisms which coexist in time; classification which, according to Simpson, separates ancestral from descendent groups and unites contemporaneous groups, or those in a similar stage of evolution, if they are derived from a common ancestry (cf. Classification, Vertical classification).

Horizontal resistance: Multigenic resistance in plants that confers broad resistance to multiple biotypes of pathogens or phytophagous arthropod (cf., vertical resistance); host-plant resistance to a broad range of pest genotypes.

Hormesis: Growth stimulation by exposure to low doses of a toxicant.

Hormoligosis: Reproductive stimulation by sublethal doses of pesticides.

Hormone: A chemical messenger that regulates some activity at a distance from the endocrine organ that produced it; a chemical secreted in one part of the body that regulates the function of some other part of the body; a chemical, secreted into the bloodstream, which produces a specific effect on some bodily organ or process; a substance produced by an organism that affects the physiology and/or growth of the individual producing it.

Hornworm: A caterpillar (larva of Sphingidae) with a dorsal spine or horn on the last abdominal segment.

Horny: Thickened or hardened.

Host: The living plant or animal that acts as a food source for pests, used in reference to parasites and parasitoids; the organism in or on which a parasite (or parasitoid) lives; the plant on which an insect feeds; organism on which the parasite lives and obtains nutrition *e.g.* tapeworm's host is man/pig; mosquitoe's host is man whose blood its female sucks.

Host discriminate: To choose between different hosts.

Host escape: A type of apparent host-plant resistance in which an unexplained lack of infestation on a susceptible plant occurs in a population of otherwise-infested plants.

Host evasion: A type of apparent host-plant resistance in which the plant passes through a susceptible stage quickly or at a time such that its exposure to potentially injurious pests is reduced.

Host-free period: The time when the host plant is not grown (e.g., celery and beet-free periods reduce the reservoir of virus available to vectors).

Host, intermediate: Not the primary but without the intervenation of which, life history of the parasitic organism would remain incomplete *e.g.* mosquito is the intermediate host of malarial parasite.

Host plant: A species of plant on which a particular insect species feeds.

Host-plant resistance: The ability of a plant to reduce the development and reproduction of a pest that is using the plant as a host. Sometimes used to describe the ability of a plant to tolerate the damage caused by a pest; the relative amount of heritable qualities possessed by a plant that reduces the degree of damage done to the plant by a pest or pests.

Host preference: To prefer one host over another.

Host race: A population of a species that shows a genetically determined preference for a particular host plant or animal species.

Host range: The organisms that a particular organism can use as food sources.

Host regulation: The ability of a parasitoid to manipulate the host's physiology.

Humectant: Material added to a spray to delay evaporation of the water carrier.

Humeral angle: The basal anterior angle or portion of the wing.

Humeral bristles: Bristles on the humeral callus (Diptera).

Humeral callus: One of the anterior lateral angles of the thoracic notum, usually more or less rounded (Diptera).

Humeral cross vein: A cross vein in the humeral portion of the wing between the costa and the subcosta.

Humeral plate: One of the articular sclerites of the neopteran wing base; *see also* **Tegula; Axillary sclerite.**

Humeral suture: The mesopleural suture (Odonata).

Humeral vein: A branch of the subcosta that serves to strengthen the humeral angle of the wing (Neuroptera; Lepidoptera).

Humeral: Blood-borne; pertaining to the shoulder; located in the anterior basal portion of the wing.

Humerus (pl., humeri): The shoulder; the lateral angle of the pronotum (Hemiptera).

Humus: Organic soil.

Hyaline: Like glass, transparent, colorless, and clear.

Hybrid: A cross between two species, or a cross between two genetic types; the offspring of parents from two different genotypes, cultivars, or species.

Hybridization: The production of individuals from genetically unlike parents (cf. Heterozygous); in taxonomy, crossing between individuals from different populations, especially different species (cf. Sympatric hybridization, Allopatric hybridization).

Hydro: Water; fluid; hydrogen.

Hydrofuge: Water repelling.

Hydrophilic: Having water-attractant characteristics; wettable.

Hydrophobic: Having water-repellent characteristics; nonwettable.

Hydroprene: A juvenile hormone analog.

Hydrostatic balance: Maintenance of body position in the water by aqatic insects.

Hydrostatic skeleton: Turgid structural support provided by fluid pressure maintained by muscle contractions on a fixed volume of liquid within especially larval insects; maintenance of body form by the pressure exerted by muscles on a fluid-filled body cavity, most important in soft-bodied larvae.

Hygroreception: Perception of moisture in the air.

Hygroreceptor: A sense organ involved in the perception of moisture.

Hymenoptera, order: (*hymen,* membrane; *ptera,* wings) The order of insects comprising bees, wasps, ants, and sawflies (Fig. 142).

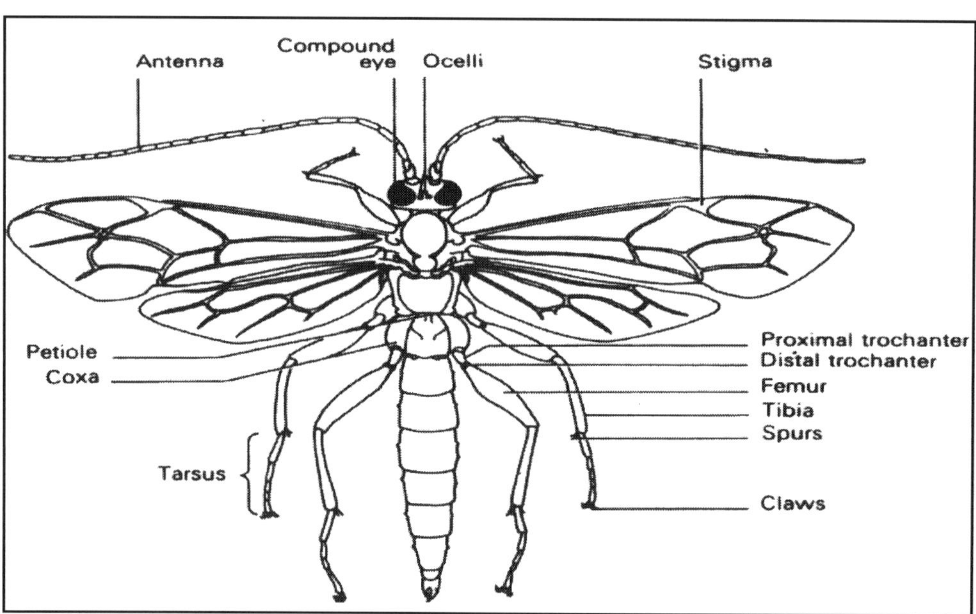

Fig. 142. External features of Hymenoptera. An ichneumon wasp (Ichneumonidae).

Hyper: Over; above; excessive.

Hypermetamorphosis: A type of complete metamorphosis in which the different larval instars represent two or more different types of larvae; a form of holometabolous development in which one or more larval instars is distinctly different from the others; insect development with successive larvae having different body forms, more or less having two metamorphoses during development (some Coleoptera, some Diptera, etc.) (Fig. 143).

Hyperparasite (adj. hyperparasitic**)** (altermative term, secondary parasite)**:** A parasite that lives upon another parasite.

Hyperparasitism: A parasite serving as a host for another species of parasite.

Hyperparasitoid: A secondary parasitoid that develops upon another parasite or parasitoid.

Hyperparasitoidism: A parasitiod serving as a host for another species of parasitoid.

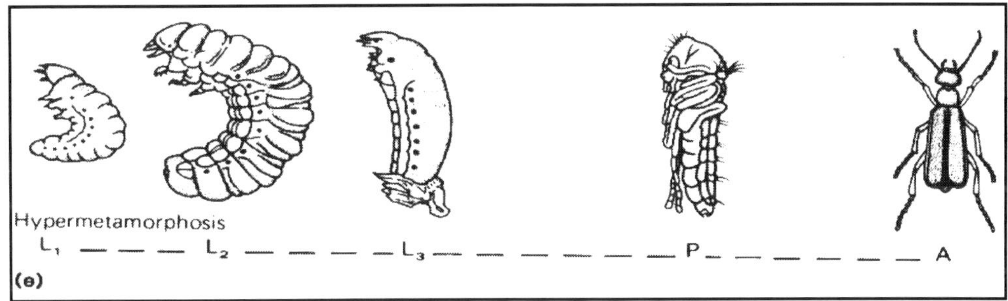

Fig. 143. Type of insect metamorphosis: Hypermetamorphosis, beetle, *Epicauta cinerea* (Coleoptera; Meloidae). (A) adult; L_1, first instar larva; L_2, second instar larva; L_3, third instar larva; P, pupa.

Hyperplasia: Proliferation of cell numbers resulting in tissue or organ size.

Hypertrophy: Abnormal increase in cell size resulting in increased tissue or organ size.

Hypha (pl. hyphae): A single thread or filament of the mycelium that forms the body of a fungus.

Hypo: Under; lower.

Hypocerebral ganglion: Ganglion located just posterior to the brain and dorsal to the foregut.

Hypocotyls: Embryo or seedling tissue located between the cotyledons and the root tip.

Hypodermis: The epidermis (see).

Hypodigm: The entire known material of a species which is available to the taxonomist.

Hypognathous: Mouthparts hung vertically from the head capsule; with the head vertical and the mouth parts located ventrally (Fig. 1 and 68).

Hypolimnion: The stagnant water at the bottom of freshwater media.

Hypopharynx: A median lobe of the preoral cavity of the mouthparts (Fig. 144); a median mouth-part structure anterior to the labium; the ducts from the salivary glands are usually associated with the hypopharynx, and in some sucking insects the hypopharynx is the mouth-part structure containing the salivary channel; a median, tonguelike mouthpart structure anterior to the labium; associated with the ducts from the salivary glands; in generalized forms, the tongue like structure that is surrounded by the externally visible mouthparts and that separates the cibarium from the salivarium; may be in the form of a stylet and channel in some sucking insects.

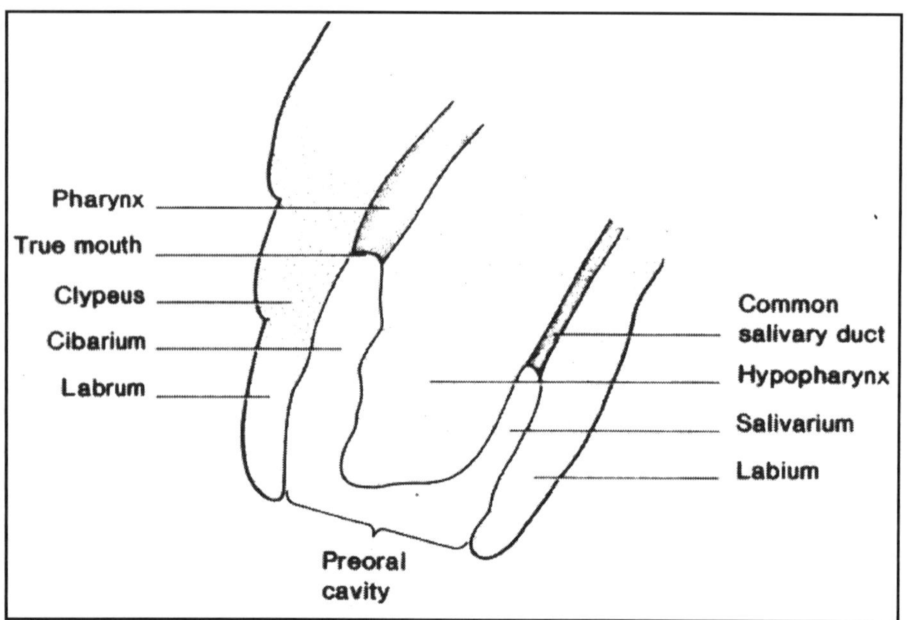

Fig. 144. Sagittal section of mouthparts region in a generalized insect: Hypopharynx.

Hypopleural bristles: A more or less vertical row of bristles on the hypopleuron, usually directly above the hind coxae (Diptera).

Hypopleuron (pl., hypopleura): The lower part of the mesepimeron; a sclerite on the thorax located just above the hind coxae (Diptera).

Hypopus: A special stage in the life cycle of some mites which is adapted for clinging to and being dispersed by other organisms.

Hyporheic: Living in the substrate beneath the bed of a waterbody.

Hypostoma: Region of mouthparts in some Diptera; part of the subgena behind the mandible.

Hypostomal bridge: Closure of the occipital foramen by fusion of the posterior region of the subgenae (Fig. 145).

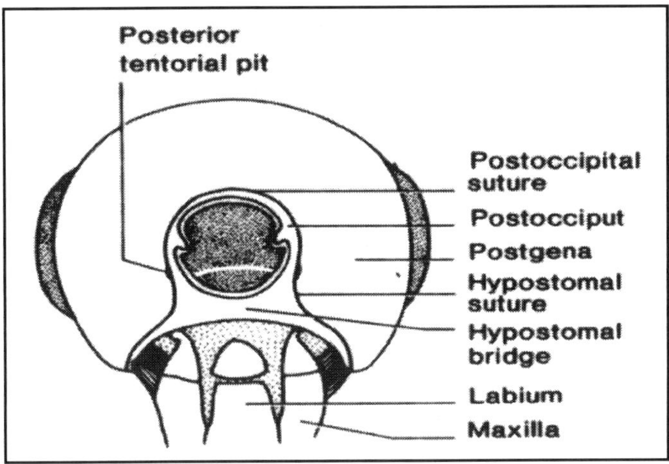

Fig. 145. Modification of the head capsule: Hypostomal bridge.

Hypovirulent: Reduced virulence in a pathogen strain that can be used to impart resistance to move virulent, destructive strains.

Hythergraph: A polygonal diagram resulting from plotting temperature means against rainfall.

I

Ichthyology: Science dealing with the study of fishes – a class of aquatic vertebrates most of whom depend for respiration exclusively on oxygen dissolved in water (Gr. *ichtys* = fish + *logos* = study).

Idiobiont: A parasitoid that prevents its host from developing any further, by paralysis or death; *see also* **Konobiont.**

IGR: See insect growth regulator

Ileum: The second section of the hindgut, preceding the colon differentiated region of the anterior intestine of the hindgut preceding the rectum in some insects (Fig. 120).

Imaginal buds: See imaginal disc.

Imaginal disc or **bud:** Latent adult structures in an immature insect, visible as groups of undifferentiated cells; mass of cells that persist in an undifferentiated state throughout the larval instars and become adult tissues (Fig. 146); a group of cells set aside in the embryo and maintained through the larval stage as a center of development of adult structures.

Imago (pl. imagoes or imagines): The adult sexually mature form of an insect.

Immature stage: Any stage of an insect prior to the imago (adult form).

Immediate effect: Of a defensive chemical, an effect that appears immediately on first encounter; *see also* **Delayed effect.**

Immigration: Movement of individuals into an area; the movement of individuals into a population.

Immune: Exempt from infection.

Immunity: Total and complete resistance of a plant to a pest or pests such that those pests are unable to use the plant for food.

Imperfect stage: The asexual portion of the life cycle of fungi during which no sexual spores are produced.

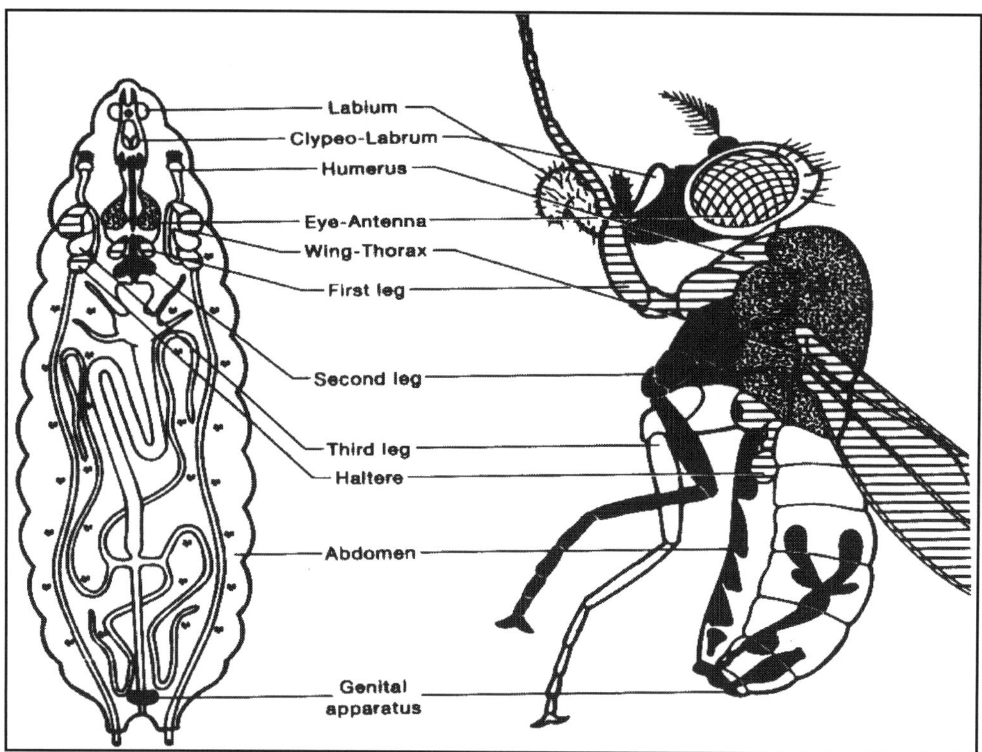

Fig. 146. Schematric representation of the imaginal discs of a fly larva and the adult organ or appendage into which they differentiate. After Ursprung and Nothiger, 1972.

Imperfectly density-dependent factor: An agent that is usually density-dependent but whose action sometimes fails; includes parasites, predators, and pathogens.

In situ: In a natural or original position.

Inclinate: Bent toward the midline of the body.

Inclusion body: A protein crystal that encloses an insect virus.

Inclusive fitness: Net genetic representation of an individual in succeeding generations, through personal reproduction and that of individuals bearing identical genes.

Incompatible: Refers to the decreased effectiveness that results if two or more pesticides are mixed; not compatible; incapable of forming a stable mixture with another chemical.

Incomplete metamorphosis: (alternative terms, partial or gradual metamorphosis) Metamorphosis in which there is a gradual, and often not very marked transformation from immature to adult stage. There is no pupal stage and immature stages are referred to as nymphs. cf. complete metamorphosis.

Incorporate: To mix or till fertilizer or pesticide into the soil.

Indeterminate growth: Without a distinctive final, adult instar - with no definitive terminal moult; *see also* **Determinate growth.**

Index: An assay used to determine if a plant has been infected by a particular pathogen, especially viruses. Suspect infected tissue to inoculated into a susceptible plant that is known to show characteristic symptoms upon infection, thereby confirming the presence of the pathogen.

Indicator: Marker.

Indigenous: Native to a region; endemic; occurring naturally in a country or area.

Indirect flight muscles: Muscles that power flight by deforming the thorax rather than directly moving the wings (Fig. 147 and 148); *see also* **Direct flight muscles.**

Indirect loss (indirect injury): Decline in quality or quantity of a marketable commodity (e.g., seeds) owing to the feeding of insects on nonmarketable portions (sometimes roots, stems, leaves) of the plant; injury to nonyield-forming organs.

Indirect pest: A pest insect that feeds on a part of the plant that is not marketed.

Indirect system: Of flight, with power for flight coming from regular deformation of the thorax by indirect flight muscles, rather than from muscular connection to the wings.

Indirect wing muscles: Muscles that contribute to insect flight but are not directly attached to the insect wings (Fig. 147 & 148).

Induced defence: A chemical change, deleterious to herbivores, induced in foliage as a result of feeding damage.

Induced pest: See *replacement.*

Induced resistance: A type of apparent host-plant resistance in which a particular plant condition or environmental state (e.g., nutrient level, soil moisture) makes a plant more resistant to pests than under other circumstances.

Inducible defenses: Defense mechanism brought into play in response to damage as, for example, in herbivore feeding on a plant.

Industrial melanism: The evolutionary development of a darker population favored by selection in the darkened surroundings of an industrial area (cf. Melanism); the phenomeon of dark **morphs** occurring with a higher than usual frequency in areas in which industrial pollution darkens tree trunks and other surfaces upon which insects may rest.

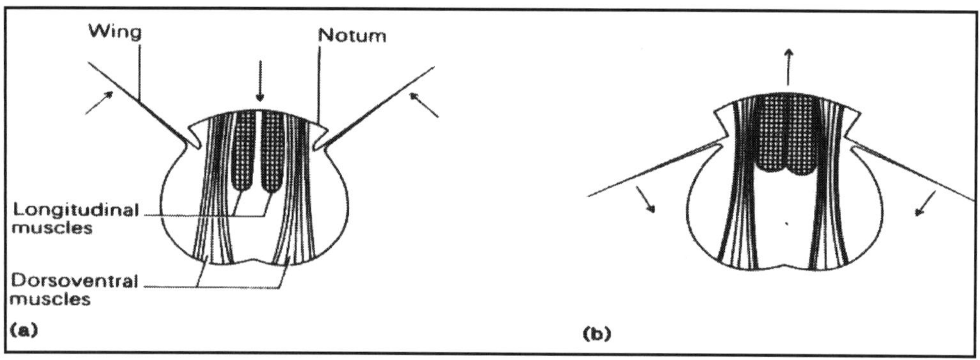

Fig. 147. Cross section of the mesothorax showing the action of the indirect flight muscles. (a) Notum depressed, wings elevated. (b) Notum arched and elevated, wings depressed. Arrows indicate direction of movement.

Inert ingredient: Biologically inactive substances that are added to pesticide formulations; auxilixary ingredient in a pesticide formulation that has no direct effect on pests.

Infect: To enter and establish a pathogenic relationship with a plant (host); to enter and persist in a carrier.

Infection site (syn. Infection court): The host location where a pathogen first becomes established.

Infection: The entrance and establishment of a pathogen in host tissues.

Inferior appendage: The lower one (Anisoptera) or two (Zygoptera) of the terminal abdominal appendages, used in grasping the female at the time of copulation (male Odonata).

Infest: To occupy and cause injury to either plant, soil or stored products.

Infestation: The physical presence of pest organisms in a location.

Inflorescence: Collective term for a terminal cluster of flowers.

Infraepisternum: A ventral subdivision of an episternum.

Infraspecific: Within the species, usually applied to categories (subspecies) and individual forms (varieties) (cf. Subspecies, Variety, Infrasubspecific form).

Infrasubspecific form: Individual and seasonal variants in asingle interbreeding population (cf. Variety, Infrasubspecific name).

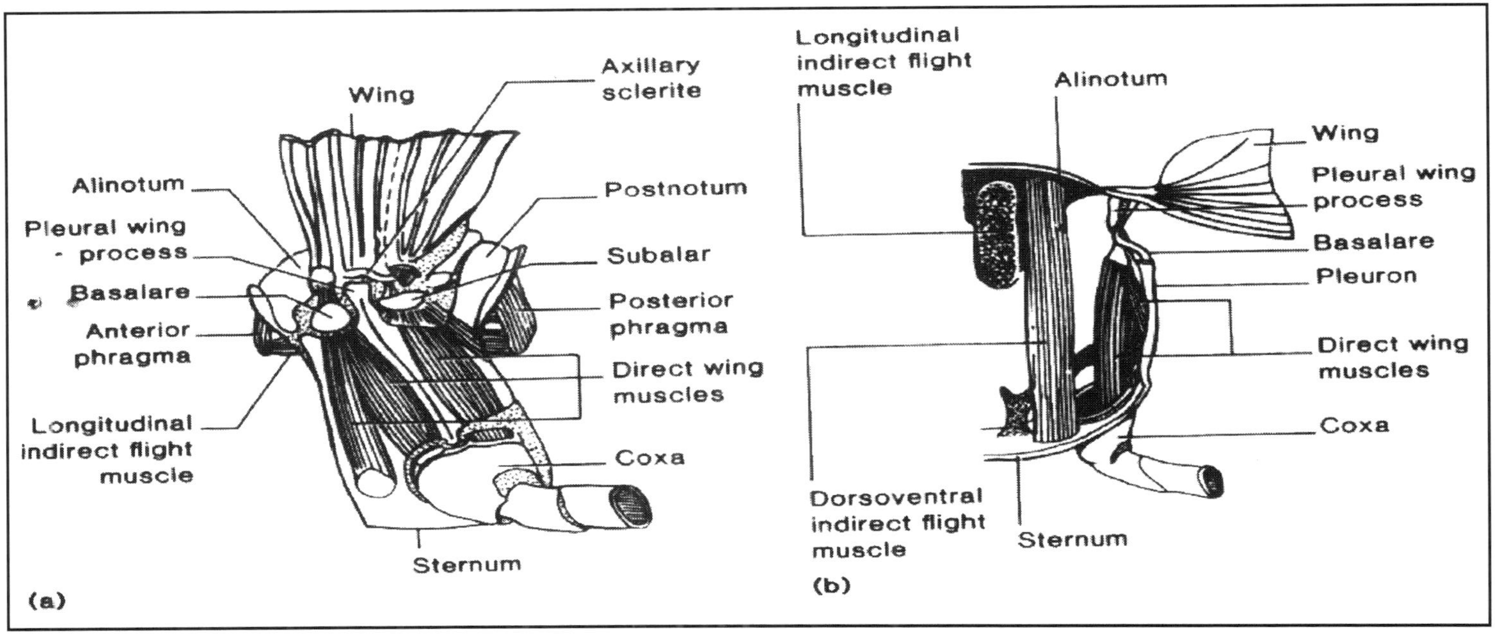

Fig. 148. Generalized wing-bearing segment. (a) Lateral view. (b) Cross-sectional view.

Infrasubspecific name: The trivial name of an infrasubspecific form (cf. Subspecific name, Infrasubspecific form) [for nomenclatural status *see Bul. Zool. Nomencl.* **4:** 89-96 (1950)].

Inhalation poison: An insecticide with a fumigant action.

Inhalation toxicity: The toxicity; of an insecticide (to higher animals) when inhaled in the form of fine droplets or vapour.

Injection: (i)Placement of chemical under tree bark using a device equipped with a reservoir of chemical. (ii) Placement of a chemical below the soil surface with a nozzle-equipped chisel or similar device.

Injector: A device for ejecting a pesticide below the soil surface, or into the transport system of a tree.

Injury: The damage caused to plant physiology by pest activity; the effects of pest activities on host physiology that are usually deleterious.

Innate behaviour: Behavior requiring no choice or learning; genetically more-or-less fixed behavior.

Innate capacity for increase in number: Maximal rate of increase attained at any particular combination of temperature, food, moisture, etc. when the quantity of food, space and other animals of the same kind are kept at an optimum level, while other organisms of different kinds are excluded.

Inner epicuticle: The innermost of the three layers of epicuticle, with the procuticle beneath it; *see also* **Outer epicuticle; Superficial layer.**

Inner vertical bristles: The more mesally located of the large bristles on the vertex, between the ocelli and the compound eyes (Diptera).

Inoculation: (alternative term, introduction)-(as applied to biological control). The introduction of new species of parasites, parasitioides, predators or disease organisms for biological control purposes.

Inoculative release: Release of natural enemies that are expected to colonize and spread throughout an area naturally.

Inoculum: Pest propagules from which new infestations or infections arise.

Inorganic pesticide: Pesticides that do not contain carbon (e.g., Bordeaux mixture, copper sulfate, sodium arsenite, and sulfur).

Inorganic: Lacking carbon atoms.

Inquiline: An organism that lives in the nest or abode or home of another and sharing its food; in entomology, used particularly of residents in the nests of social insects such as termites or social Hymenoptera, or in a hymenopterous plant gall, induced by another organism; *see also* **Integrated inquiline; Non-integrated inquiline.**

Inquiliness: Insects that are residents in the shelters or nests of other insects.

Insect equivalent: The amount of injury that could be produced by an individual pest through its complete life cycle.

Insect growth regulator (IGR, bio-rational, third-generation insecticide**):** Group name applied to chemicals which interfere with normal insect growth or development, usually by affecting growth hormone systems. Chemicals that mimic the action of insect hormones and can be used to disrupt normal insect growth and development; a substance produced by a plant that mimics or antagonizes an insect hormone; substance effective in controlling or modifying insect growth processes.

Insect pathology: The study of insect diseases.

Insect pest management: A modern approach to dealing with pest insects; control measures chosen from all available methods and applied on the basis of carefully evaluated ecological/ economic parameters.

Insecta: The class of arthropods comprising the insects; characterized by three body regions (head, thorax, and abdomen), three pairs of thoracic legs, and one pair of antennae; the taxon made up of insects; means literally "in-cut," describing the segmented appearance of insects.

Insectary crop: Plants that attract and provide food for predators and parasites, e.g., untreated alfalfa adjacent to a cotton field.

Insectary: A place where insects are reared and bred.

Insecticide: A toxin effective against insects; pesticides that kill insects; a chemical employed to kill insects; a substance that kills insects.

Insectivore (adj. insectivorous)**,** (also **entomophage**): An insect eater.

Insectoverdins: Green pigments found in insects.

Insemination: Introduction of sperm into the female reproductive tract; transfer of semen from the male reproductive system to the female reproductive system.

Insertion: Movable attachment area of skeletal muscles.

Insight learning: The ability to combine learned behavior from diverse experiences to solve a problem.

Insolution: Sunlight; solar radiation.

Instar: The form or stage of an insect between successive moults; the first instar being the stage between hatching and the first moult; the period or stage between insect molts; each developmental form of an insect's life cycle; the insect between successive ecdyses or between successive apolyses; the stage between molts in insect nymphs or larvae, numbered to designate the various stages (e.g., sixth instar).

Instinct: Behavior performed without previous experience and without interaction with other members of the species.

Instinctive behavior: Unlearned stereotyped behavior, in which the nerve pathways involved are hereditary.

Instrumental conditioning: Trial and error learning.

Insular: Living or inhabiting an island.

Integrated control: (alternative terms, pest management, **IPM)** The use of two or more different control measures together in an integrated fashion; control of pests that emphasizes use of insecticides so as to conserve natural enemies in the agroecosystem.

Integrated inquiline: An inquiline that is incorporated into the social life of the host by behavioural modification of both inquiline and host.

Integrated pest management (IPM): A system that maintains the population of any pest, or pests, at or below the level that causes damage or loss, and which minimizes adverse impacts on society and the environment; an approach to the control of pests (insects, diseases, weeds) in which all available techniques are evaluated and integrated into a unified program; or application of several different insect control methods to maintain a pest population below economically damaging levels.

Integument: The outer covering of the body that includes the epidermis and the cuticle (Fig. 84); general body covering; an outer covering such as the cuticle of insects and nematodes, or the seed coats of plants.

Intelligence: The capacity to modify behaviour as a result of experience.

Inter: Between.

***Inter*:** Between; e.g., interspecific, between two or more different species.

Interantennal suture: A suture extending between the bases of the two antennae (Siphonaptera).

Intercalary vein: An extra longitudinal vein that develops from a thickened fold in the wing, more or less midway between two pre-existing veins (Ephemeroptera).

Intercalary: Inserted between others; extra wing veins inserted between the main longitudinal veins (as in Ephemeroptera and others).

Intercostals vein: The subcosta (Hymenoptera).

Intercropping: The growing of two crops simultaneously in the same field.

Interference coloration: Coloration resulting from the reflection of light from a series of superimposed surfaces separated by distances comparable with the wavelengths of light.

Interference competition: Competition in which individuals are prevented from feeding, mating, or laying eggs as a result of the presence of other individuals.

Interference: In colours, iridescent colours produced by variable reflection of light by narrowly separated surfaces (as scales of Lepidoptera); in intra- and interspecific interactions between predators and parasitoids, a reduction in the profitability of otherwise high resource density.

Intergradation: Merging gradually through a continuous series of intermediate forms or populations.

Intermediate organ: A chordontonal organ in the subgenual organ of the foreleg of some orthopterans associated with the **tympanum** and believed to respond to sound frequencies of 2-14 kHz.

Intermoult period: see **Stadium.**

Internal parasite: A parasite that feeds from within the host body.

International code: A term sometimes applied to the *Regles Internationales de la Nomenclature Zoologique* (International Rules of Zoological Nomenclature).

International Rules of Zoological Nomenclature: Regles Internationales de la Nomenclature Zoologique (q.v.).

Interneuron: Neuron that serves as a connection between sensory and motor neurons; a nerve cell system and serving to connect other neurons; or neurons, found within ganglia, that may synapse with one another or with sensory or motor neurons; also called *association neurons* (Fig. 149).

Internode: Section of stem between two adjacent nodes.

Internuncial neurons: See interneurons.

Interplanting or intercropping: Growing more than one crop simultaneously on a piece of land.

Intersegmental groove: see **Antecostal suture.**

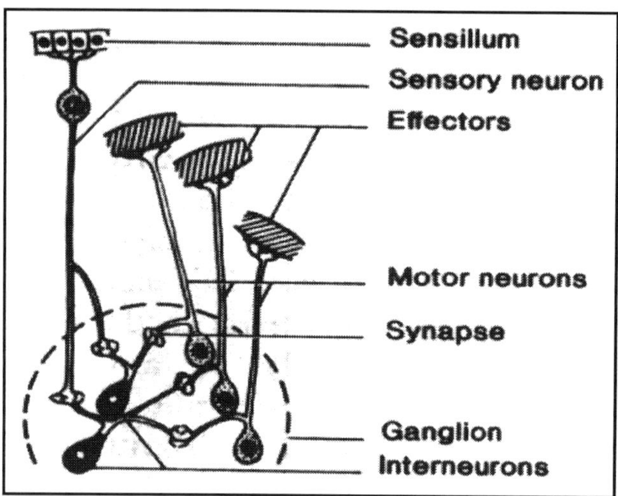

Fig. 149. Structural type of neuron: Relationship among sensory, motor and interneurons.

Intersegmental membrane (also **conjunctiva):** Membrane between segments, particularly of the abdomen (Fig. 150).

Intersegmental muscles: The principal longitudinal muscles running between successive antecostae in insects with secondary segmentation (Fig. 150).

Intersex: An individual more or less intermediate in phenotype between male and female (cf. Hermaphrodite).

Intersexual selection: Natural selection involving choices between the sexes, often on the basis of courtship displays.

Interspecific: Interactions between members of different species.

Intersternite: An intersegmental sternal plate posterior to the eusternum, known as the **spinasternum** except on the metasternum; an intersegmental sclerite on the ventral side of the thorax; the spinasternum.

Interstitial: Situated between two segments (interstitial trochanter of Coleopteran; coincident (the ends of two veins meeting).

Intestine: The anterior portion of the hindgut (Fig. 41).

Intima: The cuticular lining of the foregut and hindgut and the tracheae (Fig.).

Intra: Within; e.g., intraspecific, within a single species.

Intra-alar bristles: A row of two or three bristles situated on the mesonotum above the wing base, between the dorsocentral and the supra-alar bristles (Diptera).

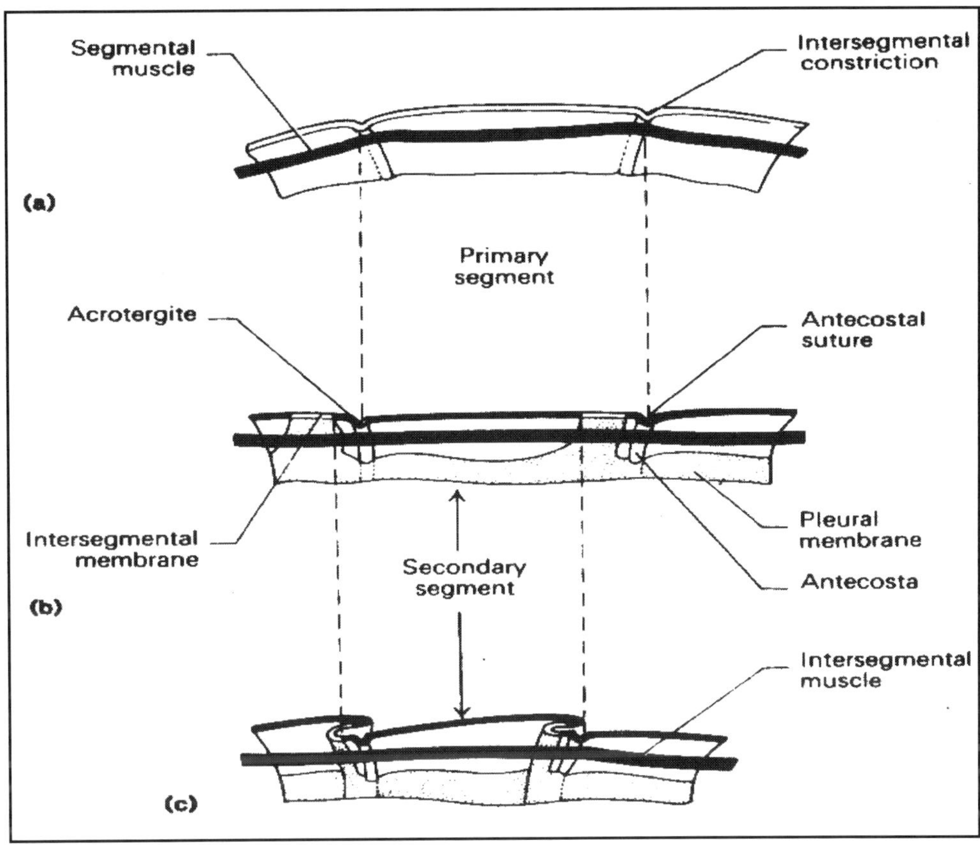

Fig.150. Types of segmentation (tergites onlt). (a) Primary. (b) and (c) Secondary (adjacent secondary segments overlapping in c).

Intrasexual selection: Natural selection involving competition among members of one sex of a species, usually for mates.

Intraspecific competition: Situation in which individuals within a population compete for a limited resource.

Intraspecific: Interactions between members of the same species.

Intrinsic muscles: Both origins and insertions entirely within a given appendage (Fig. 151).

Intrinsic: Located entirely within an organ (as contrasted to extrinsic).

Introduction (importation): A biological control practice involving identification of the natural enemies that regulate a pest in its original location, and introduction of these into the pest's new location.

Introduction: See inoculation.

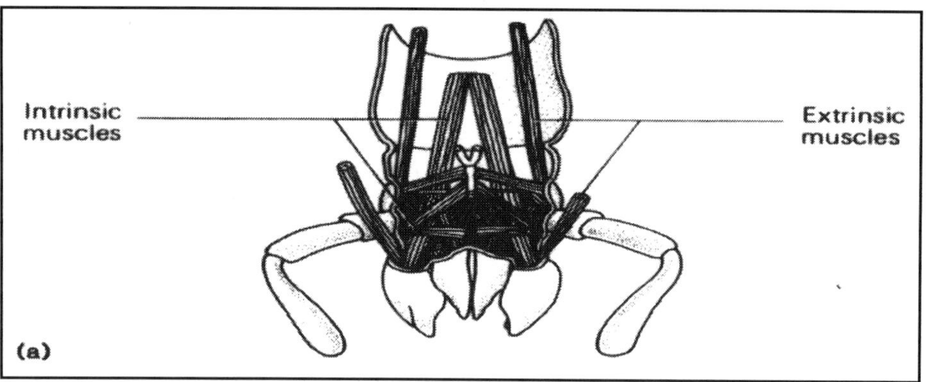

Fig. 151. Extrinsic and intrinsic skeletal muscles: Labium of a cricket.

Introgressive hybridization: The spread of one or more genes of one species into the germ plasm of another species as a result of hybridization (cf. Hybridization, Coenospecies).

Inundation: Very large number of mass-reared biological control agents released to overcome a pest population.

Inundative release: In biological control, swamping a pest with large numbers of control agents, with control deriving from the released organisms rather than any of their progeny; widespread distribution of natural enemies for the purpose of suppressing a pest population, with little or no impact expected from progeny of the released individuals.

Invagination: An infolding or inpocketing.

Invertase: An enzyme that catalyzes the breakdown of sucrose into fructose and glucose units.

Invertebrate: Animals that do not have an internal skeleton.

Invertebrate: Any animal without a backbone (spinal column).

IPM: Abbreviation of integrated pest management.

Ipsenol: An aggregation pheromone of bark beetles of the genus *Ips.*

Irreversibility: See Dollo's rule.

Irritability: A character of living beings- the ability to react to changes of surroundings.

Irruption: Local population explosion; irregularly large population increase.

Iso: Equal; uniform.

Isogonic growth: See harmonic growth.

Isolating mechanism: Any intrinsic agent which hinders the interbreeding of groups of individuals.

Isolines: Varieties of a crop plant that differ in only a single gene.

Isomer: A particular molecular arrangement of a chemical compound. Different isomers of the same substance have the same atoms within the molecule but their spatial arrangement differs.

Isophenes: Lines connecting points of equal expression of a character; lines at right angles to a cline on a map (cf. Cline).

Isoptera: The order of insects comprising the termites (Fig. 152 amd 155).

Isopteran, order: (*iso,* equal; *ptera,* wings) termites (Fig. 152 and 155).

Iteroparous: A type of life history in which the animal reproduces two or more times during its lifetime.

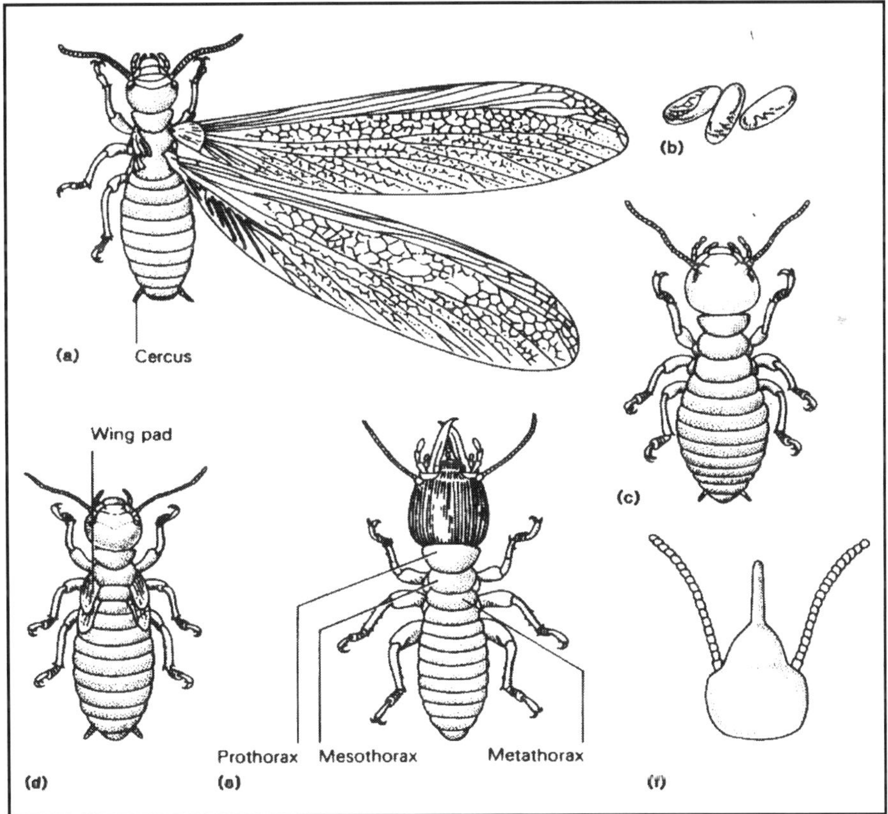

Fig. 152. Isoptera, termites. (a) Winged form, *Zootermopsis angusticollis.* (b) Eggs. (c) Third instar nymph. (d) Last instar nymph. (e) Soldier. (f) Head of *Nasutitermes* sp. After Essig, 1942.

J

Jet: Liquid emitted from a nozzle orifice (in USA = nozzle).

JH: An abbreviation of juvenile hormone.

Johnston's organ: A sense organ similar to a chordotonal organ, located in the second antennal segment of most insects; functions in sound reception in some Diptera (Fig. 153); an organ in the pedicel of the antenna, consisting of a cluster of chordotonal sensilla.

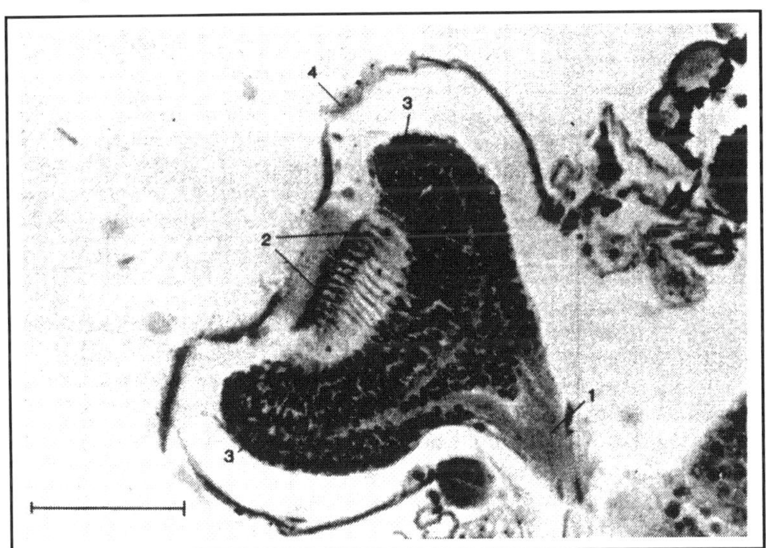

Fig. 153. Photomicrograph of a sagittal of Johnston's organ in the mosquito *Aedes triseriatus*. 1, Nerve; 2, scolopophores; 3, cellbodies of neurons; 4, pedicel. (Scale line= 50 μm).

Joint: An articulation of two successive segments or parts.

Jugal area (also **jugum,** pl. juga)**:** The posterobasal area of the wing, delimited by the jugal fold and the wing margin.

Jugal fold: A fold-line of the wing, dividing the jugal area from the anal area.

Jugal lobe: A lobe at the base of the wing, on the posterior side, proximad of the vannal lobe (Hymenoptera).

Jugal: Lobe at the base of the hindwing in some Lepidoptera and Trichoptera.

Jugate coupling: A mechanism for coupling the fore and hind wings in flight by a large jugum of the fore wing overlapping the hind wing.

Jugum: A lobelike process at the base of the front wing, which overlaps the hind wing (Lepidoptera); a sclerite in the head (Hemiptera and homoptera; lobe like wing-coupling mechanism characteristic of some Lepidoptera (Fig. 154).

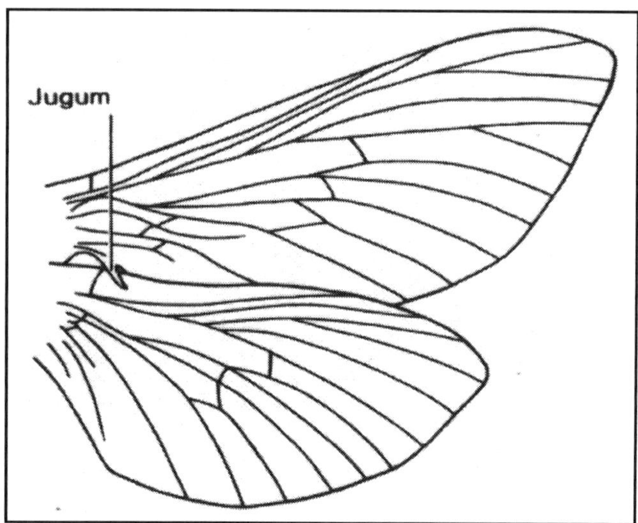

Fig. 154. Wing-coupling mechanisms: Wings of a jugate moth (Hepialidae).

Junior homonym: The more recently published of two or more identical names for the same of different taxonomic categories (cf. Homonym, Senior homonym).

Junior synonym: The more recently published of two or more available synonyms for the same taxonomic category (cf. Synonym, Senior synonym).

Juvabione: A juvenile hormone analog; juvenile hormone mimic from balsam fir; an insect growth regulator occurring in certain trees and causing abnormal development of insects feeding on the tree; a substance isolated from trees of the genus *Abies* and found to have juvenile hormone characteristics.

Juvenile: The sexually immature stages of nematodes that hatches from an egg and molts three times before becoming an adult, or the immature stages of ametabolous insects; the immature stages in primitively wingless insects with no metamorphosis.

Juvenile hormone (JH): A hormone, secreted by the insect brain that causes immature characteristic in insects as long as it is produced; a hormone secreted by the corpora allata that maintains juvenile features in immature insects and controls certain aspects of adult physiology and behavior a hormone that promotes larval development and inhibits development of adult characteristics; a 16, 17 or 18 C-chain hormone released by the corpora allata into the haemolymph, involved in many aspects of insect physiology, including modification of the expression of a moult.

Juvenile hormone mimics (also **juvenoids):** Synthetic chemicals that mimic the effect of JH on development.

K

Kairomone: A chemical produced by an individual of one species that has a beneficial effect on the recipient individual of another species; a chemical that is adaptively favourable to the receiving organism; an interspecific messenger substance that benefits the receiver but not the releaser (e.g., attractants, excitants); an interspecific chemical messenger that benefits the receiver but not the emitter; a communication chemical that benefits the receiver and is disadvantageous to the producer; *see also* **Allomone; Synomone.**

Karyological character: A character involving chromosome structure or number (cf. Taxonomic character).

Keeled: With an elevated ridge or carnia.

Keratin: The inert exoskeleton protein of vertebrate animals forming skin, hair, feathers, claws, nails, scales, etc.; similar in structure and function to the arthropod chitin.

Keratolytic: Pertaining to or promoting loosening or separation of the outer layer of the skin.

Key factor (as applied to population dynamics): A factor which is dominant in regulating the numbers of an organism.

Key pest: An important major pest species in the complex of pests attacking a crop, with a dominating effect on control practices; refers to The serious and persistent pest that routinely inflicts economic damage on commercial crops; a perennial, severe pest that causes serious and difficult crop production problems; a pest that dominates cultural activities.

Key: A tabulation of diagnostic characters of species (or genera, etc.) in dichotomous couplets facilitating rapid identification; a tabular arrangement of species, genera, orders, or other classification categories according to characters and traits that serves to identify them.

Kin selection: Natural selection that involves inclusive fitness.

Kin-, kino: Motion; activity.

Kineses: Undirected locomotor responses to stimuli.

Kinesis (pl. kineses)**:** Movement of an organism in response to a stimulus, usually restricted to response to stimulus intensity only; an undirected movement in which the speed of movement or the frequency of turning depends on the intensity of stimulation.

King: The male primary reproductive in termites (Fig. 155).

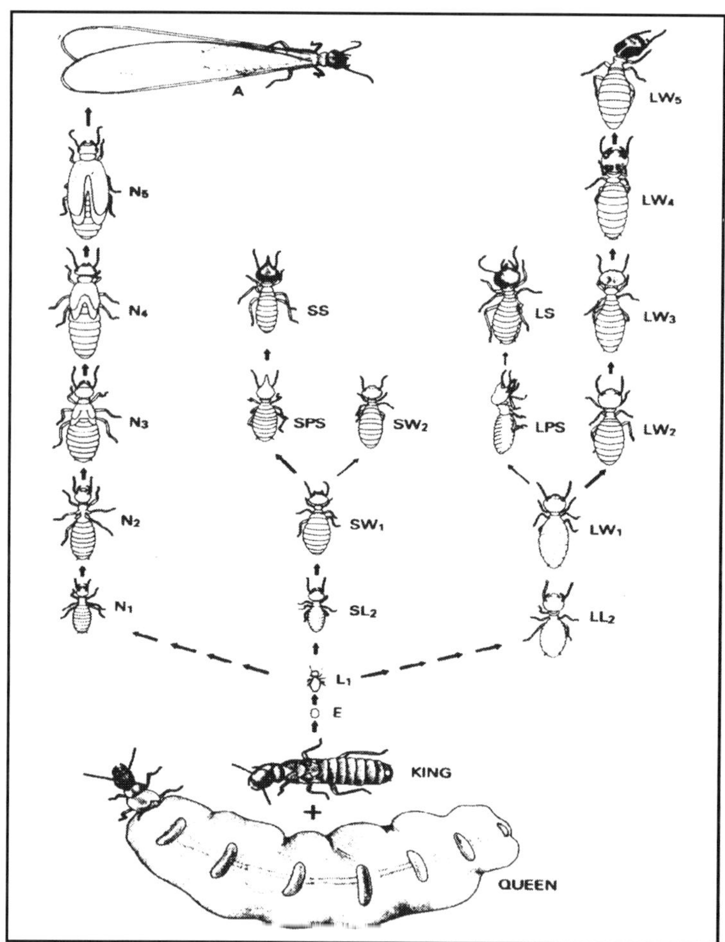

Fig. 155. Developmental pathways of the termite *Nasutitermes exitious* (Isoptera: Termitidae). Heavy arrows indicate the main lines of development, light arrows the minor lines. A, alate; E, egg; L, larva; LL, large larva; LPS, large presoldier; LS, large soldier; LW, large worker; N, nymph; SL, small larva; SPS, small presoldier; SS, small soldier; SW, small worker. The numbers indicate the stages. After Waston and Abbey, 1985.

Kinoprene: A juvenile hormone analog.

Kinship component: An indirect contribution to an individual's inclusive **fitness** derived from increased reproductive success of the individual's kin (relatives) through the altruistic assistance of the individual.

Kleptoparasite: An animal that steals its food from another species.

Klinokinesis: Movement resembling a 'random walk', in which changes in direction are made when unfavourable stimuli are encountered, with frequency of turning dependent upon the stimulus intensity.

Klinotactic orientation: A taxis characterized by moving the body back and forth across a stimulus field and moving toward or away from the region of maximum stimulation.

Klinotaxis: A movement in a definite direction relative to a stimulus, either directly towards or away from the source.

Knock-down insecticide: A poison that knocks down flying insects rapidally but may not kill them.

Knockdown: Immediate incapacitation or paralysis of an insect by a quick-acting insecticide such as pyrethrum, allethrin, or tetramethrin.

Konobiont: A parasitoid that allows its host to continue to develop; *see also* **Idiobiont.**

K-strategist: A species characterized by a low reproductive rate, increased survival mechanisms, minor tendency to disperse, and often a relatively large body size (as compared to an r-strategist); an ecological term applied to organisms which are adapted to ensure good survival. cf. r-strategist

L

Label: Technical information about a registered pesticide in the form of printed material attached to, or printed on, the pesticide container. It is illegal to use the pesticide in ways inconsistent with the label.

Labella (sing. labellum): In certain flies, paired lobes at the apex of the proboscis, derived from labial palps; apical spongelike structure on the proboscis of nonbiting muscoid flies (Fig. 156).

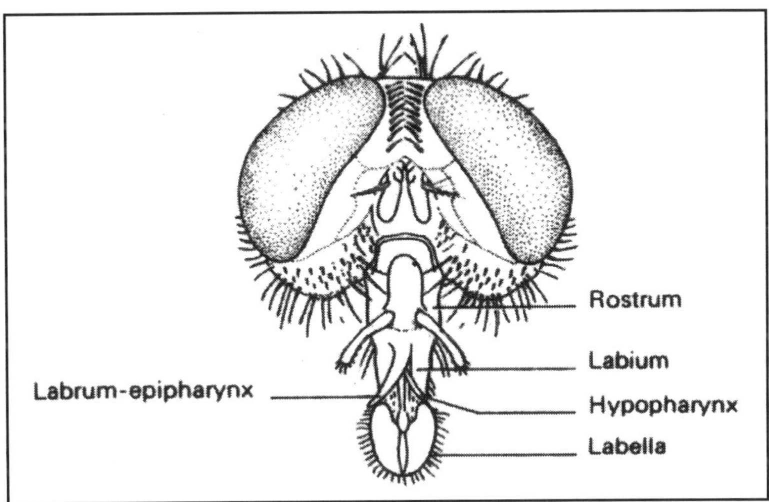

Fig.156. Examples of nonstylate mouthparts: House fly, *Musca domestica.*

Labial: Of or pertaining to the labium.

Labial glands: Glands associated with the labial segment, commonly having salivary functions.

Labial palp: A 1-5 segmented small feelerlike appendage of the **labium** (Fig. 63).

Labial palpus (pl. labial palpi): One of a pair of small feelerlike structures arising from the labium (Fig. 63).

Labial suture: Boundary between the proximal mentum, and the distal prementum, of the labium; the suture on the labium between the postmentum and the prementum (Fig. 63).

Labium (adj. labial)**:** The 'lower lip', forming the floor of the mouth, often with a pair of palps and two pairs of median lobes (Fig. 63); derived from the first head segment.

Labrum: The anterior mouthpart or upper lip, lying below the clypeus and in front of the mandibles; a flap like structure anterior to the mouthparts, below the clypeus which adjoins the clypeus in front of the mouth (Fig. 63).

Labrum-epipharynx: A mouth part representing the labrum and epipharynx.

Lac: A natural body secretion of the scale insect *Laccifer lacca* (Homoptera) that is used to make shellac, shoe polish, paint, adhesives, and many other products.

Lac: Main constituent of the commercial product known as shellac – a resinous secretion of the scale insect called lac insect, host plants several *e.g. Ficus.*

Lachrymal: Referring to tears in the Mammalia.

Lacinia (pl., laciniae): The inner lobe of the maxilla, borne by the stipes; the inner arm of the "pincers" present on the maxilla of biting insect mouthparts; the inner lobe of the insect maxilla bearing a brush of setae or spines; used for food particle handling (Fig. 63).

Lacquer: Pesticide incorporated into a lacquer or varnish to achieve slow release over a lengthy period of time.

Lamarckism: The theory advocated by Lamarck, that evolution is brought about by the inheritance of acquired characters.

Lamella (pl. lamellae): One in a series of layers or plates; a leaflike plate.

Lamellate: With or a series of leaflike platelike structures or segments; lamellate antenna.

Lamina lingualis: One of two median distal plates in the millipede gnathochilarium.

Lanceolate cell: A cell in the anal area of the wing (Hymenoptera).

Lanceolate: Spear-shaped, tapering at each end.

Lapse: A condition when air temperature is highest at ground level and decreases with elevation. Small pesticide particles can be carried aloft when warm air rises.

Lapsus calami: In nomenclature, a slip of the pen, especially an error in spelling (cf. Error, Emendation).

Larva (pl. larvae)**:** An immature insect after emerging from the egg, often restricted to insects in which there is complete metamorphosis (holometaboly) but sometimes used for any immature insect that differs strongly from its adult; *see also* **Nymph.** The ambulatory, sexually immature, wingless stages between egg and pupa of holometabolous insects; immature nematodes or first instar mites (Acarnia); an immature stage differing radically from the adult; usage of the term is commonly limited to immatures of insects with complete metamorphosis.

Larvicide: A chemical which kills the immature stages (larvae) of an insect.

Larviform: Shaped like a larva.

Larviparity: A form of viviparity in which the female gives "birth" to larvae instead of depositing eggs.

Larviparous: An insect that gives birth to living young instead of laying eggs.

Larviposition: The deposition of larvae by the adult female of some Diptera.

Latency: Dormancy, or the period within a host during which no symptoms of infection are evident even though a pathogen is present.

Latent learning: Learning without apparent reward or punishment; conditioning in which the reward occurs some time following receipt of the stimulus.

Latent period: (i) The time that passes between vector acquisition of a pathogen and when the vector becomes able to transmit the pathogen to a new host. (ii) The time between host infection and production of symptoms or inoculum by the infection.

Latent: An infection that produces no visible symptoms.

Laterad: toward the side, away from the midline of the body.

Lateral: On or pertaining to the side (that is, the right or left side) of the body.

Lateral movement: The horizontal movement of a pesticide from the application site.

Lateral ocelli: See stemmata.

Lateral oviducts: In female insects, the paired tubes leading from the ovaries to the medial common oviduct (Fig. 5).

Lateral salivary duct: Duct that carries secretion products from a salivary gland to the common salivary duct.

Laterosternite: The result of a fusion of the eusternum and a pleural sclerite.

Laterotergite: A tergal sclerite located laterally or dorsolaterally.

Lateroventral: To the side (away from the midline of the body) and below.

Law of priority: The provision in the International Rules of Zoological Nomenclature that the correct name for a genus or species can be only that name under which it was *first* designated in conformance with the requirements laid down in those rules.

Lay by application: Applications of an agricultural chemical at the last time in the growing season when it is possible to drive a tractor through the crop.

LC_{50} (also **median lethal concentration**) **:** The lethal concentration of a toxicant in the air or water which kills half of the test animals exposed to it over a 24 hour period. LC_{50} is expressed in parts per million for air or water, and in micrograms per liter for a dust or mist. It is often used as the measure of acute inhalation toxicity; or lethal concentration of toxicant required to kill 50% of a large group of individuals of one species; or the concentration of a toxicant in some medium (air, water, soil) that will kill 50 percent of the test organisms exposed; generally expressed as mg or cm^3 per animal or as parts per million or billion in the medium.

LD_{50} (also **median lethal dose**): The dose or amount of an active ingredient which, when taken orally or absorbed through the skin, kills half of the test animals exposed to it. The LD_{50} is generally expressed in milligrams per kilogram. It indicates acute oral or acute dermal toxicity; or lethal dose of toxicant required to kill 50% of a large group of individuals of one species; or the dose of a toxicant that will kill 50 percent of the test organisms to which it is administered; generally expressed as milligrams of toxicant per kilogram of body weight.

Leaching: The movement of a chemical downward through the soil in water.

Leaf area index (LAI): Ratio of leaf surface to soil surface area, relation to utilization of solar energy for photosynthesis; or the ratio of total leaf surface area of a plant to the ground surface area covered by the plant.

Leaf miner: An insect that feeds entirely within leaf mesophyll; or an insect which lives in and feeds upon the cells between the upper and lower epidermis of a leaf; these being larvae of Agromyzidae (Diptera), Lyonetiidae and Gracillariidae (Lepidoptera), Hispidae (Coleoptera), etc (Fig. 157).

Learned behavior: Behavior acquired as a result of interaction with the environment.

Lectotype: One of a series of syntypes which, subsequent to the publication of the original description, is selected and designated through publication to serve as "the type" [*see Bul. Zool. Nomencl.* **4:** 186 (1950)] (cf. Syntype).

Legal control: (see quarantine)

Leishmaniasis: Any disease caused by flagellate protozoans of the genus *Leishmania.*

Lentic habitat: Pond or lake; swamp; a region of standing water.

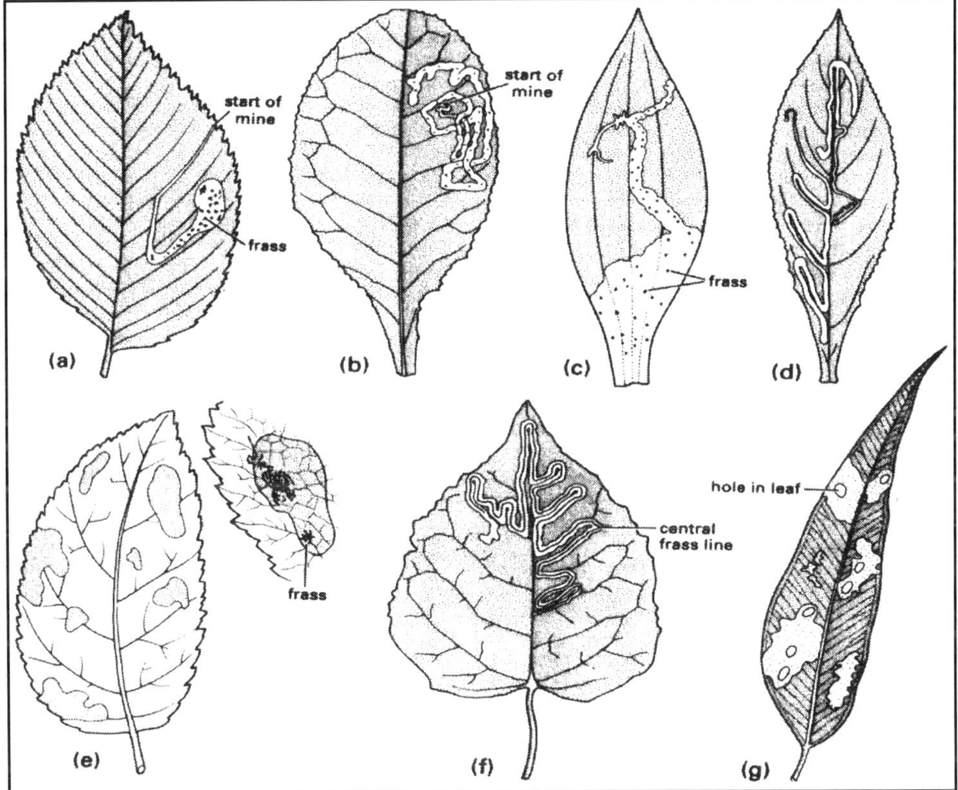

Fig. 157. Leaf mines: (a) linear-blotch mine of *Agromyza aristata* (Diptera: Agromyzidae) in leaf of an elm, *Ulmus americana* (Ulmaceae); (b) linear mine of *Chromatomyia primulae* (Agromyzidae) in leaf of a primula, *Primula vulgaris* (Primulaceae); (c) linear-blotch mine of *Chromatomyzia gentianella* (Agromyzidae) in leaf of a gentian, *Gentiana acaulis* (Gentianaceae); (d) linear mine of *Phytomyza senecionis* (Agromyzidae) in leaf of a ragwort, *Senecio nemorensis* (Asteraceae); (e) blotch mines of the apple leaf miner, *Lyonetia speculella* (Lepidoptera: Lyonetiidae), in leaf of apple, *Pyrus malus* (Rosaceae); (f) linear mine of *Phyllocnistis populiella* (Lepidoptera: Gracillariidae) in leaf of poplar, *Populus* (Salicaceae); (g) blotch mines of jarrah leaf miner, *Perthida glyphopa* (Lepidoptera: Incurvariidae) in leaf of Jarrah, *Eucalyptus marginata* (Myrtaceae). After Frost, 1959; after Spencer, 1990.

Lepidoptera, order: (*lepido-,* scale; *ptera,* wings) butterflies, moths, skippers (Fig. 158).

Lepidoptera: The order of insects comprising the butterflies, skippers, and moths (Fig. 158).

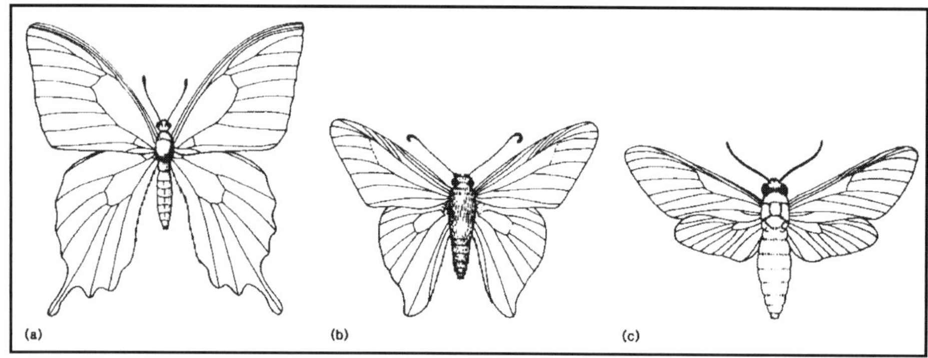

Fig. 158. (a) Butterfly. (b) Skipper. (c) Moth..

Lepidopterous: Pertaining to the order Lepidoptera, the moths and butterflies.

Lesion: A defined necrotic area or localized spot of diseased tissue.

Lieshmaniasis: Any disease caused by infection with a protozoan *Leishmania* and transmitted by sand flies.

Life cycle (life history): Habits and changes undergone by an organism from the egg stage to its death as an adult (Fig. 159); the chain or sequence of events that occurs during a lifetime of an individual organism.

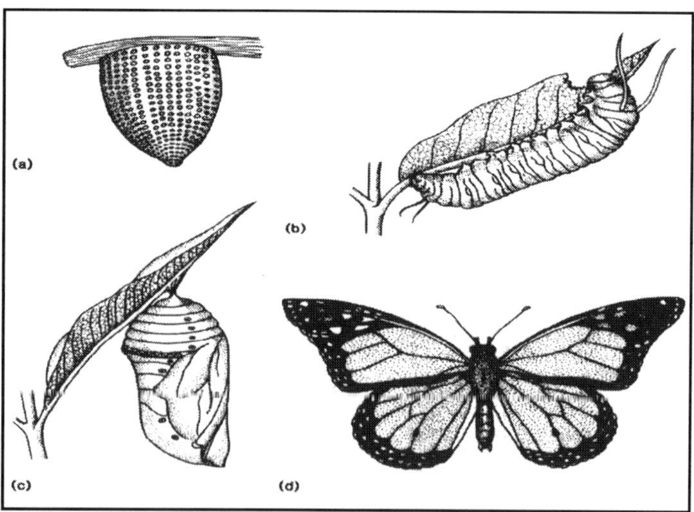

Fig. 159. Stages in the cycle of the monarch butterfly, *Danaus plexippus.* (a) Egg. (b) Larva. (c) Pupa (chrysalis). (d) Adult. (L.)

Life-form: A life-form is the common morphological features that characterize a group of organisms.

Life system: A subdivision of an ecosystem; the subject species and its effective environment; the succession of stages through which an organism passes to reach maturity.

Life table: Tabulation of birth and mortality at each life stage for a pest population. Can be used to estimate the rate of population increase; or a tabulation that accounts for age-specific deaths in a population; or a tabulation of the life stages of an insect with a cumulative record of mortality and survival.

Ligation: An experimental technique which isolates one part of the body of a living insect from another, usually by tightening a ligature.

Light compass orientation: Orientation in which a constant angle with a light source (usually the sun) is maintained.

Light compass response: see **Menotaxis.**

Lignified: Woody; plant tissues impregnated with lignin.

Ligula: The terminal lobes (or lobe) of the labium, the glossae and paralossae of the prementum of the labium, whether fused or separate.

Limnetic zone: The open-water zone to the depth of effective light penetration.

Lindane: A synthetic chlorinated hydrocarbon containing 99 percent or more of the gamma isomer of **BHC.**

Line: As a unit of measure, 1/12 in. or 2.12 mm. (cf. Millimeter).

Linear: Linelike, long and very narrow.

Lint: The fiber surrounding the seed of unginned cotton.

Lipase: Any enzyme that catalyzes the splitting of fats into glycerol and fatty acids.

Lipid: Any of a variety of compounds such as fats, oils, waxes, and steroids that are insolubie in water but soluble in alcohols and ethers.

Litter: A layer of dead vegetative matter overlying the soil.

Littoral: The shore zone between high-water and low-water, at the edge of lakes and on the seashore; shore zone; also encompasses the flora and fauna living there.

Local list: A publication giving a listing of the animals or plants recorded from a locality or district.

Local population: The individuals of a given locality which potentially form a single interbreeding community (cf. Natio, Subspecies).

Locomotion: Movement from one place to another *e.g.* in *Amoeba* by pseudopodia.

Locus: The hypothetical position of a gene in a chromosome (cf. Gene, Chromosome).

Locustol: A primer pheromone of the desert locust that triggers development from the solitary to the gregarious form.

Lodging: The failure of grain stems of remain vertical due to wind, rain, disease, or insect damage. Makes harvest difficult; and part of crop is typically lost.

Logistic curve: Sigmoid-shaped growth curve applicable to both individuals and populations of organisms plotted graphically.

Longitudinal: In the direction of the long axis of the body; lengthwise in the body or an appendage (e.g., dorsal longitudinal muscle).

Longitudinal muscles: Muscles running along the long axis of the body.

Longitudinal veins: Wing supporting structures with a basal to apical orientation (Fig. 21).

Looper: A caterpillar of the family Geometridae, with only one pair of abdominal prolegs (in addition to the terminal claspers), and which moves by looping its body; a caterpillar that moves by looping its body, that is, placing the posterior part of the abdomen next to the thorax and then extending the anterior part of the body forward a measuring worm.

Lorum (pl., lora): The cheek; a sclerite on the side of the head (Hemiptera and homoptera); the submentum in bees; a basal sclerite present in the]mouthparts of insects and well developed in bees.

Lotic: Of flowing water.

Lotic habitat: Running waters; streams and rivers.

Low volume spraying: The application of pesticides in concentrated form using about 225 to 560 litres per hectare (20-25 gals per acre).

Luciferase: An enzyme in whose presence luciferin is oxidized for the production of cold light in fire flies.

Luciferin: A chemical connected with the production of cold light in fire flies in the presence of the enzyme luciferase.

Luminescent: Producing light.

Lumper: In taxonomy, one who tends to unite related units into a single taxon; one whose criteria for determining the level to be assigned to a given taxonomic category are such that the effect of his work is to lower the rank of existing categories (as families to subfamilies, species to subspecies) (cf. Splitter).

Lunule: A crescentic mark (sclerite) on the face of some adult Diptera; a paired structure on some Hymenoptera.

Lunule, frontal: (see *frontal lunule*).

M

Macro: Large.

Macro-and microconsumers: In an ecosystem the primary consumers, feed on plants while other consumers, e.g., secondary and tertiary consumers are called macroconsumers. The decomposers, like bacteria and fungi, are called microconsumers.

Macro-and micronutrients: Some salts or elements needed in relatively large amount are termed macronutrients while some others though essential for maintaining the life-system are required in minute amounts and are the micronutrients.

Macroclimate: General climate prevailing over a large area, considered as a unit. The opposite is microclimate.

Macroevolution: Major phylogenetic patterns that develop over wide spans of geological time.

Macrofauna: The larger members of a particular fauna. Open to different levels of interpretation for they may be larger vertebrates or alternatively the larger insects of a habitat. The opposite is microfauna.

Macrogamete: The female gametocyte of a malaria parasite, *Plasmodium* spp; *see also* **Microgamete.**

Macrophage: An eater of large particles; *see also* **Microphage.**

Macropterous: With long or large wings. The opposite of apterous.

Macrotrichia: See setae.

Macrotrichium (pl. macrotrichia): A trichoid sensillum, also called a **seta** or **hair.**

Maculate: Spotted; usually refers to insect wings. Opposites are immaculate, hyaline.

Maggot: A vermiform larva; a legless larva without a distinct or well-developed head capsule Diptera and some Hymenoptera (Diptera).

Major pest: Any pest that routinely inflicts economic damage on commercial crops if not controlled, so that management strategies are focused on these pests (see also *key pest*).

Major worker: An individual of the largest-sized worker caste of termites and ant, specialized for defence; *see also* **Media worker; Minor worker.**

Malaise trap: A tent like sampling device that intercepts flying insects.

Malaria: An infectious disease fever-including of man caused by protozoa, *Plasmodium* and transmitted by mosquitoes of the genus *Anopheles.*

Male confusion technique: Method of disrupting mating in insects by artificial release of excess amounts of a sex attractant chemicals.

Mallophaga: The order of insects comprising the biting or bird lice.

Mallophaga, order: (*mallo,* wool; *phaga,* eat) The order of insects comprising the chewing lice (Fig. 160).

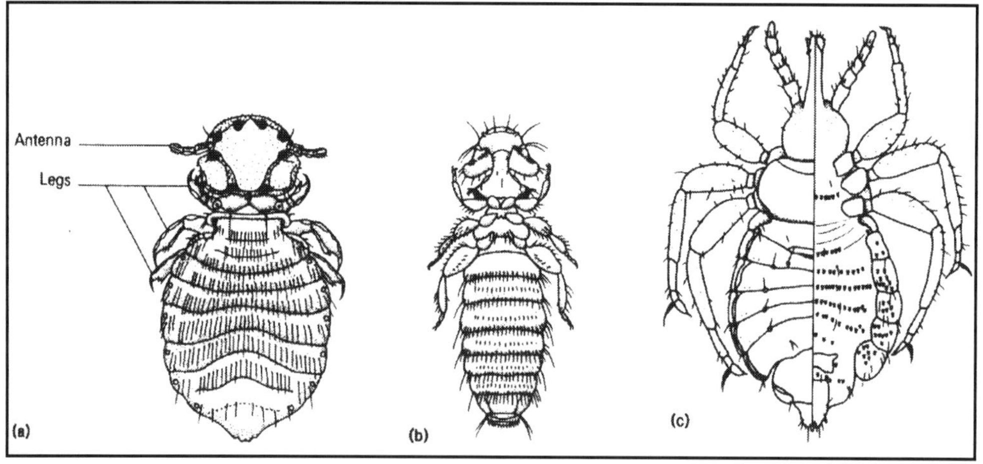

Fig. 160. Mallophaga, chewing lice. (a) Dog-biting louse, *Trichodectes canis.* (b) Chicken body louse, *Menacanthus stramineus.* (c) African bush pig chewing louse, *Haematomzyzus porci* (dorsal features to left of midline, ventral features to right). (a) After U.S. Public Health Service; (b) after Boreal Labs key card; (c) after Emerson and Price, 1988.

Mallophagan: Refers to a member of the Order Mallophaga.

Malpighian tubule: An excretory tubule, opening into the gut at the junction of the midgut and hindgut (Fig. 41); elongate thread like excretory structures of the digestive tract in most insects; located at the anterior end of the hindgut and extend into the body cavity; thin, blind -ending tubules, originating near the junction of mid and hindgut, predominantly involved in regulation of salt, water and nitrogenous wastes excretion.

Maltase: An enzyme that catalyzes the breakdown of the disaccharide maltose into glucose molecules.

Mandible (adj. mandibular): The jaws, either jaw-like stout and serrated in shape in biting and chewing (mandibulate) insects (Fig. 63 and 161), or modifies as needlelike narrow stylets in piercing and sucking insects; the first pair of jaws; derived from the fourth head segment.

Mandibular: Pertaining to the mandibles (Fig. 161).

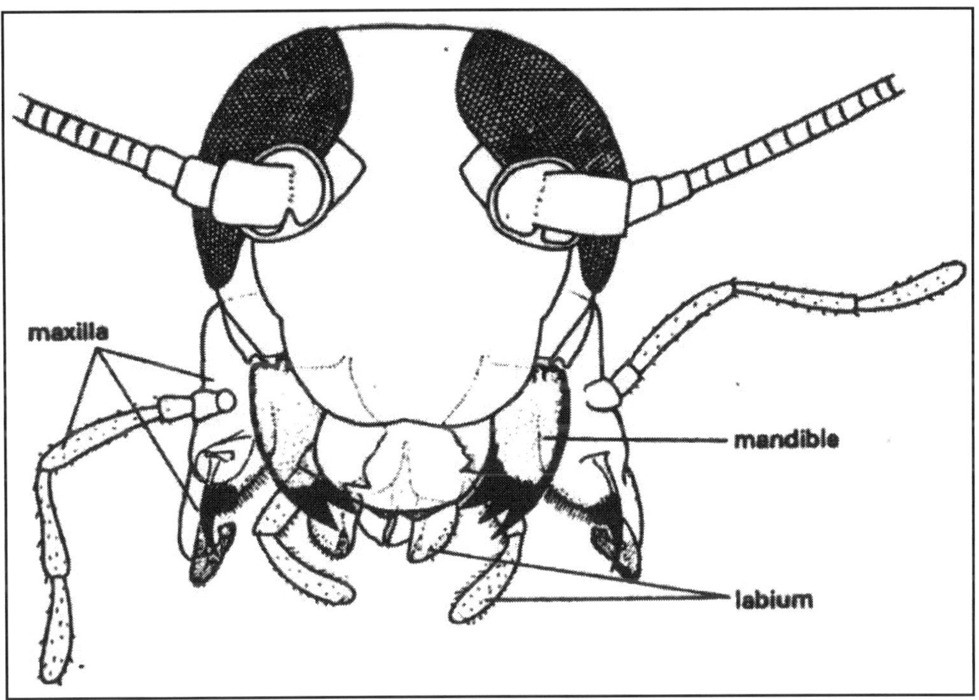

Fig. 161. Head of cockroach, frontal view. Mouthparts are shown in situ.

Mandibulate: Having primitive, jaws usually suited for chewing; mouth parts derived from mandibles (Fig. 162).

Mantle: The saddle- shaped swelling on the back of a slug.

Mantodea, order: (*mantid,* soothsayer) praying mantids (Fig. 163).

Manual: Handbook (q.v.).

Manuscript name: In nomenclature, an unpublished scientific name (cf. *Nomennudum*).

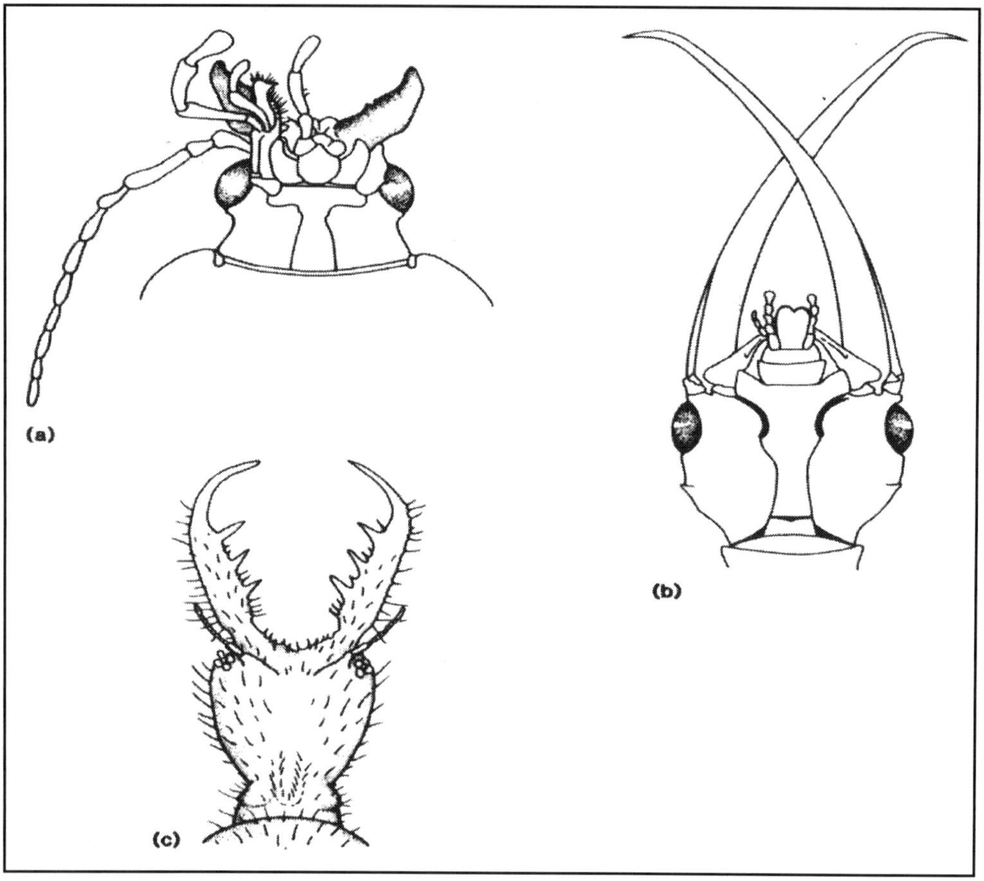

(a)

(b)

(c)

Fig. 162. Variations in form of mandibles. (a) Ground beetle, *Calosoma* sp. with left maxilla and labial palp removed. (b) Male dobsonfly, *Corydalus cornutus.* (c) Antlion larva, *Myrmeleon* sp.

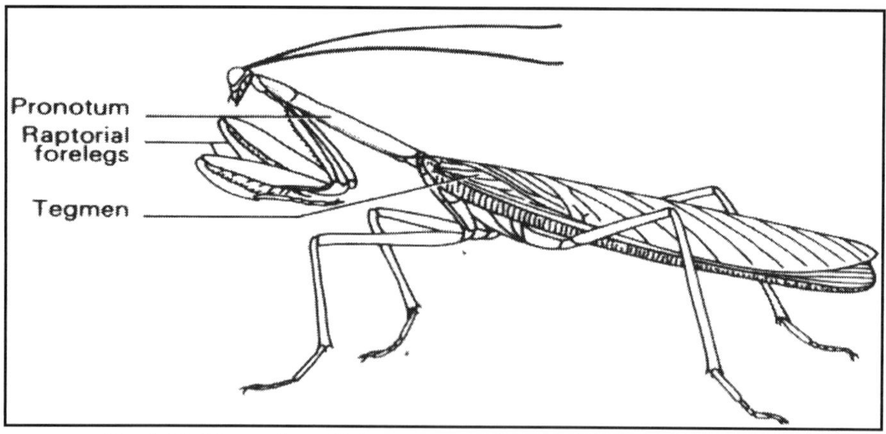

Fig. 163. Mantodea. Chinese mantid, *Tenodera aridifolia.* (VL.)

Marginal cell: A cell in the distal part of the wing bordering the costal margin (Diptera; Hymenoptera).

Marginal vein: A vein on or just within the wing margin; the vein forming the posterior side of the marginal cell (Hymenoptera).

Margined: With a sharp or keel-like lateral edge.

Mariculture: The rearing and commercial cultivation of marine organisms.

Market value: The amount of money that a seller can expect to obtain for a commodity.

Marsupial: Mammals having a pouch or marsupium formed by the fold of skin in females of the subclass Metatheria.

Marsupium: A pouch which may contain developing eggs or young. The group of animals having pouch to carry young ones is known as marsupial *e.g.* Kangaroo.

Masquerade: A form of **crypsis** in which an organism resembles a feature of its environment that is of no interest to a predator; also called **mimesis.**

Mass rearing: Rearing of the immature stages in bulk.

Material safety data sheet (MSDS): Formal description on the chemistry and toxicology of a chemical.

Material: In taxonomy, the sample available for taxonomic study (cf. series, Hypodigm).

Mating distriuption: A form of insect control in which synthetic sex pheromones (usually of the female) are maintained artificially at a higher level than the background, interfering with mate location.

Matrifilial: Of eusocial hymenoptera whose colonies consist of mothers and their daughters.

Matrix: The gelatinous substance into which eggs of some nematodes are deposited, forming an egg mass.

Maxilla (pl., maxillae): One of the paired mouth-part structures immediately posterior to the mandibles and before the labium, bearing the maxillary palpi (Fig. 63); paired, segmented, secondary jaws that aid in holding and chewing foods; located between the mandibles and the labium; stylets in many insects with piercing-sucking mouthparts; variously modified in other insects.

Maxillary palp: A small feelerlike 1-7 segmented sensory appendage borne on the **stipes** of the maxilla (Fig. 63).

Maxillary palpus (pl. maxillary palpi): A small feelerlike structure arising from the maxilla (Fig. 63).

Maxillary: Part of or belonging to the maxillae of the mouthparts.

Maxilliped: One of the appendages in Crustacea immediately posterior to the second maxillae.

Maximum dosage: The largest amount of a pesticide that can be applied safely, without excess residues or damage to the host plant. The maximum dosage printed on a pesticide label is the maximum permitted by law.

Maximum residue limit: The tolerated maximum amount of pesticide allowed to remain in food by the Food and Drug Administration (FDA).

Mean: The arithmetic average of a group of samples.

Mechanical control: Any means of pest control which involves mechanically trapping or killing insects, or providing barriers to prevent insects gaining access to plants or other materials; pest control tactics based on physical disruption of pests, with methods including cultivation, and ploughing (e.g., tillage for weed control).

Mechanical transfer: The movement of a disease organism from one host to another by passive transfer, with no biological cycle in the **vector.**

Mechanoreception: The perception of stimuli as various mechanical changes.

Mechanoreceptor: A sense organ involved in the perception of tactile stimuli and/or vibration.

Mechanosensilla: Sensory structures that respond to stimuli such as stretching, bending, or compression.

Meconium: The first excreta of a newly emerged adult following the pupal stage.

Mecoptera, order: (*meco,* long; *petra,* wing) A small order of insects comprising the scorpionflies (Fig. 164).

Media: In wing venation, the fifth longitudinal vein, lying between the radius and the cubits, with a maximum of eight branches.

Media worker: An individual of the medium-sized worker caste of termites and ants; *see also* **Major worker; Minor worker.**

Medial (or **median**): Referring to the center, usually the midline, of an animal.

Medial: Towards the middle.

Medial cross vein: A cross vein connecting two branches of the media.

Median: At or towards the middle.

Median: In the middle; lying along the midline of the body.

Median caudal filament: In Thysanura, a threadlike process at the posterior end of the abdomen that lies between the cerci.

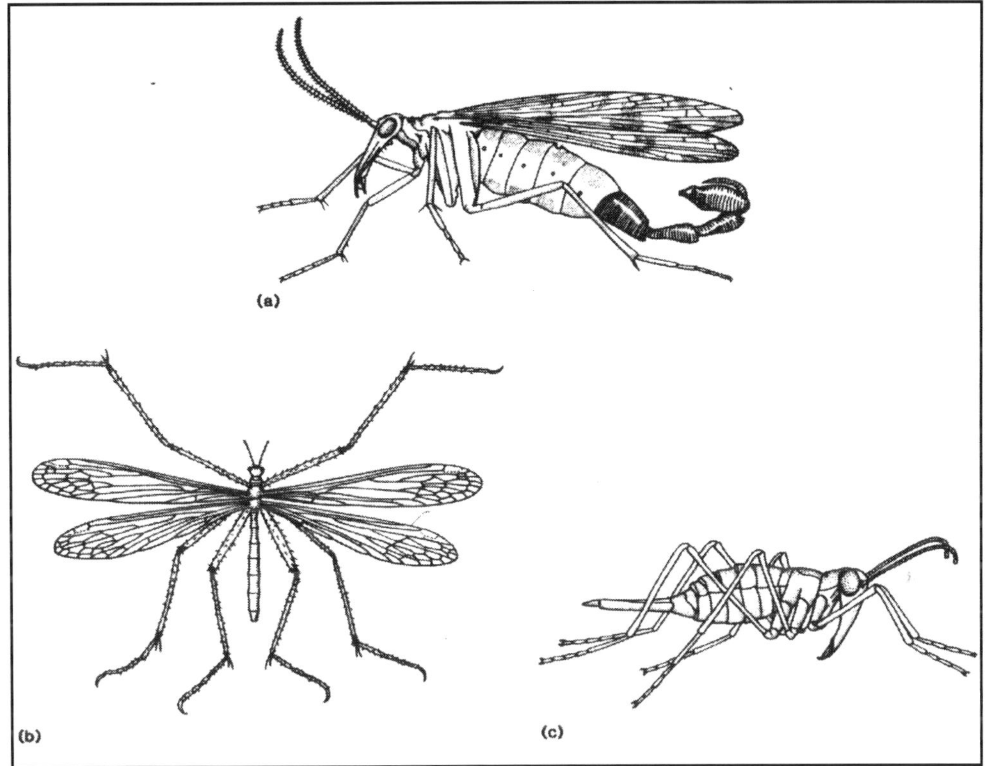

Fig. 164. Mecoptera (a) Common scorpionfly (male), *Panorpa chelata* (Panorpidae). (b) Hanging scorpionfly, *Bittacus chlorostigma* (Bittacidae). (c) Snow scorpionfly, *Boreus californicus* (Boreidae). (a) After Ross, 1962; (b) and (c) after Essig. 1942.

Median cell: A cell between the subcostal and median veins at the base of the wing (Hymenoptera).

Median flexion-line: A flod-line that runs longitudinally through the approximate middle of the wing.

Median lethal concentration: See $LC_{50.}$

Median lethal dose (MLD): See LD_{50}

Medical pest: A pest that causes a pathological condition or transmits pathogenic organisms to humans or animals.

Medio-cubital cross vein: A cross vein connecting the media and cubitus.

Medusa: Free swimming bell or umbrella-shaped form of coelenterates. Their general mode of reproduction is sexual *e.g.* Jellyfishes.

Mega: Large.

Meiofauna: Tiny animals that move through the substratum without displacing the particles; in sand usually species 50 μ -3 mm in length.

Meiosis: Cell division that produces cells with half the number of chromosomes, called gametes; or the two successive divisions of a cell's nucleus in which the chromosome number is reduced by one-half; or a process in the maturation of a sex cells involving the halving of the number of chromosomes (Gr. *meioun* = to make small).

Meiotic drive: The unequal recovery of homologous chromosomes during meiosis.

Melanic: Darkened.

Melanin: A pigment that produces black, amber, and dark brown colors in the cuticle.

Melanins: Insect pigments of yellow, black, or brown.

Melanism: An unusual darkening of color owing to increased amounts of black pigment; sometimes a racial character, sometimes, as in cases of polymorphism, restricted to a certain percentage of individuals within a population (cf. Industrial melanism, Albinism).

Melanization: Darkening.

Melittophily: Pollination by bees.

Membrane, undulating: A protoplasmic membrane between body and port of flagellum in flagellates *e.g.* in *Trypanosoma.*

Membrane: A thin film of tissue, usually transparent; that part of the wing surface between the veins; the thin apical part of a hemelytron (Hemiptera); or the thin sheet consisting of cells or of the material secreted by cells covering the organs, or lining the cavities/canals of body.

Membranous: Like a membrane; thin and more or less transparent (wings); thin and pliable (cuticle).

Menotaxis: (including **light compass response**) Movement that maintains a constant orientation relative to a light source.

Mental setae: Setae on the mentum (Odonata); distal portion of the postlabium; the distal part of the labium, which bears the palps and ligula; a median, more or less triangular plate in the millipede gnathochilarium.

Mentum: The distal ventral fused plate derived from the labium which bears the palpi.

Meristem: Undifferentiated plant tissue at shoot and root tips and in buds that is capable of division.

Meroblastic cleavage: Method of blastoderm formation characteristic of most insects and associated with eggs containing a large amount of yolk; the fusion nucleus and associated cytoplasm proliferate mitotically, and the daughter nuclei migrate to the periphery of the egg and form the blastoderm (Fig. 165).

Meroistic ovarioles: Polytrophic and telotrophic ovarioles taken together.

Meroistic ovary: An ovary possessing nurse cells that are connected to or accompany the oocytes.

Meron: The posterior portion of the coxa.

Meropleuron (pl., meropleura): A sclerite consisting of the meron (basal part) of the coxa and the lower part of the epimeron.

Merozoite: The third stage in the asexual cycle of a malaria parasite (*Plasmodium* spp.), derived from division of the **schizont.**

Mesad: Toward the midline of the body.

Mesal: At or near the midline of the body, also **medial.**

Mesenteron: The midgut, or middle portion of the alimentary tract; primary site of digestion and absorption.

Mesentoderm: A longitudinal band of tissue in the insect embryo that is surrounded by the ectoderm.

Mesepimeron (pl., mesepimera): The epimeron of the mesothorax.

Mesepisternum (pl., mesepisterna): The episternum of the mesothorax.

Mesian neurosecretory cells: Cells in the pars intercerebralis region of the brain that secrete hormones, including brain and eclosion hormones.

Mesintfaepisternum: A ventral subdivision of the mesepisternum.

Meso: middle.

Mesoclimate: Climate of a localized area, such as a valley or hillside, that differs from the macroclimate.

Mesocuticle: A procuticular layer lying between the exocuticle and endocuticle.

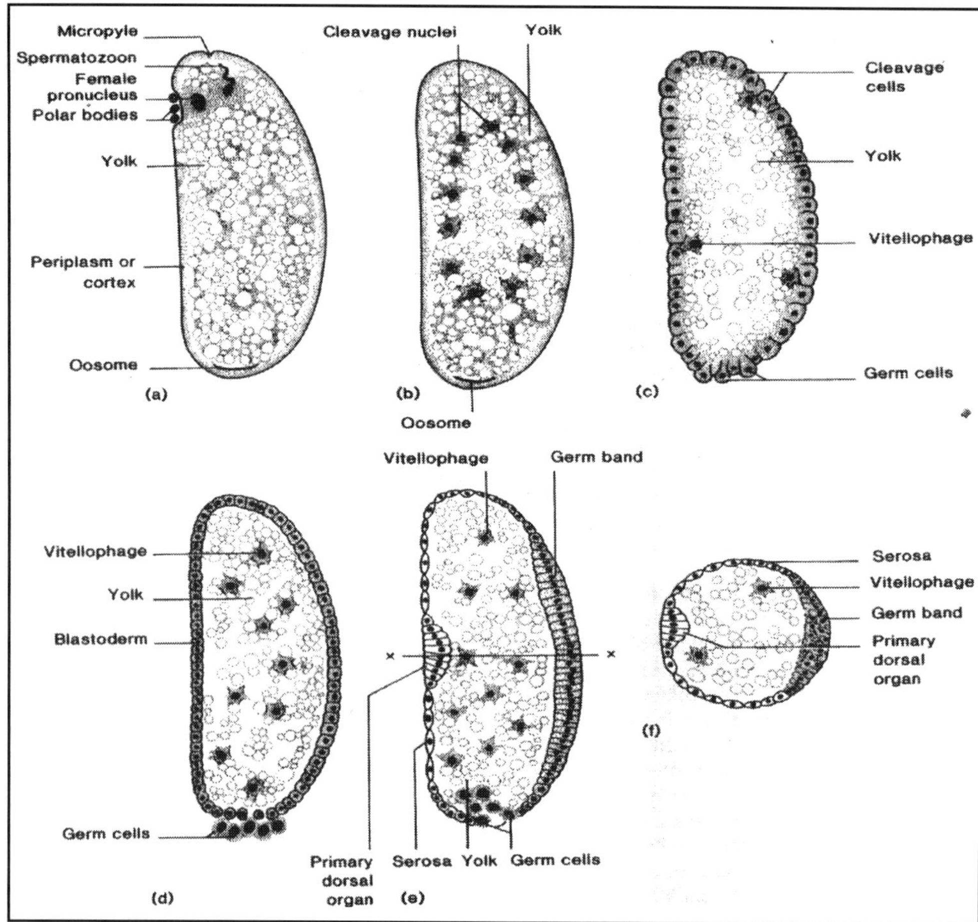

Fig. 165. Fertilization – germ band formation (chorion omitted; (a-e) sagittal sections). (a) just prior to fertilization (at the stage of fusion of spermatozoon and female pronucleus). (b) Cleavage. (c) Blastoderm forming and germ cells differentiating. (d) Blastoderm formation complete. (e) Germ band and primary dorsal organ formataion. (f) Cross section at same as (e). The location of the section is indicated in (e) by the line x-x. After Johannsen and Butt, 1941.

Mesoderm: The middle germ layer (Fig. 166); One of the three primitive germ layers in the developing embryo; gives rise to muscles, fat body, heart, blood cells, and reproductive organs; the third germ layer developed between the ectodermal and endodermal layers in triploblastic metazons *e.g.* Ctenophora, Platyhelminthes and animals higher than these in evolutionary scale (Gr. *mesos* = middle + *derma* = skin).

Mesogloea: Jelly like layer between ecto and endodermal layers in diploblastic organisms fore runner of the mesoderm of the triploblastic animals. (Coelenterata) in *e.g.* sea anemone, *Hydra,* Jellyfishes.

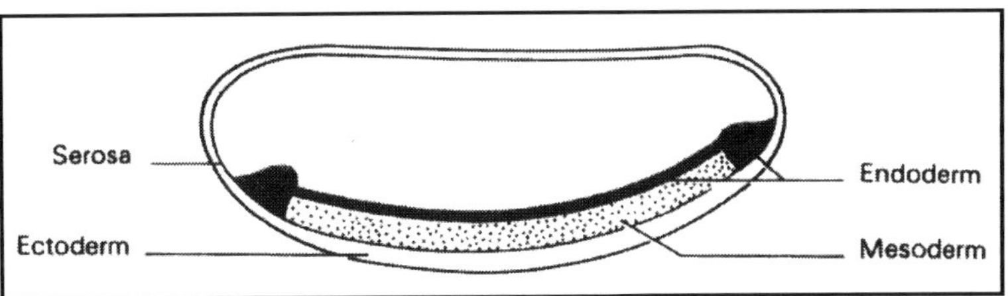

Fig. 166. Sagittal section of an embryo.

Meson: The midline of the body, or an imaginary plane dividing the body into right and left halves.

Mesonotum: Notum or dorsal plate of the mesothorax.

Mesopleural bristles: Bristles on the mesopleuron (Diptera).

Mesopleuron (pl., mesopleura): The lateral sclerire(s) of the mesothorax; the upper part of the episternum of the mesothorax (Diptera).

Mesoscutellum: The scutellum of the mesothorax, usually called simply the scutellum.

Mesoscutum: The scutum of the mesothorax.

Mesosoma: The middle of the three major divisions (**tagmata**) of the insect body, equivalent to the thorax but in apocritan Hymenoptera including the **propodeum**; called the **alitrunk** in adult ants.

Mesosternum: The sternum, or ventral sclerite, of the mesothorax.

Mesothorax: The middle and second segment of the thorax, bearing the forewings and middle legs.

Meta: Later; latter; changing.

Metabola: (alternative term, Pterygota) The major subclass of insects which undergo metamorphosis whether complete or incomplete and which usually possess wings in the adult state. cf. Ametabola.

Metabolic resistance: The ability to avoid harm by biochemical detoxification of an insecticide.

Metabolism: The chemical changes which go on within the living organisms. rate of metabolism is directly proportional to the activity of the organism *e.g.* metabolism in frog during inactive winter hibernation is negligible (Gr. *metabolos* = changeable).

Metabolite: The component parts that result from the breakdown of compound, such as a pesticide, through the physiological processes of an organism.

Metamere: A primary body segment (usually referring to the embryo); essentially identical primary body segments found in embryos and thought to be characteristic of the primitive arthropod; also called somites.

Metamorphosis: A marked and abrupt change in morphology through which a holometabolous insect passes in developing from larva to adult (meaning literally "to change form"); a change in body form by an animal during the course of its development and the adult formed is different in appearance from the one hatched out of the egg; the change in body form during insect development; either complete (drastic change) or incomplete (slight change).

Metanotum: Tergum (notum; dorsal plate) of the metathorax.

Metapneustic: Respiration when the larva has only the last pair of abdominal spiracles open; usually refers to Diptera.

Metascutellum: The scutellum of the metathorax.

Metasoma: In apocritan Hymenoptera, the **petiole** plus **gaster**; last six somites of the abdomen of scorpions where last/terminal one bears a sharp, curved poison claw, which is actually the sting.

Metasternum: The sternum, or ventral sclerite, of the metathorax.

Metatarsus (pl., metatarsi): The basal segment of the tarsus.

Metathorax: The third, or most posterior, of the three segments of the thorax, bearing the hindwings and posterior pair of legs.

Metatype: A specimen compared by the author of the species with the holotype and determined by him as conspecific with it.

Metazonite: The posterior portion of a millipede tergum, when the tergum is divided by a transverse groove.

Metepimeron (pl., metepimera): The epimeron of the metathorax.

Metepisternum (pl., metepisterna): The episternum of the metathorax.

Methoprene: A juvenile hormone analog.

Metinfraepisternum: A ventral subdivision of the metepisternum.

Metric system: A decimal system of measures (with the meter as base) and weights (with the gram as base); the universal system for reporting measures and weights in the scientific field.

Mg/kg: Milligrams of pesticide (or any toxicant) per kilogram of animal body weight.

MH (alternative terms, moulting hormone, ecdysone)**:** An abbreviation of moulting hormone.

Micro: Small.

Microbial control: The use of pathogenic micro organisms to control pest species.

Microbial insecticide: A biological preparation of viruses or microorganisms or their products, applied and used in ways similar to conventional chemical insecticides.

Microbial pesticide: Pesticides that have a living bacterium, virus, fungus, protozoa, or nematode as the active ingredient.

Microbiology: The science which deals with the study of microorganisms; the biological relationship of microorganisms.

Microclimate: The climate of a localized area, such as within a plant canopy; climate of a microhabitat; typically more constant and less variable than either the general climate of an area or a mesoclimate.

Microcoryphia: The order of insects comprising the jumping bristletails.

Microencapsulation: A pesticide formulation technique in which pesticide-containing microscopic spheres or capsules are created to permit pesticide release at a slow, steady, and effective rate.

Microevolution: Evolutionary changes that occur in populations, e.g., industrial melanism and development of resistance to pesticides.

Microfauna: The fauna of a microhabitat or used as a general term for microscopic animals.

Microfilariae: A group of very small, thin, parasitic nematodes, usually blood-inhabiting; often transmitted among the vertebrate hosts by blood-sucking Diptera.

Microgamete: The male gametocyte of a malaria parasite, *plasmodium* spp., flagellate initially, after which flagella are lost; *see also* **Macrogamete.**

Microgeographic race: A local race, restricted to a very small aarea.

Microhabitat: The specific habitat of an organism composed of its immediate surroundings and the microclimate.

Microlethical: An egg lacking large yolk reserves.

Micro-organism: An organism of microscopic size, such as a bacterium, virus, fungus, viroid, or mycroclimate; a group term applied to living organisms that are in most stages too small to see with the naked eye; includes bacteria, viruses, fungi and protozoa.

Microphage: A feeder on small particles, such as spores.

Micropyle: A minute opening or pore (or openings or pores) in the chorion of an insect egg, through which sperm enter the egg; channel through which spermatozoa enter an egg.

Microsclerotium (pl. microsclerotia): Very small sclerotia, such as those produced by the *Verticillium* wilt and rice stem rot fungi.

Microtrichium (pl. microtrichia): Minute fixed hairs that lack the basal articulation characteristic of setae; tiny cuticular setae on the surface of the wings, espically in adult Diptera; a subcellular cuticular extension, usually several to very many per cell.

Microvilli: Numerous, finger like processes that serve to increase absorptive surface area, for example, in midgut epithelial cells.

Midgut: Middle region of the insect alimentary canal; site of most digestion and absorption of food; also called mesenteron; the middle section of the gut, extending from the end of the proventriculus to the start of the ileum.

Migrant population: With reference to insects, a population that has left its place of origin in search of food or shelter, or for other reasons.

Migrantes: Adult, winged, parthenogenetic viviparous female aphids, developing on the primary host and then dispersing to secondary hosts.

Migration: Mass movement, usually in search of more appropriate conditions; long-distance movement of organisms in relation to seasons; a form of dispersal involving long-distance movements under at least partial control of the insects; movements over great distances, during which locomotion is not inhibited by food, mates or oviposition sites; an adaptation to periodically transport insects beyond the boundaries of their old reproductive sites and into new ones.

Migratory ectoparasitic nematode: A nematode that feeds from the exterior of the host and remains capable of movement to other locations on the host or to other hosts.

Migratory endoparasitic nematode: A nematode that enters host-plant tissue, feeds internally, and continues to move through the tissue as it feeds and reproduces.

Mildew: A fungus disease in which the mycelium grows over the surface of the plant causing a whitish discoloration.

Milk glands: Specialized accessory glands in certain **adenotrophically viviparous** files (Hippoboscidae and *Glossina*) that produce secretions fed upon by larvae.

Milky disease: A bacterial disease of Japanese beetle caused by *Bacillus popilliae* and *Bacillus lentimorbus*..

Millimeter (mm): 1/1,000 m., or 0.03937 in., approximately 1/25 in. (cf. Metric system).

Millimeter: 0.001 meter, or 0.03937 inch (about 1/25 inch).

Millipede: Arthropod of the class Diplopoda, having two pairs of legs on most body segments. They inhabit debris and are not usually agricultural pests.

Mimesis: Resemblance to an inedible object in the environment; *see also* **Masquerade.**

Mimetic polymorphism: The occurrence within a population of several (often strikingly different) forms, each of which resembles closely a different sympatric species, in butterflies often restricted to females.

Mimic (adj. mimetic): One of the three components of a mimicry system, the emitter of false signal(s) received by an **observer;** an individual, population or species that resembles a **model,** usually another species or part thereof; *see also* **Automimic; Batesian mimicry; Mullerian mimicry.**

Mimicry: The resemblance of a **mimic** to a **model,** by which the mimic derives protection from predation provided to the model (e.g. by unpalatability); close resemblance in form, color, pattern, and so on, between one species and another that is interpreted as conferring a benefit on one or both of these species; presence of a pattern in a palatable species that closely resembles the pattern of an unpalatable species (= Batesian mimicry) (see *also* Mullerian mimicry).

Mine: Tunnel in plant parts produced by the feeding of immature insects.

Minimim tillage: Techniques of sowing crops which involve little or no disturbance of the soil.

Minor pest: Pests that occasionally cause economic damage on commercial crops.

Minor use pesticide: A pesticide with small potential market due to the small acreages of the commodity it is used on.

Minor worker: An individual of the smaller-sized worker caste of termites and ants; *see also* **Major worker; Media worker.**

Minute: Very small; an insect a few millimeters in length or less would be considered minute.

Miscible liquid (m.l.): A formulation in which the technical product is dissolved in an organic solvent which is then, on dilution, dissolved in the water carrier.

Miscible: Liquids that can be mixed to yield a uniform solution.

Mist blower (alternative term, air blast sprayer)**:** A sprayer which uses a stream of air to carry spray droplets on to the object being sprayed.

Mist blower: See *air-blast sprayer;* sprayer producing a fine-carried spray.

Mite: Tiny arthropod of the order Acarina. Closely related to spiders with eight jointed legs, no antennae, and no wings.

Miticide (alternative term, acaricide)**:** A pesticide that kills mites and ticks; preferably called Acaricide.

Mitigation: The act of reducing pest numbers.

Mitochondrion (pl. mitochondria): An organelle that is the principal site of energy synthesis in the cell.

Mitosis: Nuclear division characterized by a complex movement of chromosomes, which results in the formation of new nuclei having the same number of chromosomes as the original nucleus; indirect cell division on account of which the cell divides into two daughter-nuclei each having the same number and kind of chromosomes (Gr. *Mitos* = thread).

Mixed-function oxidase (MFO): An enzyme that metabolizes many foreign substances.

Mode of action (as applied to insecticides)**:** The mechanism by which an insecticide kills an insect.

Mode of action: The physiological mechanism by which a pesticide exerts a toxic effect on the target organism; means by which a toxin affects the anatomy, physiology, or biochemistry of an organism.

Model (mimicry system): One of the three components of a mimicry system, the emitter of signal (s) received by the **observer;** the organism resembled by a **mimic,** protected from predation, for example, by distastefulness.

Model: A simplified depiction of reality. May be a mathematical representation of the elements of a population or crop system.

Moderately toxic: Poisons with LD_{50} **equal** to 50-500 mg/kg.

Molar area: The grinding surface of the mandible.

Mold: Fungus with conspicuous mycelial or spore masses growing on damp or decaying matter.

Mollicute: A class of microorganisms that includes mycoplasmas and spiroplasmas. They lack cell walls and are variable in shape. Spiroplasmas canbe cultured but mycoplasmas cannot.

Mollusca: The phylum of animals to which slugs and snails and their relatives belong.

Molluscicide: A toxicant or pesticide that kills slugs and snails.

Mollusk: Snails and slugs (snails without external shells) of the phylum Mollusca, class Gastropoda.

Molt: A process of shedding the cuticle (ecdysis) in case of insects, mites, and nematodes; the shedding of the exoskeleton or cuticle before entering the next life stage.

Molting: The process of digesting the old cuticle, secreting the new cuticle, and shedding the old cuticle (Fig. 167).

Molting fluid: A substance secreted by the epidermal cells; contains chitinase and protease and digests old endocuticle (Fig. 167).

Molting hormone (MH): Ecdysone; hormone that stimulates molting.

Monecious: Possessing both male and female sex organs, hermaphroditic.

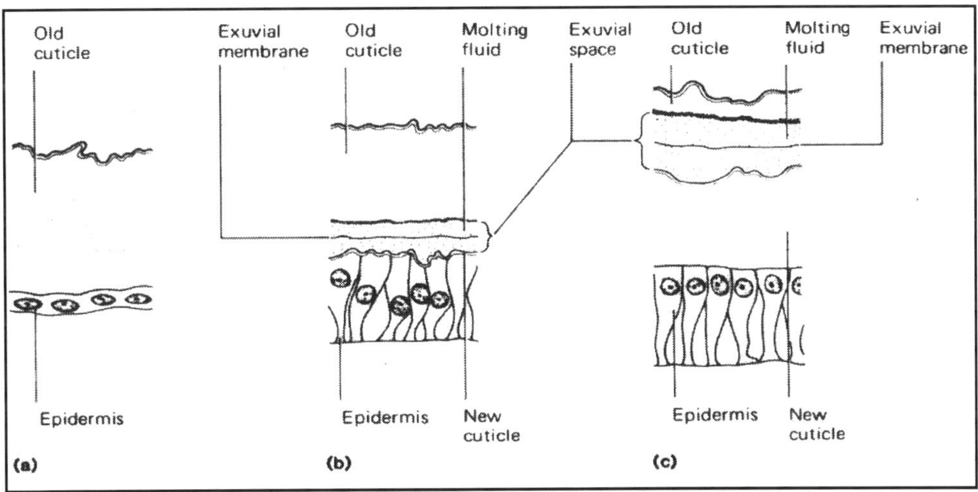

Fig. 167. Molting process. (a) Section of integument prior to molting. (b) After apolysis, digestion of old endocuticle in progress and new cuticle being deposited. (c) Just prior to ecdysis.

Moniliform: Beadlike, with rounded segments; moniliform antenna e.g. termite.

Monitoring system: A system to track of (i) whether pesticides are escaping into the environment, or (ii) the numbers, life stages, evidences of feeding, damage, and locations of pests in a crop.

Monocondylar: An articulation (as of a mandible) with one point of articulation (condyle).

Monocondylic (articulation): Joint containing a single condyle, allowing considerable movement (Fig. 168).

Monocropping (alternative term, monoculture)**:** The practice of growing the same crop (and often the same cultivar) on the same land year after year. In some situations it may involve the same crop plant over large areas.

Monoculture: A single species or single varietal planting; the cultivation of a single crop to the exclusion of all other plants, the uniform extensive cultivation of a single species of crop plant.

Monocyclic: A disease with one reproductive cycle per year (cf., polycyclic).

Monoecious: Having only one set of sexual organ in an individual, *e.g.* male or female.

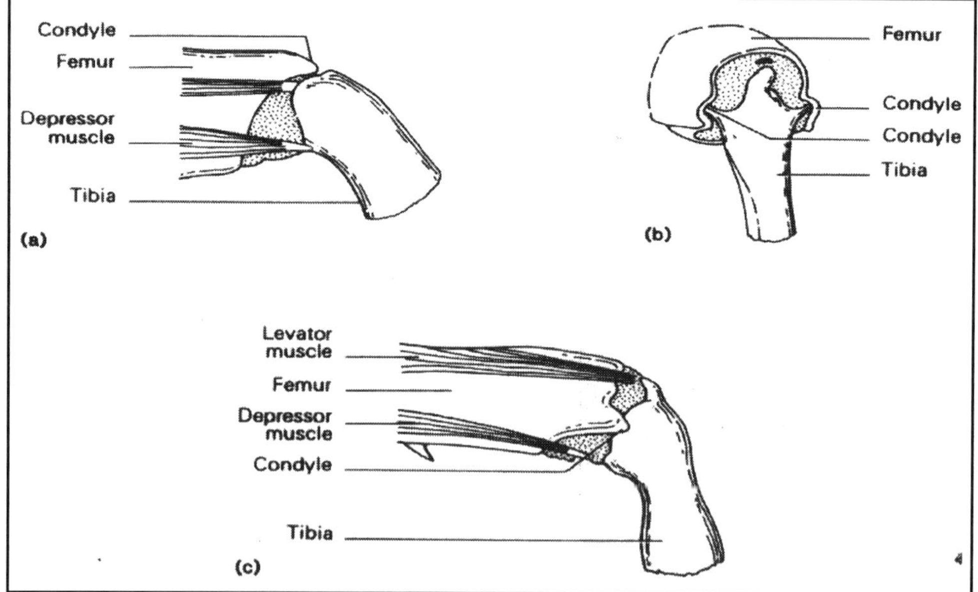

Fig. 168. Types of joints. (a) Monocondylic. (b) Dicondylic (end view). (c) Dicondylic (lateral view).

Monogamy: Mating with a single individual.

Monogenic resistance: Resistance resulting through expression of a single gene.

Monogenic: A characteristic (such as resistance) controlled by a single gene.

Monograph: In taxonomy, an exhaustive treatment of a phylogenetic group in terms of all available information pertinent to taxonomic interpretation; usually involving full systematic treatment of all included taxonomic units in terms of comparative anatomy, biology, ecology, and detailed distributional analyses (cf. Revision, Synopsis).

Monogyny: A colony of eusocial insects dominated by one queen.

Monolectic: Describes pollinating insects that visit a single plant taxon; compare with polylectric.

Monolectic: Refers to bees that use pollen from only a single plant species.

Monophage (adj. monophagous): An eater of only one kind of food, used particularly of specialized **phytophages.**

Monophagous: Feeding on only one type of food; monotropic; feeding on a single species of plant (or animal) or at least a group of closely related species. cf. polyphagous, oligophagous insects feeding on only one genus or species of plants or animals species.

Monophyletic: A term applied to a taxonomic category in which the contained units are all part of a single immediate line of descent; evolutionarily derived from a single ancestor, recognized by the joint possession of shared derived feature(s); evolving from a single taxon. (cf. Polyphyletic).

Monotypic: A category containing but one immediately subordinate zoological unit, as a genus containing but one species, or a species containing but one (the nominate or nominotypical) subspecies [for nomenclatural meaning, *see Bul. Zool. Nomencl.*, **4:** 153 (1950)] (cf. Polytypic).

Monoxene (adj. monoxenous): A parasite restricted to one host.

Morph-, morpho: Form.

Morph: A genetic form or variant.

Morphogenesis: All events occurring between the formation of the zygote and the emergence of a sexually mature adult.

Morphology: The science of form or structure; the study of the bodily form of living organisms.

Mortality factor: The cause of death of an organism.

Mortality: Population decrease factor; death rate or numbers dying per unit of time.

Mosaic theory: Model of insect vision in which each ommatidium of the compound eye "sees" only a small portion of the external environment, the perception of the total environment appearing as a mosaic of the contribution from each ommatidium.

Mosaic: Virus disease symptoms of variegated yellow and green patterns in leaves.

Motile: Capable of movement.

Motor neuron: A unipolar neuron that stimulates muscles; also called efferent neuron; a nerve cell with an axon that transmits stimuli from an interneurone to muscles.

Moult increment: The increase in size between successive instars.

Moult, moulting (alternative term edysis)**:** The process of shedding the skin (exoskeleton) undergone periodically by insects and other arthropods.

Moulting hormone (alternative terms, **MH,** ecdysone)**:** The hormone in insects which initiates the moulting process.

Moulting: The formation of new cuticle followed by **ecdysis**.

Mouth hooks: The paired hardened portions of the mouthparts of fly larvae which protrude from the mouth and enable them to shred their food; the head skeleton of the maggot larva of higher flies (Fig. 169).

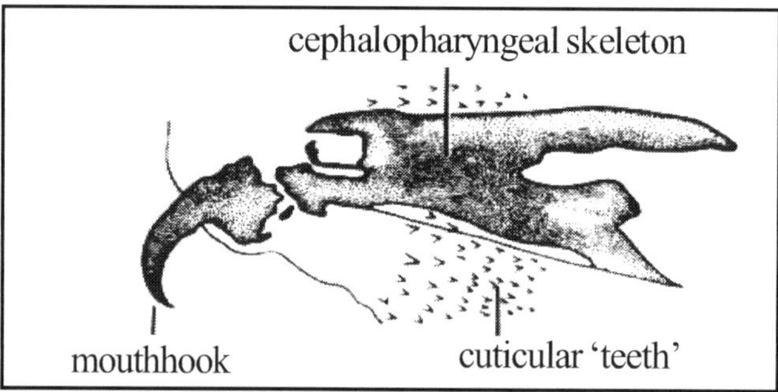

Fig. 169. Old World screw-worm fly maggot, *Chrysomya bezziana*.

Mouthparts: Organs of ingestion.

Metapleuron (pl., metapleura): The lateral sclerite(s) of the metathorax.

Mulch: A layer of material laid over the soil to prevent weed growth, and reduce water loss.

Mullerian mimicry: A mimetic system in which two or more unpalatable species obtain protection from predation by resembling each other; a number of unpalatable or distasteful species that share a similar color pattern; presence of a similar aposematic pattern in unrelated, distasteful or poisonous species. *See also* **Batesian mimicry.**

Multiple factors: Two or more pairs of genes with a complementary or cumulative effect (cf. Blending inheritance).

Multiple feeding bait: Rodent bait that contains anticoagulant toxins (e.g., warfarin) and requires several feedings to be effective.

Multiple parasitism: When the parasites compete with each other for the same host population.

Multiple resistance: The concurrent existence in a single insect population of two or more defence mechanisms against insecticide.

Multipolar neurons: Neurons with an axon and several branched dendrites.

Multiporous: Having several small openings.

Multiterminal innervation: Pattern of insect motor axons that branch out and form terminals at several regularly spaced points along a muscle fiber.

Multivoltine: Any insect that has several generations per year (cf., univoltine).

Multivoltine: Having several generations in one year; *see also* **Bivoltine; Univoltine.**

Mummy: (i) Unharvested fruit or nut remaining on the tree (also called sticktight). (ii) Remains of an aphid whose body contents have been consumed by a parasite.

Muscle fiber: Basic structural unit of a muscle (Fig. 170).

Muscle unit: A group of 10 to 20 muscle fibers organized as a discrete morphological entity and enveloped by a tracheolated membrane (Fig. 171).

Mushroom bodies: See corpora pedunculata.

Mushroom body: A cluster of seminal vesicles and accessory gland tubules into a single mushroom-shaped structure, in certain orthopteroid / blattoid insects; a complex fiber tract in the anterior part of the brain, often suggesting the shape of a mushroom, associated primarily with the integration of sensory information.

Mutagen: A compound that can cause mutations in the genome; a material that induces genetic changes.

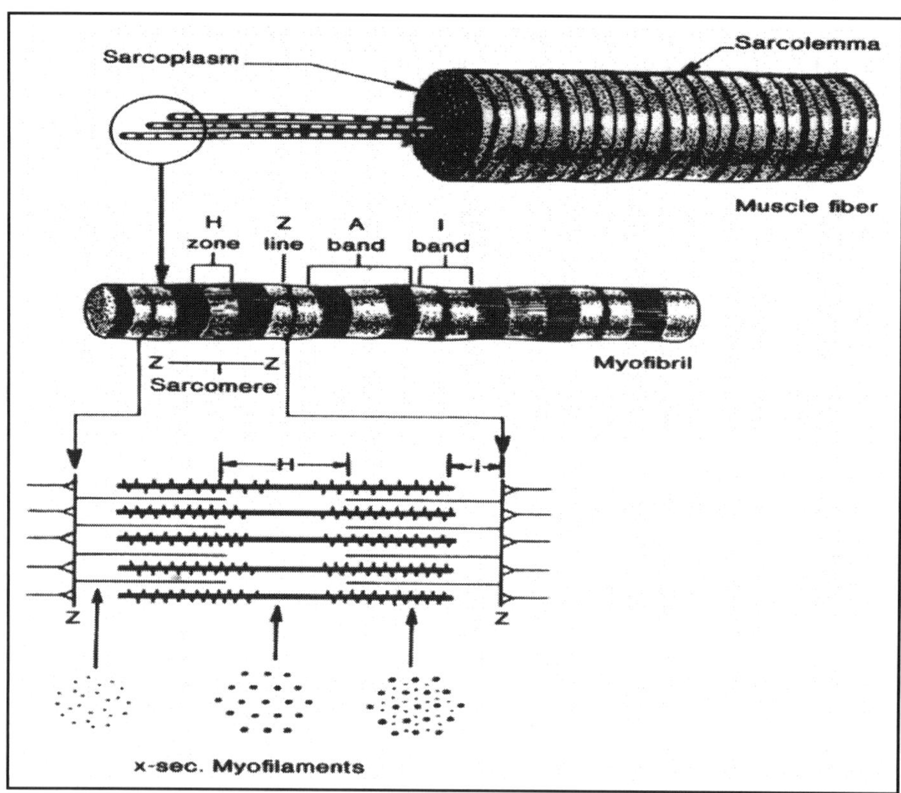

Fig. 170. Structure of generalized muscle fiber. After Herried, 1977.

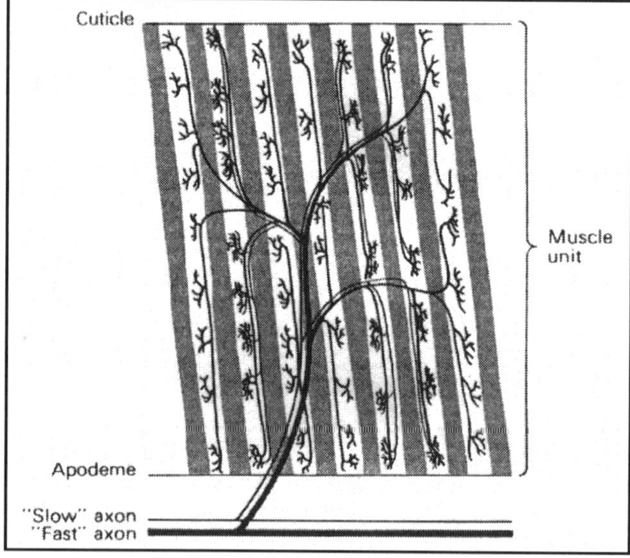

Fig. 171. Innervation of a typical muscle unit showing fast and slow axons and multiterminal endings.

Mutation: Any relatively stable heritable change in the genetic material (genome); in genetics, a discontinous change of a genetic factor; in paleontology, a sudden change in a phyletic series of fossils.

Mutualism: The symbiotic relationship between two organisms in which both parties derive benefit; or intimate relationships between two or more kinds of animals or between plants and animals that are advantageous to both.

Mycangia: Pocket-shaped receptacles used to carry symbiotic fungi (for example, Coleoptera, Scolytidae).

Mycangium (pl. mycangia)**:** An invaginated cuticular pouch in which fungal spores are carried.

Mycelium (pl. mycelia): The vegetative body of a fungus that consists of a mass of slender filaments called hyphae. *e.g.* in *Mucor.*

Mycetocyte (s): A cell containing symbiotic microorganisms, scattered throughout the body, particularly within the fat body, or aggregated in organs called **mycetomes;** cells specialized to house microbes essential to the normal growth and development of certain insects.

Mycetome: An organ containing aggregations of **mycetocytes,** usually located in the fat body or gonads; a specialized internal organ that houses symbiotic microorganisms; the ovary follicle cells through which some aphid eggs are infected by symbiotic fungi; tissues composed of mycytes and associated with the gut. Fat body, or gonads.

Mycetophagous: Fungivorous; feeding on fungi, on fungal mycelium.

Mycoherbicide: A living fungus applied to kill plants.

Mycophage (adj. mycophagous): An eater of fungi; *see also* **Fungivore.**

Mycoplasma: Pleomorphic prokaryotic microorganisms that do not have a cell wall; several are plant pathogens; a pathogenic microorganism that passes through bacterial filters but has certain features in common with bacteria.

Mycoplasma-like organism (MLO): See *phytoplasma.*

Mycorrhiza (pl. mycorrhizae): The symbiotic association of fungi with plant roots, wherein fungus and plant exchange nutrients in a mutually beneficial association.

Mycotoxin: Toxins produced by fungi. Such toxins may be present in infected seeds, feeds, or foods, and may cause illness or death in animals that consume the contaminated produce.

Myemecophilous: "Ant-loving"; refers to organisms, other than ants, that interact with ants symbiotically.

Myiasis: A condition caused by the invasion of open wounds or cavities by certain dipterous fly larvae (maggots); a disease caused by the infestation of the maggots of flies.

Myo-: Muscle.

Myofibril (s): A contractile element of a muscles fiber; unit of muscle fiber composed of myofilaments (Fig. 170); contractile fibres that run the length of a muscles fibre, comprising actin sandwiched between myosin fibers.

Myogenic flight muscles: Flight muscles that contract repeatedly as a result of mechanical stretch and do not require a nervous impulse for each contraction.

Myogenic rhythms: Rhythms produced by spontaneously active muscles.

Myophily: Plant containing by flies.

Myriapoda: A class of arthropods which includes millipedes and centipedes.

Myrmecochory: The collection and dispersal of seeds by ants.

Myrmecodomatia: Structures in higher plants that seem to have evolved, through mutualistic association, to serve as dwelling places for ants.

Myrmecophile: A symbiont found in the colonies of ants, usually living at the expense of the food in the nest; an insect that lives in the nest of ants; a form of inquiline.

Myrmecophily: Plants containing **domatia** that house ants; 'ant-plants.'

Myrmecophyte: A plant that has special cavities in which ants live.

Myrmecotrophy: The feeding of plants by ants, notably through the waste products of an ant colony.

Mythical name: A name proposed for hypothetical or mythical forms; without status in nomenclature.

N

Naiad: The immature stage in insects with incomplete metamorphosis that feed and develop in water; an aquatic, gill-breathing, immature instar of hemimetabolous insect.

Narcotic: A chemical that causes reversible depression of the nervous system.

Narrow spectrum (as applied to insecticides)**:** Having an effect on only a narrow range of insects.

Nasus: A nose, the snout of certain termite soldiers (**nasutes**).

Nasute (pl. nasuti): A type of soldier caste in certain termites; this form bears a median frontal rostrum or snout through which it ejects a defensive fluid; the jaws are small or vestigial.

Nasutus (pl., nasuti): An individual of a termite caste in which the head narrows anteriorly into a snoutlike projection.

Natality: Population 'increase' factor; birth rate; often measured as total number of eggs or eggs per female per unit of time.

Natatorial (legs): Legs adapted for swimming (Fig. 172).

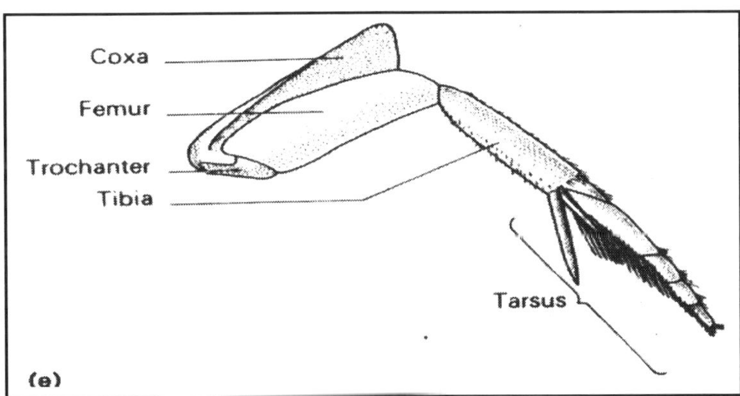

Fig. 172. Modification of insect leg: Natatorial (swimming) leg of a water beetle.

Natio: Local populations within a subspecies (cf. Population, Local population, Subspecies).

Natural classification: As currently used, classification based on characters or groups of characters which indicate phylogenetic relationship (cf. Clasification, Phylogeny, Artificial classification).

Natural control: The suppression of pest populations by naturally occurring biological and other environmental agents; the maintenance of a population at nonoutbreak levels by natural environmental factors, biotic and abiotic; the natural regulation of populations of living organisms that takes place without human interference.

Natural enemies: Living organisms found in nature that kill insects outright, weaken them, or reduce their reproductive potential; the predators and parasites that prey on or live in pest organisms; any living organism which is harmful to a species by way of its predatory, parasitic or disease inducing habit.

Natural group: A group of organisms having a common ancestor.

Natural selection: An increase in the frequency of occurrence of some genes or gene combinations and a decrease in the frequency of others because of the selective action of the environment; the process by which the environment eliminates the less welladapted members of a population or causes a differential reproductive success of different genotypes; the "survival of the fittest."

Natural system: The arrangement of taxonomic categories in a hierarchy based on an evolution of all their known characters.

Necrophage (adj, necrophagous): An eater of dead and / or decaying animals.

Necrophagous: Feeding on dead animal matter.

Necrosis: Death of a defined area of plant or animal tissue.

Nectar guide: A streak on a flower that guides insects to nectar sources.

Nectar: A sugary secretion produced by flowering plants which is highly attractive to insects.

Nectary: The floral gland from which nectar is produced; normally located within the flowers but sometimes external to them.

Neem *Azadirachta indica* (Meliaceae): The neem tree.

Negative binomial model: A mathematical model used to describe a clumped, or contagious, dispersion of organisms.

Negligible residue: A pesticide residue tolerance set for a food or feed crop.

Nekton, Necton: Organisms that swim actively in water.

Nemathelminthes: Systematic interpretation which places the Platyhelminthes and Nematoda into a single Phylum.

Nematicide: A substance or a pesticide that kills or incapacitates nematodes.

Nematocyst: Minute round capsule filled with poisonous fluid and having coiled thread, an organ of offence in coelenterates *e.g.* in jellyfishes (Gr. *nema* = a thread + *kystis* = a bladder).

Nematode (alternative terms, eelworm, roundworm)**:** Unsegmented smooth bodied worm belonging in the phylum Nematode; mostly minute in size; unsegmented roundworms that are parasites in plants, animals, and humans. Some species live in soil and water feeding on bacteria and fungi. Sometimes called eel worms.

Nematology: The study of nematodes.

Nematophagous: Feeding on nematodes, as in nematophagous fungi.

Neo: New.

Neontology: The systematics of recent organisms (cf. Paleontology).

Neoptera: The infraclass of Pterygota comprising insects capable of wing flexion.

Neopterous: Possessing a wing-flexion mechanism; having the ability to fold the wings backwards over the abdomen.

Neotenic: In termites, supplementary reproductives, arrested in their development, which have the potential to take on the reproductive role should the primary reproductives be lost.

Neoteny (adj. neotenous): The retention of juvenile features into the adult stage.

Neoteny: Attainment of sexual maturity in an immature or larval stage.

Neotype: A specimen selected as type subsequent to the original description in cases where the primary types are definitely known to be destroyed.

Nephr: Kidney.

Nephridium (pl, nephridia): Organ of excretion present in several invertebrates like Annelida, Platyhelminthes *etc.*

Nephrocytes (also **pericardial cells**): Cells that sieve the haemolymph for products that they metabolize.

Nerve: A bundle of neiron fibers; a bundle of axons that provides connections among ganglia and other parts of the nervous system.

Net reproductive rate: Rate of multiplication in one generation. It is also expressed as the ratio of the total female births in two successive generations.

Neural lamella: Outer connective tissue layer of a ganglion; a fibrous, noncellular layer that surrounds and supports a ganglion.

Neuroendocrine cells: see **Neurosecretory cells.**

Neurogenic flight muscles: Flight muscles that contract each time a nerve impulse is received.

Neurogenic rhythms: Rhythms maintained by spontaneously active neurons.

Neurohaemal organ (s): Structures that serve as storage and release sites for neurosecretory material such as corpora cardiaca; or an organ associated with the nervous system that stores and releases hormone.

Neurohormone: The largest class of insect hormones,also known as neuropeptides, being small proteins secreted within defferent parts of the nervous system (Fig. 173).

Neuron: A nerve cell; the basic unit of nerve impulse transmission; a basic functional unit of the nervous system; an elongated, excitable cell that carries information in the form of electrical charges.

Neuron: The nerve cell, comprising a cell body, **dendrite** and **axon**.

Neuropeptide: see **Neurohormone.**

Neuropile: Central region of a ganglion, consisting of intermingling, synapsing axons encapsulated by glial cell processes; the mass of closely packed nerve cell processes comprising the central part of a ganglion.

Neuroptera: (*neuro,* nerve; *ptera,* wing) the order of insects comprising the lacewings, antlions, dobsonflies and their relatives (Fig. 174).

Neurosecretory cell: A cell of the nervous system that is specialized for the production and release of hormones; modified neurons that found throughout the nervous system, producing insect hormones except ecdysteroids and juvenile hormones.

Neurotoxin: A substance that is poisonous or destructive to nerve tissues.

Neurotransmitter (s): A substance released from an axon terminal that either excites or inhibits a target cell; molecules, such as acetylcholine, that are secreted at the terminal arborizations of one neuron and that initiate an action potential via the dendrites or axon of another neuron.

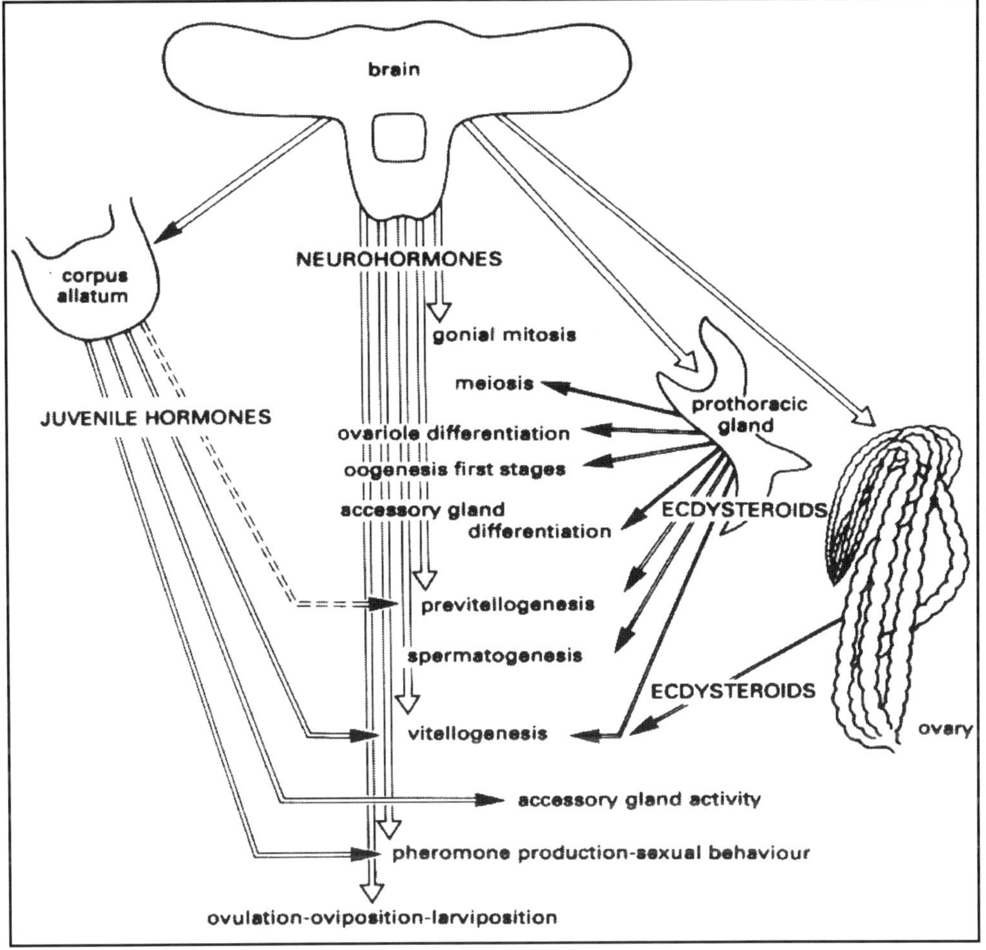

Fig. 173. A schematic diagram of the hormonal regulation of reproductive events in insects. After Raabe, 1986.

Neuston (adj. neustic): The water surface.

Neuston: Organisms that live on the surface film of water bodies.

Neutral term: A taxonomic term of convenience, such as form or group, which may be employed without reference to the formula taxonomic hierarcy of categories, and which has no nomenclatural significance.

Niche (ecological): The precise constellation of environmental factors into which a species fits or which is required by a species; the role that a species occupies in nature; that is, its precise habitat plus its behavior in that habitat.

New name: A new name for a preoccupied name (cf. Substitute name).

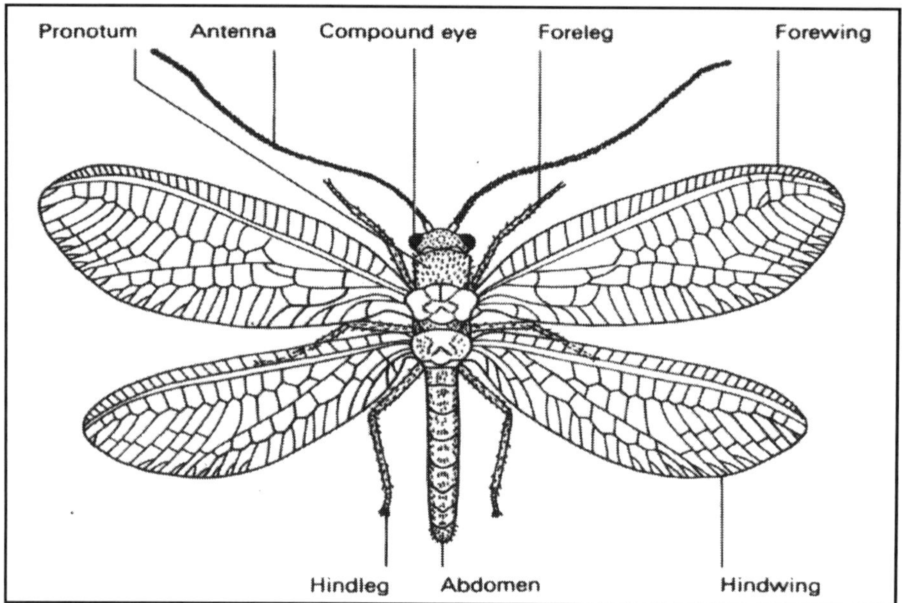

Fig. 174. Neuroptera, Green lacewing (Chrysopidae), *Chrysopa perla*.

Nicotine: A botanical alkaloid insecticide derived from the tobacco plant; secondary plant substance in the common tobacco plant that is toxic to insects.

Nidi: Groups of regenerative cells located at the bases of midgut epithelial cells.

Nit: The egg of a louse, often attached to a feather or hair.

No observable effect level (NOEL): The highest dose of a chemical that does not produce observable effects when administered to test animals.

Nocturnal: Animals active, and food seeking, at night only; *see also* **Crepuscular; Diurnal.**

Node: A knoblike or knotlike swelling; joint on a plant stem; origin of leaves and branches.

Nodiform: In the form of a knob or knot.

Nodule: A small swelling on the roots of certain legumes that contain nitrogen-fixing bacteria.

Nodus: A strong cross vein near the middle of the costal border of the wing (Odonata).

Nomen dubium: The name of a nominal species for which available evidence is insufficient to permit recognition of the taxonomic species to which it was applied [for nomenclatural status *see Bul. Zool. Nomencl.,* **4:** 76 (1950)].

Nomen novum: New name (cf. Substitute name).

Nomen nudum: A published scientific name which does not meet the requirements for availability defined in Art.25 of the International Rules of Zoological Nomenclature (cf. Valid name).

Nomen rejectum: Rejected name (q.v.).

Nomen specificum: Specific name (q.v.).

Nomen triviale: Trivial name (q.v.).

Nomen vanum: An indeterminate name (cf. *Nomen dubium*).

Nomenclator: A book containing lists of scientific names assembled for nomenclatural, rather than taxonomic, purposes (cf. Catalogue).

Nomenclature binaire: See Binary nomenclature.

Nomenclature: A system of names (cf. International Rules of Zoological Nomenclature); the naming of organisms; the science of naming (of plants and animals).

Nomina conservanda: Names whose usage has been preserved by agreement or decision in spite of actual or potential conflict with established rules of nomenclature; often applied to the Official Lists established by the International Commission on Zoological Nomenclature (cf. Official List of Generic Names in Zoology).

Nominal genus: "The concept denoted by a given generic name" as contrasted with the concept represented by a taxonomically accepted genus (International Commission).

Nominal species: "The concept denoted by a given specific name," as contrasted with the concept represented by a taxonomically accepted species (International Commission).

Nominate subspecies: Nominotypical subspecies (q.v.).

Nominotypical (or nominate) subspecies: That subspecies of a polytypic species which shares with its species the name type specimen and the same name; the subspecies of species with the earliest valid name (*e.g., X-us albus albus* in contrast to *X-us albus niger*).

Nominotypical subgenus: That subgenus of a polytypic genus which shares with its genus the same type species and the same name [*e.g., X-us (X-us) albus* in contrast to *X-us (Y-us)rufus*] (cf. Subgeneric name).

Nonaccumulative: Refers to pesticides that do accumulated in the environment because they break down rapidly.

Non-integrated inquiline: An **inquiline** that is adapted ecologically to the nest of the host, but does not interact socially with the host; *see also* **Integrated inquiline.**

Nonpathogenic: Does not cause disease.

Nonpersistent pesticide: Pesticides that decompose to nontoxic by-products in less than 1 to 3 days.

Nonpersistent virus: A virus carried on the mouthparts of its insect vector, but the virus is lost after the vector feeds a few times; a virus that does not undergo part of its development within the body of its vector; a stylet-borne virus.

Non-preference: A mechanism of plant resistance to the pests whereby the pest chooses not to feed (or oviposit) on the resistant plant in comparison with susceptible ones.

Nonresonating muscles: Muscles that require a nervous impulse for each contraction (compare with resonating muscles); also called synchronous muscles.

Nonselective pesticide: A pesticide that is toxic to a range of organisms.

Non-target organism: Any organism other than that against which control measures (especially chemicals) are applied.

Nontarget species: An organism that is not the intended object of a management tactic.

Nontoxic: An economic poison with an LD_{50} greater than 5,000 mg/kg. Not poisonous to humans.

Nosema: Cause of disease in silkworm larvae and honey bee. *N. Bombycis* is causative organism of pebrine disease in silkworm larvae and *N. apis* causative organism of nosema disease of honey bee.

Notaulix (pl., notaulices): Parapsidal suture (see).

No-till: Growing crops without using tillage.

Notopleural bristles: Bristles on the notopleuron (Diptera).

Notopleural suture: A suture between the notum and the pleural sclerites.

Notopleuron (pl., notopleura): An area on the thoracic dorsum, at the lateral end of the transverse suture (Diptera).

Notum (pl. nota): A thoracic **tergum;** the dorsal surface of a body segment (usually used when speaking of the thoracic segments); a dorsal sclerite of the insect thorax.

Noxious weed: A weed designated by law as particularly undesirable and difficult to control.

Nozzle: Device that regulates drop size and controls volume flow rate, uniformity or thoroughness of coverage and safety of liquid pesticide spray applications. 1. **Air blast:** nozzle using high velocity air to break up the spray liquid supplied at low pressure. 2. **Anvil:** nozzle in which the spray liquid jet strikes a smooth, solid surface at a high angle of incidence. 3. **Cone** (or swirl)**:** nozzle in which the liquid emerges from the orifice with tangential velocity imparted by passage through one or more tangential or helical channels in the swirl chamber. 4. **Hollow cone:** nozzle in which spray jet has a core of air breaking to give drops in an annular pattern. 5. **Fan nozzle:** the aperture is an elongate horizontal slit, producing a fan-shaped spray pattern. 6. **Deflector:** nozzle in which a fan-shaped sheet of spray is formed by directing the liquid over a sharply inwardly curving surface.

Nuclear cytoplasm: Cytoplasm around the nucleus of an insect egg (Fig.175).

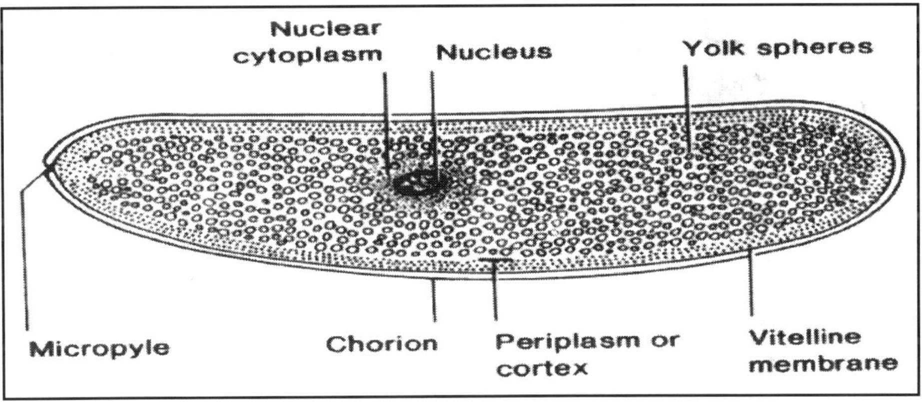

Fig.175. Sagittal section of a typical insect egg: Nuclear cytoplasm.

Nuclear Polyhedrosis Virus (NPV): A lethal virus disease of insects, particularly larvae of certain Lepidoptera and Hymenoptera; a virus that multiples in cell nuclei, chiefly in the epidermis, fat body, and blood cells.

Nucleus: A differentiated structure of specialized protoplasm in a cell controlling metabolic activities in all cells except of bacteria (L. demunitive of *nux* = nut).

Nulliparous: A female that has laid no eggs.

Nuptial flight: Mating flight of a large swarm of males following one or more females prior to copulation; e.g., in bees and ants.

Nurse cells: Nutritive cells in polytrophic and telotrophic ovarioles; also called trophocytes.

Nutrient: A mineral element necessary for plant or animal growth.

Nutrition: Act of nourishing to promote growth of bodies (L. *nutritive* = to nourish).

Nymph: The immature stage (following hatching) of an insect that does not have a distinct pupal stage; also the immature stages of Acarina that have eight legs; the wingless immature stage of an insect having incomplete metamorphosis, e.g., Hemiptera, Orthoptera; the immature stage in insects with gradual metamorphosis; Immature instars of ametabolous and hemimetabolous insects.

Objective synonym: An absolute or nomenclatural synonym resulting from the proposal either of a replacement name for a supposedly preoccupied name or of names based on the same specimen, illustration, or taxonomic entity (cf. Synonym, Subjective synonym).

Obligate parasite: A parasite that requires a living host to survive and reproduce and which cannot exist in any other way.

Obligatory: Compulsory, or exclusive; e.g. obligatory diapause is a resting stage that occurs in every individual of each individual of each generation of a univoltine insect.

Oblique sternals: Lateral abdominal muscles in other than longitudinal or dorsal-ventral orientation (Fig. 176).

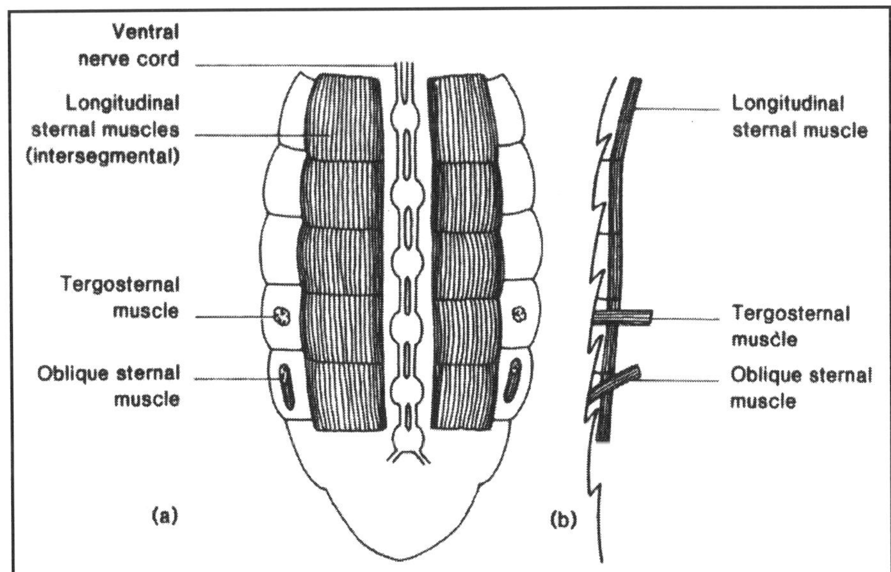

Fig. 176. Skeletal musculature of the ventral portion of the abdomen. (a) Dorsal view. (b) Lateral view.

Oblique vein: A slanting cross vein; in Odonata, where Rs crosses M_{1+2}.

Observer: One of the three components of mimicry systems, the receiver of the signal(s) emitted by the **model** and **mimic.**

Obtect pupa: A pupa in which the appendages are more or less fused (cemented) to the body surface; commonly covered by a cocoon (Fig. 207 b) as in the Lepidoptera (moth).

Obtect: Concealed; covered; refers to pupa where wings and legs are appressed to the body, abdomen immovable.

Occasional pest: A species normally present, but infrequently causing damage; a pest with a general equilibrium position substantially below the economic-injury level, highest pest population fluctuations occasionally and sporadically exceed the economic-injury level; one which reaches significant levels only occasionally.

Occipital foramen: The opening of the back of the head.

Occipital suture: A line that runs from the posterior end of the coronal suture to just above the mandibles on either side of the head; a transverse suture in the posterior part of the head which separates the vertex from the occiput dorsally and the genae from the postgenae laterally (Fig. 123).

Occipital: Refers to the back part of the head, the occiput.

Occiput: The dorsal posterior part of the head, between the occipital and postoccipital sutures (Fig. 123).

Occupational Safety and Health Administration (OSHA): Fedral agency that regulates matters that affect the health and safety of workers.

Ocellar bristles: Bristles arising close to the ocelli (Diptera).

Ocellar triangle: A slightly raised triangular area in which the ocelli are located (Diptera).

Ocellus (pl. ocelli): The 'simple' eye of adult and nymphal insects incorporating a single lens, typically three in a triangle on the vertex, with one median and two lateral ocelli; the **stemma** of some holometabolous larvae; also a colored spot on the wing of Lepidoptera.

Odonata: The order of insects comprising the dragonflies and damselflies.

Oenocyte: A cell of the haemocoel, epidermis or especially the fat body, probably with many functions, most of which are unclear, but including synthesis of cuticle paraffins and/ or haemoglobin.

Oesophagus: The foregut that lies posterior to the pharynx, anterior to the crop.

Official Index of Rejected and Invalid Generic Names in Zoology: A record of generic names suppressed by the International Commission on Zoological Nomenclature under the plenary powers or declared by the Commission to be invalid or nonexistent [*see Bul. Zool. Nomencl.,* **4:** 334 (1950)].

Official Index of Rejected and Invalid Specific Trivial Names in Zoology: A record of specific trivial names suppressed by the International Commission on Zoological Nomenclature under the plenary powers or declared by the Commission to be invalid or non existent [*see Bul. Zool. Nomencl.,* **4:** 334 (1950)].

Official List of Generic Names in Zoology: A record of generic names (and the type species of each) which have been validated, conserved, or stabilized by the International Commission on Zoological Nomenclature through use of the plenary powers or the rendering of an Opinion [*see Bul. Zool. Nomencl.,* **4:** 267-268 (1950)].

Official List of Specific Trivial names in Zoology: A record of trivial names of species or subspecies which have been validated, conserved, or stabilized by the International Commission on Zoological Nomenclature through the use of the plenary powers or the rendering of an Opinion [*see Bul. Zool. Nomencl.,* **4:** 269-271 (1950)].

Oigophagous: Feeding on a limited range of plants (or animals), usually within a single family. cf. monophagous, polyphagous.

Olfaction: Distance chemoreception; the sense of smell.

Olfactory lobes: See antennal lobes.

Oligogenic resistance (major-gene resistance): Host-plant resistance conferred by one or a few genes.

Oligolectic: Bees that gather pollen from only a few closely related species of flowers; utilizing a very limited number of plant species as sources of pollen (said chiefly of bees).

Oligophage (adj. oligophagous): An eater of few kinds of food, e.g. several plant species within one genus or one family; used particularly of phytophages.

Oligophagous (=Stenophagous**):** An animal feeding upon only a few, closely related host plant or animal species; having a restricted range of food; phytophagous insects that feed on only one family of plants.

Oligopneustic: A respiratory system with one to two functional spiracles on each side of the body; *see also* **Apneustic; Polypneustic.**

Oligopod larva: A larva with legs on the thorax and not on the abdomen.

Oligoxene (adj. oligoxenous)**:** A parasite restricted too few hosts.

Ommatidium (pl., ommatidia): A single functional element or unit or visual section of a compound eye expressed externally as a facet (Fig. 177).

Ommochromes: Red, yellow, and brown pigments commonly found as insect eye pigments.

Omnivore (s): An animal that eats a variety of foods including both plant and animal matter; insects or other animals that fall into two or more of the following categories: detritivores, herbivores, mycetophagous, or carnivores.

Omnivorous: Feeding on a variety of foods; eating all kinds of food of animal and / or plant origin (L. *omnis* = all + *voro* = eat).

Onchocerciasis: Infection by filarial worms of the genus *Onchocera* transmitted by some species of black flies.

Oncogen: A substance that causes the formation of tumors in animals; may be carcinogenic.

Onisciform larva: A flattened platyform larvae, like a wood-louse in appearance.

Onisciform: Insect larva resembling a woodlouse, being depressed and broadly ovate.

Onomatophore: "Name bearer" = type (Simpson) (cf. Type).

Ontogeny: The development from egg to adult; the developmental history of an individual organism from egg to adult.

Oocyst: A stage in the sexual cycle of a malaria parasite (*Plasmodium* spp.) within the **ookinete** and which undergoes **sporogeny** (asexual reproduction).

Oocytes: An immature egg cell formed from the **oogonium** within the ovariole; germ cells in females.

Oogonia: Primary female germ cells located in the germarium of an ovariole.

Oogonium: The first stage in the development in the **germarium** of an egg from a female germ cell.

Ookinete: An active sexual stage in the cycle of a malaria parasite (*Plasmodium* spp.) formed from the zygote and which penetrates the body cavity within the mosquito.

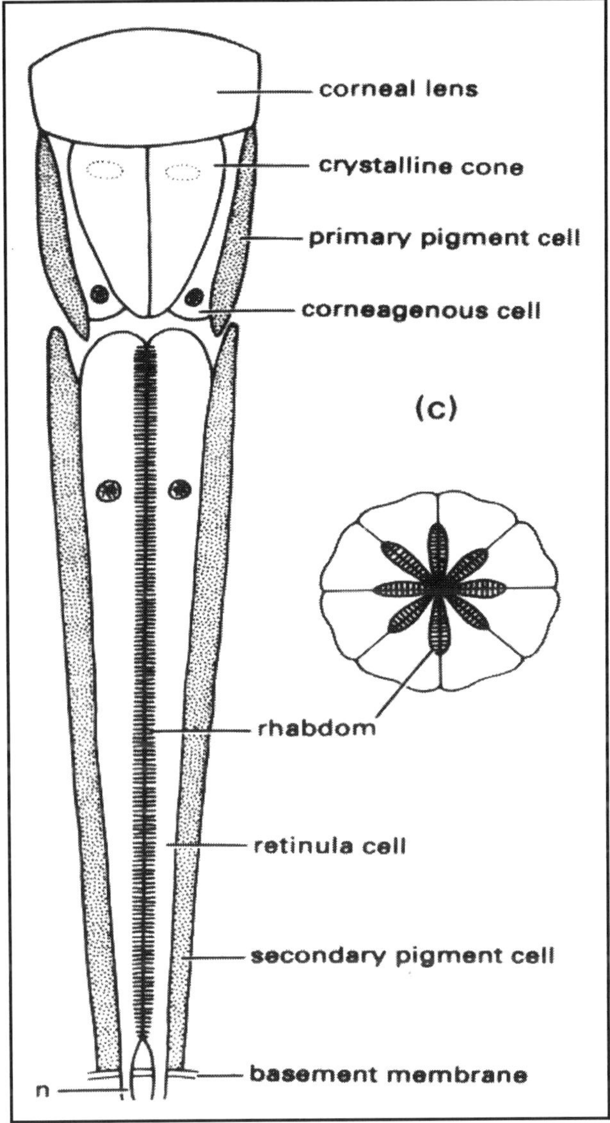

Fig. 177. An ommatidium from a compound eye, with enlargement showing a transverse section.

Oosome: Specialized region of the egg associated with differentiation of germ cells in the embryo (Fig. 178).

Oospore: Resting fungus spore; the result of fusion of gametes.

Oostatic hormone: A peptide hormone produced either by the ovary or associated neurosecretory tissue, with inhibitory activities on one or more ovarian functions contains several to many eggs.

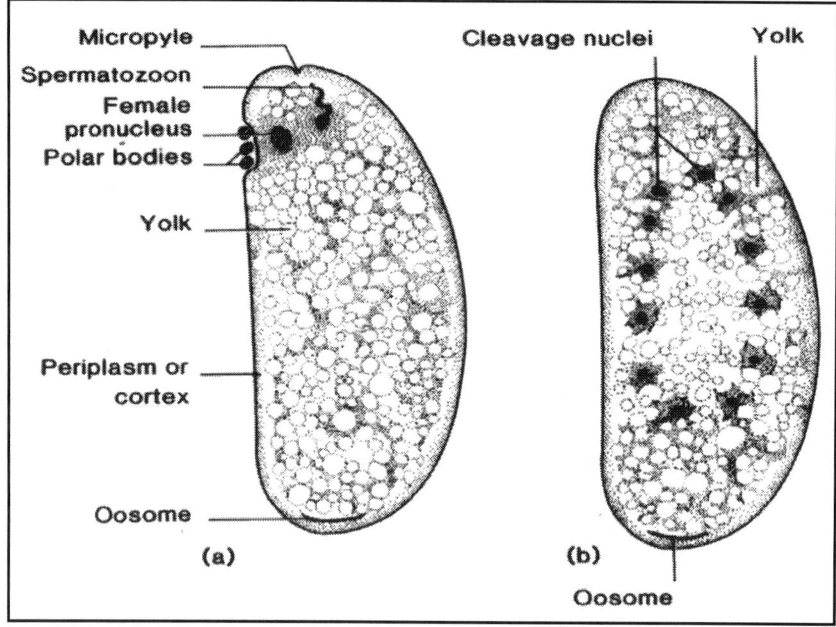

Fig. 178. Fertilazation – germ band formation. (a) Just prior to fertilization (at the stage of fusion of spermatozoon and female pronucleus). (b) Cleavage.

Ootheca (pl., oothecae): An egg purse or egg capsule of distinct form; as produced by cockroaches for example; eggcase; a mass of eggs enclosed in a container secreted by the female accessory glands, best shown in the Dictyoptera (Fig. 179).

OP: Abbreviation of organophosphate.

Open cell: A wing cell extending to the wing margin, not entirely surrounded by veins; an area of the wing membrane partially bounded by veins but including part of the wing margin.

Open coxal cavity: One bounded posteriorly by a sclerite of the next segment (front coxal cavities, Coleoptera, or one touched by one or more of the pleural sclerites (middle coxal cavities, Coleptera).

Open tracheal system: A gas exchange system comprising tracheae and tracheoles and with spiracular contact with the atmosphere; *see also* **Closed tracheal system.**

Operculum (pl., opercula): A lid or cover like structure. Many different specific uses in the Class insecta.

Opisthognathous mouthparts: Mouthparts directed ventroposteriorly relative to the head capsule; head positioned obliquely and mouthparts pointed backward (Fig. 180).

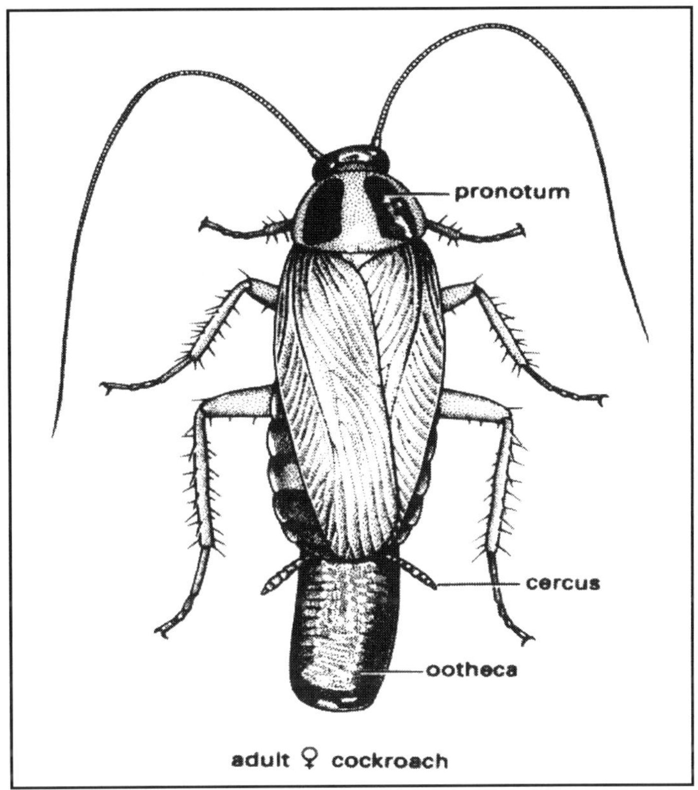

Fig. 179. Blattodea (Blattaria; Cockroaches, Roaches): Ootheca.

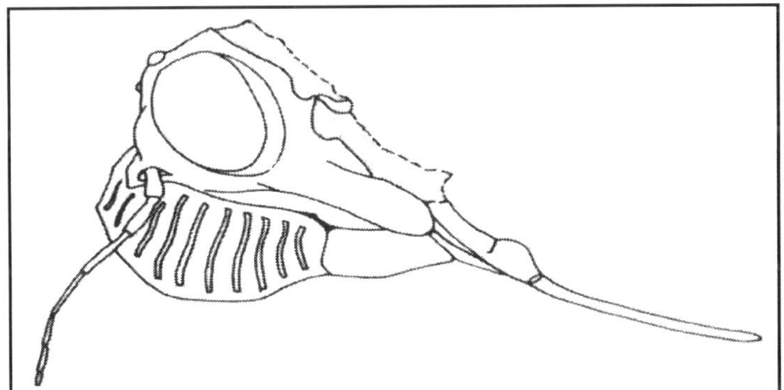

Fig. 180. Position of mouthparts relative to the head capsule: Opisthagnathous.

Opisthognathous: With the head deflexed backward such that the mouthparts are directed posteriorly, as in many Hemiptera; *see also* ***Hypognathous; Prognathous.***

Oposthosoma: The hind body region of arachnids (spiders, mites, ticks).

Optic lobes: Bilateral portions of the brain that receive sensory input from the compound eyes.

Optic: Pertaining to the eye or sense of sight (Gr. *opticos* = seen).

Optomotor responses: Compensating adjustments based on visual input during flight.

Oral (or stomodeal) trophallaxis: In social and subsocial insects, the transfer of nutritive fluid from the mouth to another individual; may be mutual or unidirectional.

Oral toxicity: The toxicity of an insecticide (to higher animals) when ingested by mouth.

Oral vibrissae: A pair of stout bristles, one on each side of the face near or just above the oral margin, and larger than the outer bristles on the vibrissal ridge (Diptera).

Oral: Pertaining to or near the mouth; via the mouth (L. *os* = mouth).

Orbital plate: A sclerite on the head next to the compound eye (Diptera), also called genovertical plate.

Order: A category of classification; one of the principal divisions of the class Insecta, based largely on wing structure and usually ending in -ptera (e.g., Lepidoptera); a subdivision of a class or subclass, containing a group of related families. The insect orders are the main insect types, e.g. order Coleoptera, beetles; order Lepidoptera, butterflies and moths.

Organ: Any part of an animal performing some definite function; a group of cells or tissues acting as a unit for some special purpose (Gr. *organon* = instrument).

Organelle: A specialized part in a protozoan that performs some special function (like an organ in a metazoan). Gr. *organon* = instrument.

Organic matter: The component of soil that is composed of decaying plants and animals.

Organic pesticide: Pesticides made from carbon-containing chemicals. The two major groups are petroleum oils and synthetic organic pesticides.

Organic produce: Commodoties produced from plants or animals which are grown without the use of synthetic fertilizers or pesticides.

Organic: Possessing carbon atoms.

Organism: A single plant or animal; one that functions as a unit.

Organochlorine (insecticides) (alternative term, chlorinated hydrocarbon): A group name applied to synthetic organic insecticide containing carbon, hydrogen, and chlorine. Many are persistent in the environment (e.g., DDT, chlordane, and aldrin).

Organophosphate (insecticide): A group name applied to synthetic organic chemical insecticides which have a phosphorous atom at the core of the molecule; malathion and parathion are example.

Organosulfur: A sulfur-containing organic compound used as an acaricide (e.g., tetradifon).

Organotin: A tin-containing organic compound used as an acaricide (e.g., cyhexatin) and/or fungicide.

Orifice (nozzle) velocity: Velocity at which the spray leaves the nozzle orifice.

Origin: Stationary attachment area of skeletal muscle.

Original description: The summary of characters accompanying the proposal of a name for a new taxonomic entity in conformance with Art.25 of the International Rules of Zoological Nomenclature.

Ornithologist: One studying ornithology, or making a special study of birds.

Ornithology: Science dealing with the study of the vertebrate feathered class Aves (birds) (Gr. *Ornis, ornithos* = a bird + *logos* = study).

Orthogenesis: A term usually applied either to a tendency to evolve consistently in the same direction or to the concept of "predetermined" evolution toward a definite goal.

Orthokinesis: A response to a stimulus in which the rate of response (such as speed of movement) is positively proportional to the intensity of the stimulus.

Orthoptera, order: (*ortho-*, straight; *ptera,* wing) The order of insects comprising the grasshoppers, locusts, crickets, katydids, and related forms (Fig. 181).

Osculum: Main opening of the sponge acting as outlet of water from the spongocoel - the main opening acting as outlet instead of inlet *e.g.* in sponges.

Osmeterium (pl. osmeteria): A fleshy, tubular, eversible, usually Y-shaped scent gland at the anterior end of certain caterpillars.

Osmeterium (pl. someteria): An eversible tubular pouch, usually **Y**-shaped scent gland on the prothorax of some larval swallow tail butterflies (Lepidoptera: Papilionidae) (Fig. 182), used to disseminate volatile toxic, defensive compounds; a fleshy, tubular structure, eversible through a prothoracic slit found on the larvae of some Papilionidae and producing a strong repungnatorial substance; an eversible gland on the thorax of the larvae of swallowtail butterflies that secretes allomones used for defense.

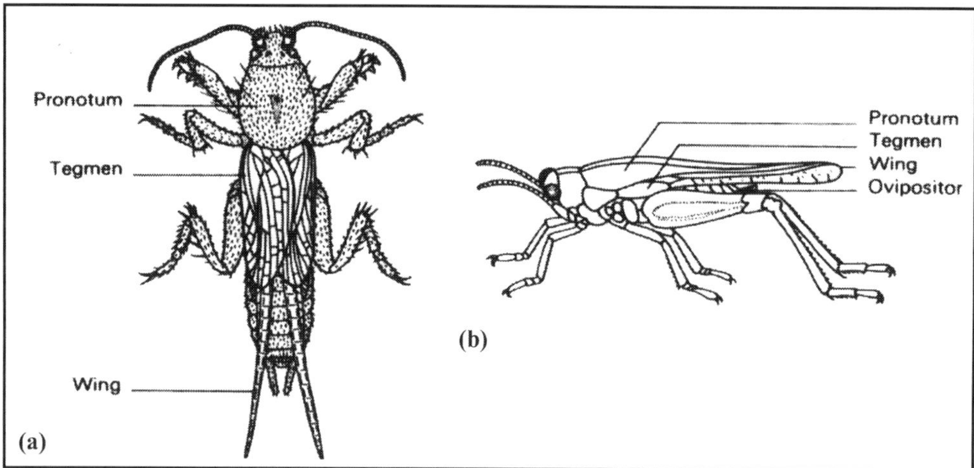

Fig. 181. Representative Orthoptera. **(a)** Mole cricket, *Gryllotalpa hexadactyla* (Gryllotalpidae). **(b)** Pygmy grasshopper, *Telmatettix Hesperus* (Tetrigidae).

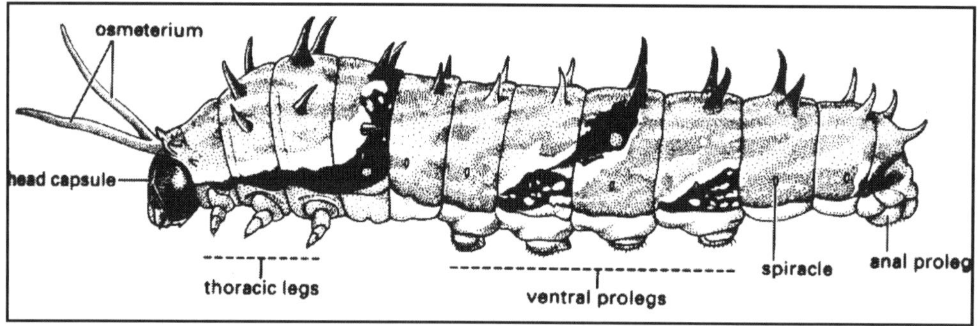

Fig. 182. A caterpillar of the orchad butterfly, Papilio aegeus (Lepidoptera: Papilionidae), with the osmeterium everted its head. Eversion of this glistening, bifid organ occurs when the larva is disturbed; it is accompanied by a pungent smell.

Osmoregulation: The regulation of water balance, maintaining the homeostasis of osmotic and ionic content of the body fluids.

Ostia (s. ostium): Valves by which blood enters the insect heart; segmentally arranged inlet pores in the walls of the heart.

Ostiole: A small opening or hole; the external opening of certain glands (as in many Heteroptera).

Ostium (pl. ostia): A slit-like opening in the dorsal vessel ('heart') present usually on each thoracic and the first nine abdominal segments, each ostium having a one-way valve that permits flow of haemolymph from the pericardial sinus into the dorsal vessel (L. os = mouth).

Outbreak: High numbers of a particular pest species.

Outer epicuticle: The middle of the three layers of epicuticle, with the inner epicuticle beneath it; *see also* **Inner epicuticle; Superficial layer.**

Outer fork: The fork of R_{4+5} (Ephemeroptera).

Outer vertical bristles: The more laterally located of the large bristles on the vertex, between the ocrlli and the compound eyes (Diptera).

Ov-, ovi: Egg.

Ovarian cycle: The length of time between successive ovipositions.

Ovaries: Paired gonads in the female reproductive system (Fig. 5).

Ovariole sheath: Cellular layer that surrounds an ovariole.

Ovariole: A more or less tubular division of an ovary in which the eggs are formed; one of several ovarian tubes that form the ovary, each consisting of a **germarium,** a **vitellarium** and a stalk or **pedicel** (Fig. 5 and 183).

Ovary: The egg-producing organ of the female in which the egg cells multiply and are nourished. one of the paired gonads of female insects, each comprising several **ovarioles** (Fig. 5). (L. *ovum* = egg).

Overwinter: Survival of adverse winter conditions until the next growing season.

Ovicide: A pesticide that kills the eggs of insects and mites.

Oviduct: The tube leading away from the ovary through which eggs pass (Fig. 5 and 183); the tube by which eggs are conveyed from the ovary to the uterus or to the exterior. (L. *ovum* = egg + *duco* = lead).

Oviparous: Reproduction by laying eggs; deposition of chorionated egg; egg-laying; producing eggs that hatch outside the mother's body. (L. *ovum* = egg + *pario* = produce) *e.g.* in insects in general.

Oviposit: To lay or deposit eggs.

Oviposition: The act of egg laying.

Ovipositor: The egg-laying apparatus (Fig. 183); the external genitalia of the female; the organ used for laying eggs typically composed of two sets of valves or a tubular extension of the abdomen; *see also* **Appendicular ovipositor; Substitutional ovipositor.**

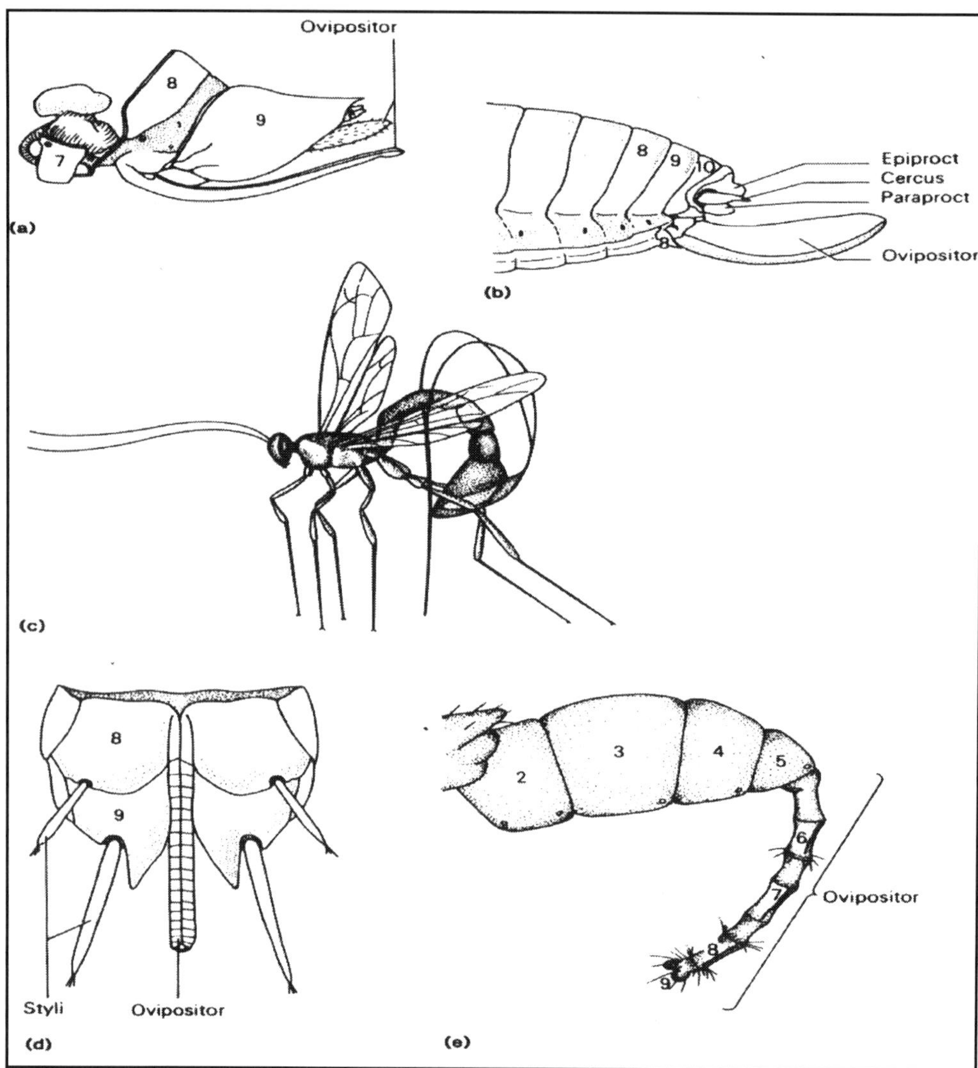

Fig. 183. Modifications of the ovipositor. (a) Cicada, *Magicicada septendecium.* (b) Katydid, *Scudderia* sp. (c) Ichneumon wasp, *Megarhyssa lunator.* (d) Firebrat, *Thermobia domestica.* (e) Telescopping terminal abdominal segments serving as an ovipositor in the house fly, *Musca domestica.* Numerals indicate abdominal segments. (a), (b), and (d) after Snodgrass, 1935; (c) after Riley, 1888; (e) after West, 1951.

Oviscapt: The abdomen when used as an "ovipositor" (Fig. 183 e) functional ovipositor produced by elongation of terminal abdominal segments rather than the appendages; applies to some Diptera and Trichoptera.

Ovisorption: Resorption of eggs prior to the time of oviposition.

Ovitubus: See oviscapt.

Ovoviparous: Forms producing an egg with a definite shell hatching inside the maternal body.

Ovoviviparity: A condition in which a fully developed egg with a chorion hatches within the parent and a larva is "born" instead of an egg deposited; producing active young immediately following the hatching of the eggs within the female; producing eggs that are incubated and hatched within the parent's body, as with some fishes, reptiles and invertebrates (L. *Ovum* = egg + *verus* = alive + *pario* = produce); retention of the developing fertilized egg within the mother, considered to be a form of **viviparity** but in which there is no nutrition of the hatched young; *see also* **Oviparity.**

Ovoviviparous: The production of living young immediately following the hatching of the eggs within the female.

Ovulation: The movement of an oocytes from the ovariole.

Ovum: An egg, the sex cell of a female (L.).

P

Pacemaker: A region from which waves of contraction in the heartbeat cycle are propagated.

Paedogenesis: Reproduction by the juvenile form; the production of eggs or young by an immature or larval stage of an animal; reproduction by larviform individuals.

Paeudotrachea: A ridged groove on the ventral surface of the labellum of some higher Diptera, used to take up liquid food.

Page precedence: The principle that when two or more conflicting homonyms or synonyms are published in the same work (or portion of a work) and, as a consequence, are of the same date, the names shall have priority according to the sequence in which they first appear in the work (or portion) concerned.

Paint gun: Type of small, hand-carried, air-blast machine.

Pair-wise coevolution: see **Specific coevolution.**

Paleontology: The science that deals with the life of past geological periods (cf. Neontology).

Paleoptera: The infraclass of Pterygota comprising insects incapable of wing flexion.

Paleopterous: Wing flexion mechanism absent; lacking the ability to position the wings backward over the abdomen.

Palp (also **palpus,** pl. palpi**):** Finger-like, sensory feeler in the mouthprts of insects. There are usually two pairs of palps; one pair (maxillary palps) attached to the maxilla, and one pair (labial palps) attachéd to the labium.

Palpifer: Lateral sclerite located on the stipes of a maxilla and bearing the maxillary palpus; the lobe of the maxillary stipes which bears the palp.

Palpiger (s): The lobe of the mentum of the labium which bears the palp; lateral sclerites located bilaterally on the prelabium that bear the labial palps.

Palpus (pl. palpi): A sensory organ associated with mouthpart structures; a paired, segmented appendage arising on the maxilla or labium and serving sensory functions associated with food ingestion.

Panoistic ovariole: An ovariole that lacks nurse cells; *see also* **Telotrophic ovariole.**

Panoistic ovary: An ovary in which the ovarioles lack nurse cells.

Panoistic: Type of ovariole that lacks trophocytes.

Pantropical: A species occurring widely throughout the tropical and subtropical parts of the world.

Papilla: Any nipple-like structure or elevation, larger or small; small, raised, finger-like process usually of minute size.

Para-: Alongside of.

Parabiosis: Joining the hemolymph circulation of one individual with that of another.

Paraglossa (pl. paraglossae)**:** One of a pair of lobes distolateral on the prementum of the labium, lying outside the glossae, but mesal to the labial palp; one of a pair of lobes at the apex of the labium, lateral of the glossae; two outer lobes found on the prementum of the labium. The outer part of the extremity of the labium in insect mouthparts.

Paramere: One of a pair of lobes lying lateral to the penis, forming part of the **Aedeagus** (Fig. 8).

Paranota (sing. Paranotum, adj. paranotal)**:** Postulated lobes of the thoracic terga from which, it has been argued, the wings derive.

Paraphyletic: A group (grade) evolutionarily derived from a single ancestor, but which does not contain all descendants, recognized by the joint possession of shared primitive character(s); rejected in **cladistics** but often accepted in **evolutionary systematics;** *see also* **Monophyletic; Polypheletic.**

Paraproct (s): One of a pair of lobes adjacent to the anus lateroventrally; plates of the eleventh abdominal segment located on either side of the anus. Ventral relics of segment 11(Fig. 184).

Parapsidal suture: A longitudinal suture of the mesonotum separating the median area from the lateral area (Hymenoptera). (Also called parapsidal furrow or notaulix).

Parapsis: (see *scapula*).

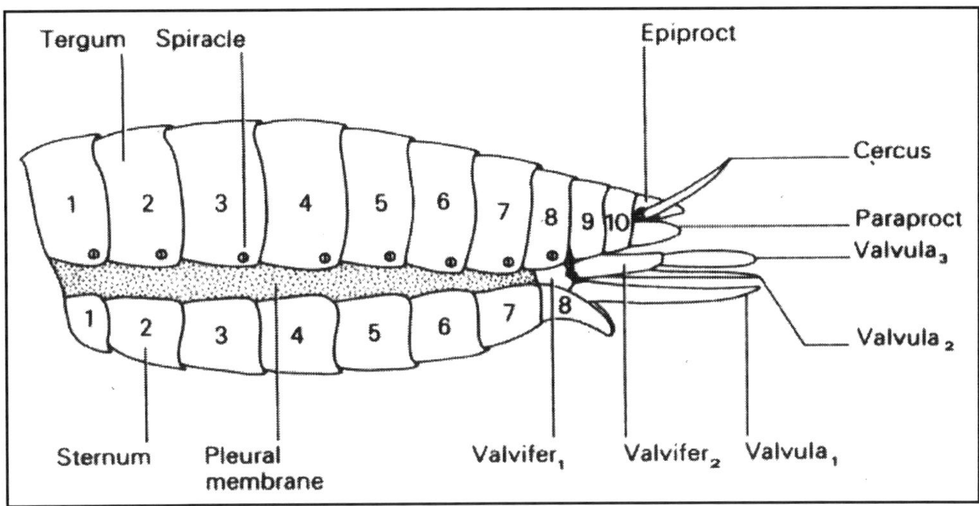

Fig. 184. Generalized insect abdomen.

Parasite guild: A group of parasite species that use a particular host or host stage upon or within which to develop.

Parasite, ecto: Which feeds upon the animal from outside *e.g.* ratflea.

Parasite, endo: Which feeds from within the body of the animal *e.g. Taenia,* liverfluke.

Parasite, obligate: (L. *obligare* = bound) Which cannot exist independently of the host *e.g.* plant parasitic nematodes.

Parasite: An animal or plant that lives for most of its time in or on another organism, the host, from which it obtains nourishment, and to which it causes some damage; usually not lethal to its host; an organism living in intimate association with a living organism (plant or animal) from which it derives material essential for its existence while conferring no benefit in return; an animal that lives in or on the body of another living animal (its host), at least during some portion of its life cycle; a parasite usually does not kill its host or consume a large proportion of its tissues; an animal that completes its development on or in another organism (host) from which it obtains nutrition, may live and feed upon the host, externally or internally but does not normally kill it.

Parasitic: Living as a parasite.

Parasitism: An intimate relationship in which one individual (the parasite) benefits at the expense of another (the host); the relationship between parasitoid or parasite and host.

Parasitization: The condition of being parasitized, by either a parasitoid or a parasite.

Parasitized: The state of a host that supports a parasitoid or parasite.

Parasitoid: An organism alternately parasitic and free-living; most parasitic Hymenoptera and Diptera fall into this category as usually only the larvae are parasitic; or an arthropod parasite that kills its host. Parasitiod adults are free-living, the immature stages are parasitic; an animal that feeds in or on another living animal for a relatively long time, consuming all or most of its tissues, and eventually killing it. (Also used as an adjective, describing the mode of life of a parasitoid.); or an arthropod parasite that parasitizes and kills an arthropod host; is parasitic in its immature stages but is free-living as an adult; an insect that lives in its immature stages in or on another insect, which it kills after completing its own feeding; insects that resemble both parasite and predator; they begin life as small parasites but may outgrow and eventually devour the host.

Paratype: A specimen other than the holotype which was before the author at the time of preparation of the original description and was so designated or indicated by the original author (cf. Allotype).

Parazoa, Porifera, sponges: Said to be evolved separately from other metazoans. Beginning of tissue formation but no organized tissue formation yet.

Parenchyma: Loose tissue of vacuolated, irregularly arranged cells constituting a large part of body of Platyhelminthes; soft and thin-walled plant cells which form the pith of plant stems, leaf mesophyll, and other plant parts.

Parental investment: Behavior of a parent that increases the probability of offspring survival at the cost of the parent's ability to produce more offspring.

Parous: A female that has laid at least one egg.

Pars intercerebralis: Cell mass, located in the middorsal region of the brain above the protocerebral bridge, that contains the median neurosecretory cells (Fig. 185).

Parthenocarpy: The development of fruit without fertilization and seed.

Parthenogenesis: Reproduction without fertilization; usually through eggs but sometimes through viviparity; the development of eggs without fertilization by spermatozoa. Occurs in some nematodes, aphids, beetles, parasitic wasps, and other insects; production of individuals from unfertilized eggs, asexual reproduction; development of a new individual from an unfertilized egg, as in rotifers, plant lice *etc.* (Gr. *parthenos* = virgin + *genesis* = origin); *see also* **Amphitoky; Arrhenotokous parthenogenesis; Paedogenesis; Thelytoky.**

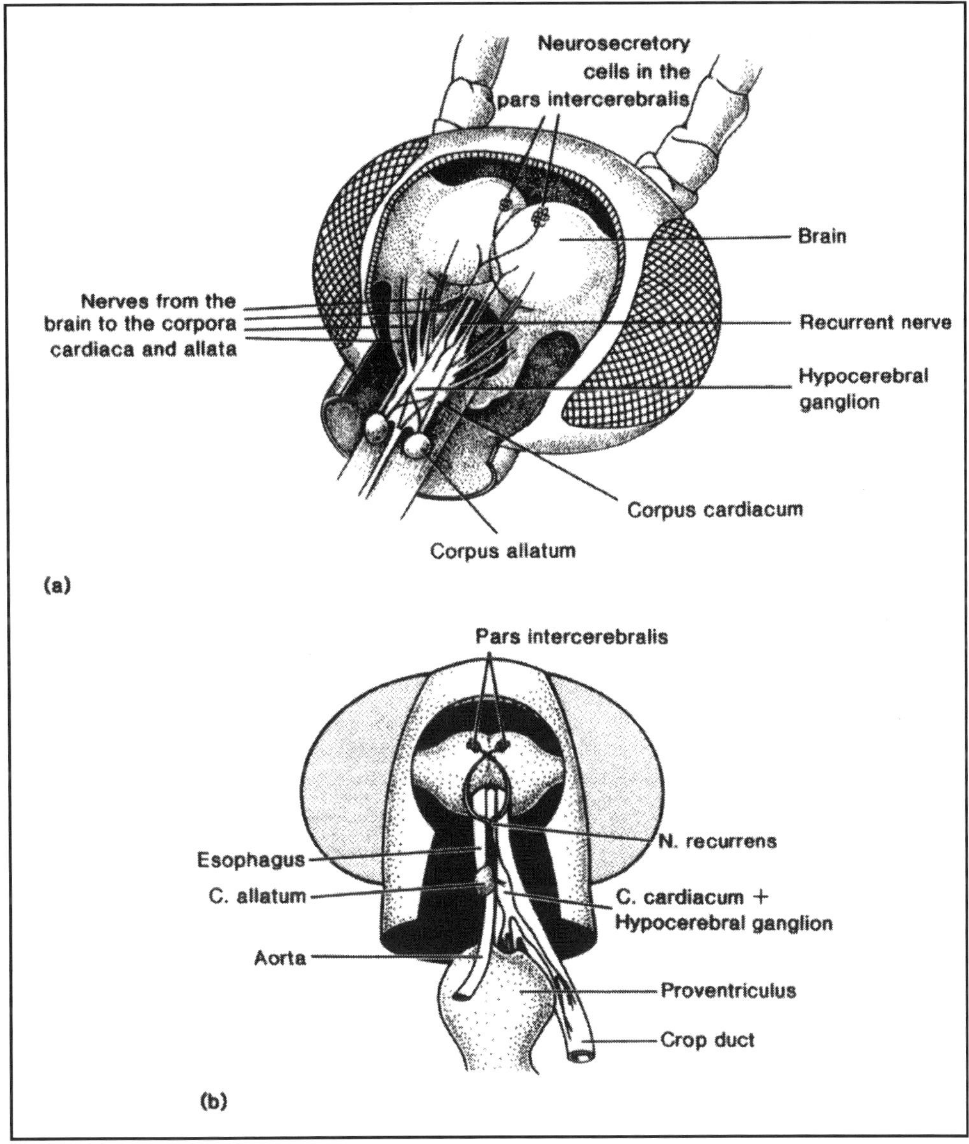

Fig. 185. Stomatogastric nervous system and endocrine tissues; (a) an orthopteran species. (b) A diptera species. After Engelmann, 1970.

Parthenogenetic: The production of offspring from unfertilized eggs.

Partial metamorphosis (alternative terms, incomplete metamorphosis, gradual metamorphosis)**:** Metamorphosis in which there is a gradual and often not very marked transformation from immature to adult stage; there is no pupal stage and immature stages are referred to as nymphs.

Parts per million (ppm): Proportion of toxicant present in relation to that of plant material on which it has beend deposited. Usually applied in connection with the edible portion of a crop and its suitability for consumption.

Passive suction ventilation: Sucking in of air through nearly closed or fluttering spiracles as a result of negative pressure in the tracheae.

Passive ventilation: Simple diffusion of gases into the tracheae and tracheoles without the aid of pumping or other movements.

Pasture spelling: Leaving a pasture idle for a period of time by removing livestock from it.

Patch: A discrete area of microhabitat.

Patenty: In the course of the malaria disease, the first appearance of parasites in the red blood cells.

Pathogen reservoir: A host or group of hosts harboring a disease-causing organism.

Pathogen: A parasite that causes disease; a disease-causing organism.

Pathogen: A parasitic organism capable of causing disease in a plant or animal.

Pathogenic: The ability of a pathogen causing or producing disease. (Gr. *pathos* = suffering + genesis = origin).

Pathovar: In bacteria, a subspecific designation for isolates that infect only certain plants within a genus or species.

Patronymic: In nomenclature, a dedicatory name, a name based on that of a person or persons.

Paurometabolous: Insects that go through three stages of growth, egg, nymph, and adult, with nymphs resembling adults except for the absence of wings. Grasshoppers, termites, thrips, true bugs, planthoppers, and aphids are examples; or undergoing a gradual metamorphosis, passing through immature and adult instars in essentially the same environment.

Pearl, artificial: Looking like pearls though made out of only certain synthetic materials like glass.

Pearl: Formed by the secretion of mantle of certain bivalve molluscs.

Pebrine: A disease caused by the protozoan *Nosema bombycis* in the larvae of the mulberry silk worm, *Bombyx mori,* responsible sometimes for a fair amount of loss to the silk industry.

Pecten: A comblike or rakelike structure; any comb-like structure or organ; can refer to many different structures in different groups of insects.

Pectinate: With branches or processes like the teeth of a comb; pectinate antenna; pectinate tarsal claw; comb like; that is, having a series of slender projections from an elongate shaft.

Pectine: Tactile comb-like organ on abdomen in arachnids.

Pedicel: Second antennal segment, between the scape and the flagellum; the constricted region between the thorax and abdomen of Hymenoptera, Apocrita, which is also called a petiole; the second segment of the antenna (Fig. 15 and 19); the stem of the abdomen, between the thorax and the gaster (ants); the base of the ovariole.

Pediculosis: Infestation with lice.

Pedipalp (pl. pedipalpi): An important part of the mouth part of the arachnids being outer in position to the chelicerae, variously modified in various arachnids depending upon their food habits *e.g.* in scorpion.

Pedipalps: The second pair of appendages of an arachnid.

Pedogenesis: Reproduction by immature insects, for example, the beetle *Micromalthus debilis.*

Peduncle: Stalk or petiole; a stalk-like structure supporting another structure, such as a flower stalk.

Pelagic: Inhabiting the open sea; ocean-dwelling.

Pelagic: Living in the open sea or the open water of large freshwater lakes.

Pellet: Seed coated with inert material, often incorporating pesticides, to ensure size and shape for precision drilling.

Pellets: A dry pesticide formulation with particles over 10 cubic millimeters in size. See *granules.*

Penetrant: Oil added to a spray to enable it to penetrate the waxy insect cuticle more effectively; or a wetting agent that enhances liquid absorption to a surface.

Penis (pl. penes), also **phallus:** The median intromittent organ (Fig. 8), variously derived in different insect orders; The copulatory organ of the male; also called aedeagus; the copulatory organ of a male for conveying sperm to the genital tract of a female. (L).

Pentadactyl: (Gr. *pente* = five + *daktylos* = finger) Having 5 fingers, toes, or digits.

Penultimate: Next to the last.

Perennial pest: A pest with a general equilibrium position below but close to the economic-injury level; highest pest population fluctuations regularly reach the economic-injury level.

Perennial: A plant that lives for three or more years.

Perfect stage: Part of fungus life cycle during which gametes and sexual spores are produced.

Perfectly density-dependent factor: A factor that never fails to control the increase in a population's numbers (i.e., intraspecific competition).

Peri: Surrounding.

Pericardial cells: Large, multinucleate athrocytes found in association with the heart, alary muscles, and surrounding connective tissue (Fig. 186).

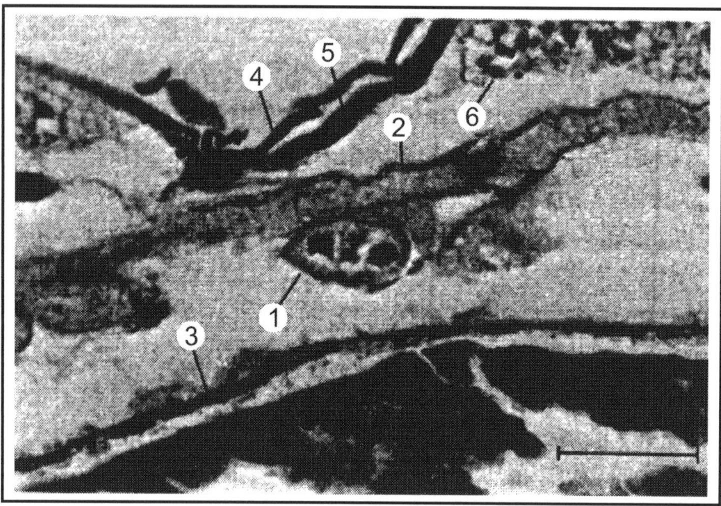

Fig. 186. Photomocrographs of tissues associated with the circulatorysystem in adult mosquito, *Aedes triseriatus*: Pericardial cell. 1, pericardial cell; 2, heart; 3, midgut epithelium; 4, cuticle; 5, muscle; 6, fat body. (Scale lines = 50 μm).

Pericardial septum: See dorsal diaphragm.

Pericardial sinus: Body cavity region above the dorsal diaphragm; a space around the heart, limited below by the dorsal diaphragm; the body compartment that contains the **dorsal vessel** ('heart').

Pericardial: Surrounding the heart.

Pericardium: (Gr. *peri* = around + *kardia* = heart); The membranes lining the cavity and covering the heart; also the cavity enclosing the heart.

Perikaryon: Cell body of a neuron; also called the soma or cell body.

Perineural sinus: The ventral body compartment that contains the nerve cord, separated from the perivisceral sinus by the **ventral diaphragm**.

Perineural: Surrounding nervous tissue.

Perineurium: Cellular layer located beneath the neural lamella of a ganglion that probably secretes the neural lamella; or the layer of cells surrounding a ganglion, which secretes the neural lamella.

Periodic release: The regular release of biological control agents that are effective in control but unable to establish permanently.

Periodicity: Recurrence of particular behavior(s) or physiological events in a regular pattern.

Peripheral : (Gr. *peri* = around + *stalsis* = constriction) Rhythmic involuntary muscular contractions passing along a hollow organ, espically of the digestive tract.

Peripheral nervous system: All the nerves emanating from the ganglia of the central and visceral nervous systems; the network of nerve fibers and cells associated with the muscles.

Periphyton: Organisms in water attached to or clinging to rocks, twigs, and other submerged structures.

Periplasm: Cytoplasm located around the periphery of the yolk of an insect egg (Fig. 187).

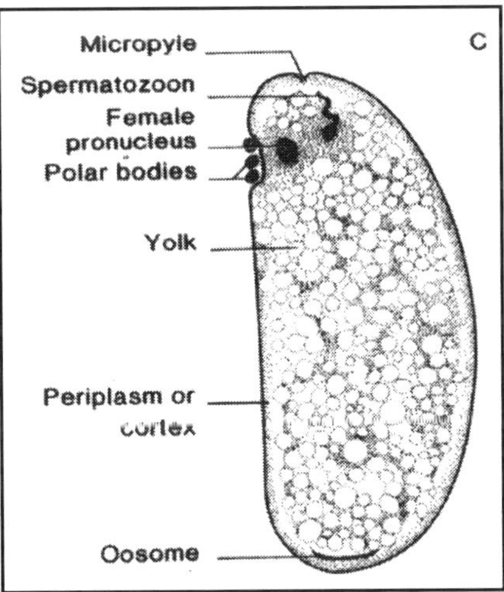

Fig. 187. Fertilization – germ band formation (chorion omitted; (a-e) sagittal sections): Just prior to fertilization (at the stage of fusion of spermatozoon and female pronucleus).

Peripneustic: Respiratory system with one metathoracic and eight abdominal pairs of spiracles; most insect larvae have this type.

Periproct: Posterior somite of a segmented animal, in which the anus is located; also called the telson (Fig. 188).

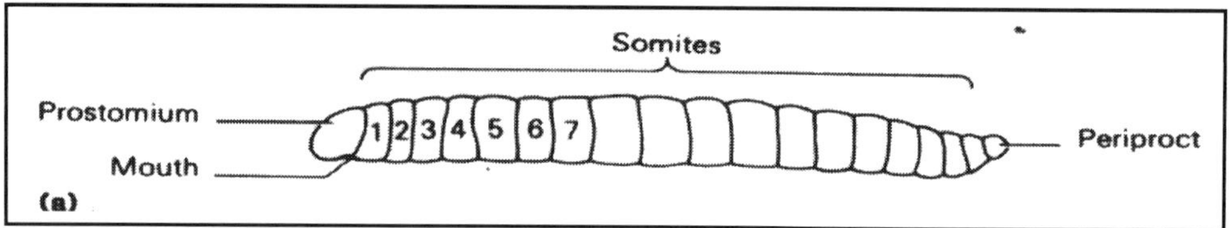

Fig. 188. Hypothetical stage in the evolution of the insect form: Periproct.

Peristaltic contractions: Rhythmic contractions of longitudinal and circular muscles surrounding or composing a tubular organ such as the alimentary canal.

Peristome: The ventral margin of the head, bordering the mouth.

Periterme: A sclerotized plate surrounding an orifice, notably around a spiracle.

Perithecium: Flask shaped structure of some ascomycete fungi containing asci and ascospores.

Peritoneum: The thin serous membrane (mesodermal) that lines the body cavity and covers the organs therein in many animals. (Gr. *peri* = around + *teino* = stretch).

Peritrophic membrance: A chitinous membrane secreted by cells lining the midgut; forms a semipermeable envelope around the food; a chitinous, noncellular membrane that surrounds the food bolus in most insects; a membrane secreted by the cells lining the midgut; this membrane is secreted when food is present, and forms an envelope around the food; it usually pulls loose from the midgut, remains around the food, and passes out with the feces; or the delicate, tubular sheath that surrounds the food within the midgut; a thin sheath lining the midgut epithelium of many insects; a membrane produced within the midgut of insects and which encloses the food as it passes along the alimentary canal.

Perivisceral sinus: Body cavity region below the dorsal diaphragm and above the ventral diaphragm (if present); the central body compartment, delimited by the **ventral diaphragm** and **dorsal diaphragm.**

Permit: A document required in some states to allow application of restricted use pesticides.

Persistence: (i) The term applied to chemicals that remain active for a long period of time after application. (ii) The amount of time after application that a pesticide remains effective and stable in the environment. (iii) A virus that infects its insect vector so that the vector can transmit for the remainder of its life.

Persistence: The degree to which the activity of an insecticide remains after it is applied.

Persistent virus: A virus that persists within the body of a vector; circulative virus.

Personal protective equipment (PPE): Clothes, materials, or devices that offer protection from pesticides, especially important when handling or applying highly toxic pesticides. Such equipment can prevent injury or death, and is specified on the label.

Pest categories: Include pathogens, weeds nematodes, arthropods, mollusks, and vertebrates.

Pest control advisor (PCA): A person who makes agricultural pest control recommendations, or any person who holds her/himself out as an authority or general advisor on any agricultural use, or solicits services or sales.

Pest control operator (PCO): A person or firm that is officially licensed to apply pesticides, or use any method or device for hire, to control pests or prevent, destroy, repel, mitigate, or correct any pest infestation or disorder.

Pest control: The actions taken to eradicate, inhibit, or limit pests and to prevent them from interfering with crop profitability or agricultural operations; or the application of technology, in the context of biological knowledge, to achieve satisfactory reduction of pest numbers or effects.

Pest density: The population level at which a pest species causes economic damage.

Pest management (alternative terms, integrated control, **IPM**): An approach to pest control that blends biological and chemical control methods and involves application of insecticides according to needs, rather than according to a set progamme.

Pest management (integrated pest management, IPM): A comprehensive approach to dealing with pests that strives to reduce pest status to tolerable levels by using methods that are effective, economically sound, and ecologically compatible.

Pest management strategy: A overall plan for eliminating or alleviating a pest problem.

Pest management tactic: A method of controlling a pest or its damage; a method of implementing a pest management strategy.

Pest resurgence: The rapid increase in numbers of a pest following cessation of control measures or resulting from development of **resistance** and/ or elimination of natural enemies.

Pest spectrum: The complete range of pests attacking a particular crop.

Pest status: The ranking of a pest relative to the economics of dealing with the species.

Pest status: Loss of 5 per cent or more in the yield of a crop.

Pest: An animal or plant causing damage to man's crops, animals or possessions; or any organism that interferes with the activities and desires of humans, including pathogens, weeds, nematodes, mollusks, arthropods, and vertebrates.; or a species that interferes with human activities, property, or health or is objectionable.

Pesticide dealer: Any distributor or retailer that: sells pesticides for agricultural use, methods and devices for the control of agricultural pests, or soliciting pesticide sales.

Pesticide kill: The death of nontarget organisms due to careless or improper pesticide use.

Pesticide resistance: See *resistance.*

Pesticide tolerance: The amount of pesticide residue which may legally remain on or in food at the time of sale. The EPA is responsible for establishing federal residue tolerances.

Pesticide: A chemical which by virtue of its toxicity (poisonous properties) is used to kill pest organisms. A term of wide application which includes all the more specific applications- insecticide, acaricide, bactericide, fungicide, herbicide, weedicide, molluscicide, nematicide, rodenticide, etc; refers to any economic poison intended to control pests. See *FIFRA;* or any substance for controlling, preventing, destroying, disabling, or repelling a pest.

Petiolate: Attached by a narrow stalk or stem; some Diptera and Hymenoptera Apocrita have slender segments between the thorax and abdomen.

Petiole: A stalk or stem; the narrow stalk or stem by which the abdomen is attached to the thorax (Hymenoptera); the constriction following the first abdominal segment of Hymenoptera, Apocrita; also called a pedicel; the waistlike construction in apocritan; hymenoptera, the narrow second (and sometimes third) abdominal segments that precede the gaster, forming the 'waist' (Fig. 189).

Petroleum oil: Refined spray oil classed as paraffin, naphthalene, aromatic, or unsaturated oil based on the hydrocarbons it contains.

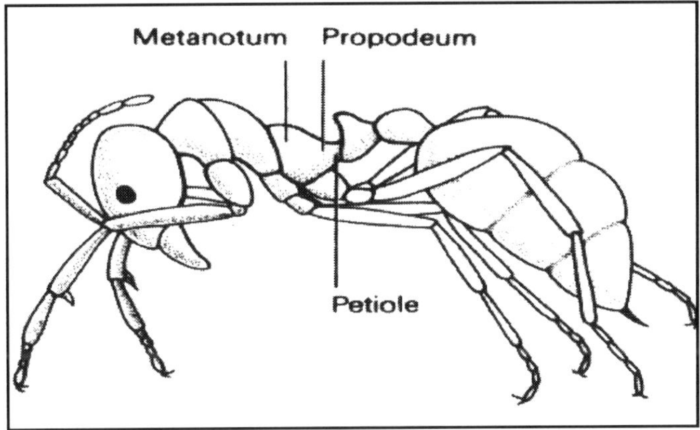

Fig. 189. Modifications of the abdomen: Fire ant.

pH (PH): A measure of relative acidity or alkalinity of aqueous solutions. A pH of 7 is equal to neutrality; above 7 is basic and below 7 is acidic.

Phagocyte: (Gr. *phagein* = to eat + *kytos* = hollow vessel) A blood cell capable of degrading tissues and foreign matter; or a white blood cell that engulfs and digests bacteria and other foreign materials.

Phagostimulant: A natural plant substance that induces feeding.

Phalaenophily: Plant pollination by moths.

Phallobase: In male genitalia, the support for the **aedeagus**.

Phallomere: A lobe lateral to the penis.

Phallus: see **Penis.** The male copulatory organ, including any processes that may be present at its base.

Pharate stage: A stage in which molting has occurred but the insect has not cast off the old cuticle.

Pharate: Period of time between apolysis and ecdysis, while the insect is still within the confines of the old cuticle; cloaked; living inside the cuticle of a previous instar; usually occurs as a pharate pupa or pharate adult insect.

Pharynx: Foregut structure into which the true mouth opens; often, a well-developed pump; the anterior part of the foregut, between the mouth and the oesophagus; the region of the digestive tract between the mouth cavity and the oesophagus; often muscular and sometimes with teeth in invertebrates; the gill region of many aquatic vertebrates. (Gr. throat).

Pharynx: The anterior part of the foregut, anterior to the oesophagus.

Phasic: Characterized by short-duration stimulation; phasic receptors.

Phasmida, order: (*phasm,* phantom) The order of insects comprising stick and leaf insects (Fig. 190).

Phenetic: A classification system in which overall resemblance between organisms is the criterion for grouping; *see also* **Cladistics; Evolutionary systematics.**

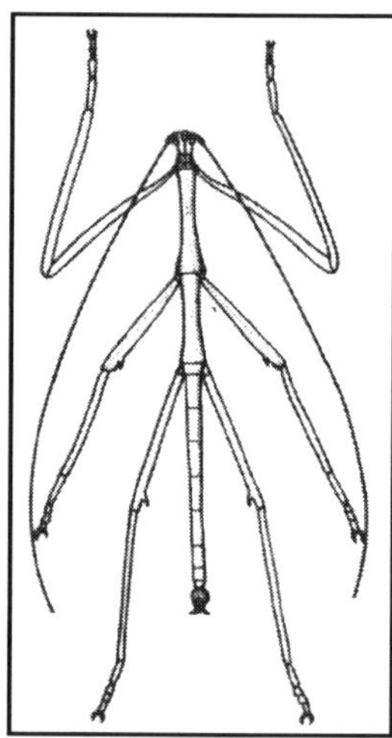

Fig. 190. Phasmida. A stick insect, *Diapheromera femorata.*

Phenetics: A school of systematic thought based on the assumption that phylogeny cannot be deduced with certainty and therefore classification should be based on as many different characteristics as possible (compare with cladistics).

Phenological asynchrony: Lack of synchrony between the life cycle of a pest and the appropriate stage of its host plant.

Phenology: The growth stages that an organism passes through over time; the periodicity of biological phenomena.

Phenotype: The class in which an individual falls on the basis of visible characters, as the result of an interaction between genotype and environment (cf. Genotype). The observable structural and functional attributes of an organism; the sum total of the observable or measurable characteristics of an organism without reference to its genetic nature; the external appearance of an individual without regard to its genetic or hereditary constitution. Compare Genotype. Phenotype is the physical manifestation of a genetic trait. (Gr. *phaino* = show + *typos* = impression, type).

Phenyl: An organic radical, C_6H_5, derived from benzene via the removal of a hydrogen atom.

Pheromone mass trapping: The use of pheromones to lure pest insects, which are then killed.

Pheromone: (= Ectohormone) A chemical released by one organism that influences the behaviour of another of the same species; a substance secreted by one animal to the exterior that causes a specific reaction on reception by another animal of the same species (e.g., sex pheromone); an external chemical messenger that passes between individuals of the same species and controls intraspecific interactions.

Phloem: The tissue that translocates the products of photosynthesis from the foliage to other plant parts.

Phloeophagy: Feeding on the phloem tissues in the inner bark of tree trunks.

Phore: Carrier.

Phoresy: One organism being transported on the body of an individual of another species; a commensalistic relationship in which one kind of animal attaches to another and thereby gains a mode of transportation without harming that individual.

Phorids: Flies belonging to the family Phoridae.

Photo: Light.

Photoperiod: The duration of the light (and therefore also dark) part of the 24-hour daily cycle.

Photoperiodism: A response by an organism to the duration or pattern of light and dark materials.

Photoreception: Reception of stimuli in the form of light.

Photoreceptor: A sense organ involved in the perception of light.

Photosynthate: The products of photosynthesis used to support plant growth, respiration, and fruit production.

Photosynthesis: The process by which green plants use energy in sunlight to convert water and CO_2 into carbohydrates; the formation of carbohydrates from carbon dioxide by chloroplasts of the chlorophyll of green plants or flagellate protozoans in the presence of light (Gr. *phos* = light + synthesis).

Phototaxis: Directed response relative to light.

Phragma (pl., phragmata): A platelike apodeme or invagination of the dorsal wall of the thorax; plate like inflections of the terga (nota) that provide surface area for the attachment of the dorsal longitudinal wing muscles (Fig. 191).

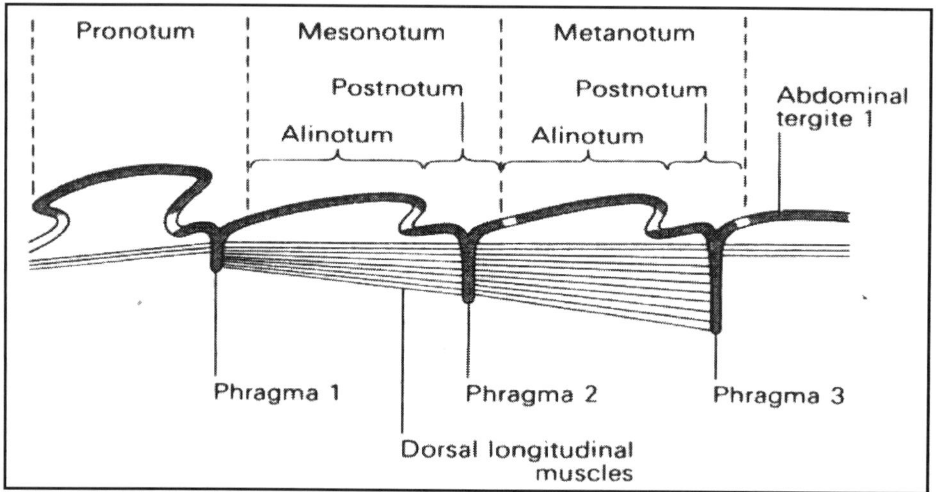

Fig. 191. Sagittal section of dorsum of generalized pterygote insect thorax. After Snodgrass, 1935.

Phragmosis: The closing of a nest opening with part of the body.

Phyletic tracking: Strict coevolution in which the phylogenies of each taxon (e.g. host and parasite, plant and pollinator) match precisely.

Phyletic: Pertaining to a line of descent (cf. Phylogeny).

Phyll: Leaf.

Phylogenetic tree: A diagrammatic presentation of assumed lines of descent, based on paleontological, morphological, or other evidence.

Phylogenetic: Relating to the evolutionary history of a group of organisms.

Phylogeny (adj. phylogenetic): Evolutionary history (of a taxon); the study of the historical development of the line or lines of evolution in a group of organisms; the origin and evolution of higher categories (cf. Classification). (Gr. *peylo* = race + *gen* = became).

Phylum (pl. phyla): Major division of the animal kingdom corresponding to the main animal types, e.g. Phylum: Arthropoda – arthropods, phylum: Mollusca-molluscs.

Physical control: Pest control which involves either modification of some physical factor in the environment, (e.g., temperature) or use of a physical factor (e.g., light) to attract insects.

Physical gill: A bubble or packet of air that adheres to the body of an aquatic insect and is continuous with the tracheal air space.

Physiological disorder: A disorder caused by factors other than a pathogen; abiotic disorder.

Physiological race: See Biological race.

Physiological resistance: The ability of an insect strain to tolerate insecticides or other factors through physiological change (e.g., via modification of cuticle structure).

Physiological species: See Sibling species.

Physiological time: A measure of development time based upon the amount of heat required rather than calendar time elapsed.

Physogastry: Enlargement of the abdomen of mature female due to increased size of reproductive system, e.g., termites, ants and bees.

Phytoalexins: Defense compounds produced by plants as a response to pathogen arthropod attacks; phenolic compounds produced by plants when they become diseased or injured by insects; enable plants, once damaged, to resist further attack by insects.

Phytoecdysone (s): A plant product that mimics ecdysone; compounds found in certain plants that mimic the action of ecdysone.

Phytopathology: The study of plant disease.

Phytophagous: Herbivorous; feeding on or in plants, or on vegetable matter.

Phytophagy: The eating of plants.

Phytoplasma: Mollicutes that infect plants and cause disease.

Phytosanitation: Measures requiring the removal or destruction of infected or infested plant material likely to form a source of reinfection or reinfestation. (See Crop Hygiene.) International phytosanitation refers to inspection of plants and seeds to prevent noxious pests and diseases from being brought into a country; often involves plant quarantine.

Phytotelemata: Plant structures inhabited by or used by insects, ranging from pithers, special galls, to flooded leaf axils.

Phytotoxemia: A disease like plant condition produced by the injection of toxic substances by insects.

Phytotoxic: A chemical liable to damage or kill plants (especially the higher plants), or plant parts; toxicant that causes damage or death to plants.

Phytotoxicity: Poisonous or chemically injurious to plants.

Pictured: With spots or bands (wings; for example).

Pigment: Colouring matter.

Pilifer: One of a pair of lateral projections on the labrum (Lepidoptera).

Pilose: Covered with hair.

Pioneer communities: In the process of succession, the first community or organisms to become established are the pioneer communities.

Piroplasmosis: Infection with tick-transmitted piroplasma protozoans, *Babesia* species, that attack the red blood cells of cattle, dog, and other animals.

Pitfall trap: A sampling device designed to capture insects moving along the ground.

Placenta: The organ by which the fetus (embryo) of higher mammals is attached in the uterus of the mother and through which diffusible substances pass for the metabolism of the fetus. See Chorion. (L. flat cake).

Placoid sensilla: Plate like cuticular sensory structure (Fig. 192).

Placoid: Platelike.

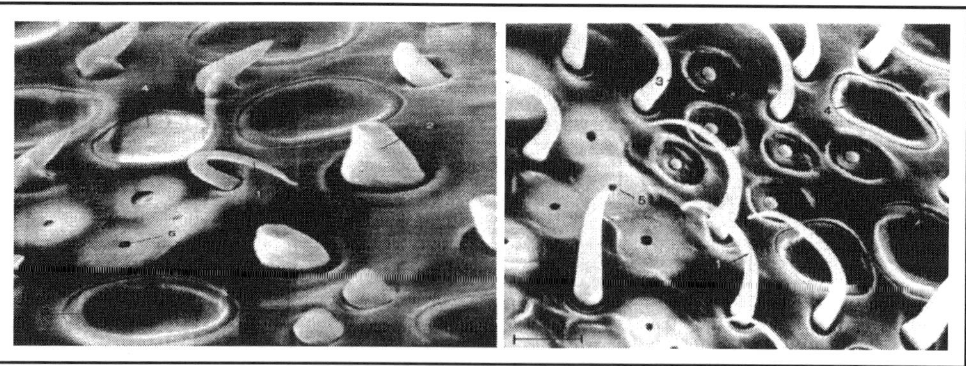

Fig. 192. Scanning electron micrographs of sensilla on the antennal flagellum of a worker honey bee. 1, trichoid; 2, basiconic; 3, campaniform; 4, placoid; 5, coeloconic or ampullaceous. (Scale lines = 10 μm.)

Plague (bubonic plague, black death): An infectious disease caused by the bacterium *Yersinia pestis* and spread via the oriental rat flea *Xenopsylla cheopis.*

Planarium: Flatworm; a free-living member of the Platyhelminthes, Turbellaria; either terrestrial or aquatic in habits.

Planidium larva: A type of first instar larva in certain Diptera and Hymenoptera which undergo hypermetamorphosis; a larva that is legless and somewhat flattened.

Plankton: Tiny marine or freshwater organisms floating and drifting with the surrounding water.

Plant cover: The degree to which plants within a sprayed area receive a deposit of pesticide.

Plant growth regulator (PGR): Chemicals that affect plant growth.

Plant host range: The species of plants on which an insect species feeds.

Plant pathogen: A parasite that causes disease in plants.

Plant resistance (to pests): The ability of a plant to grow and produce economically despite the presence of a pest.

Plant resistance: A range of mechanisms by which plants resist insect attack; *see also* **Antibiosis; Antizenosis; Tolerance.**

Plantigrade: Walking on the whole sole of the foot, as a man or bear. (L. *planta* = sole + *gradior* = walk).

Plasm-, plasmo-, -plasm: Something molded or modeled.

Plasma: The aqueous noncellular fluid portion of the blood hemolymph (Gr. a thing molded).

Plasmid: A self-replicating, extrachromosomal circular DNA found in certain bacteria and fungi.

Plasmodium (pl. plasmodia): The protozoans that may cause malaria.

Plastron: A film of air over the body of certain aquatic insects that serves as a physical gill for respiration; fine hydrofuge hairs or other fine cuticular networks which hold a thin layer of gas permanently in place; acts as a permanent "physical gill;" a framework of stiff, water-repellent hairs or cuticular structures on the bodies of aqatic insects, containing a film of air into which oxygen diffuses from the water which is used in respiration.

Platyform larva: A larva that is extremely flattened, as the larva of the Psephenidae. (Also called onisciform larva.)

Plecoptera, order: (*pleco,* pleat; *petra,* wing) The order of insects comprising the stone flies (Fig. 193).

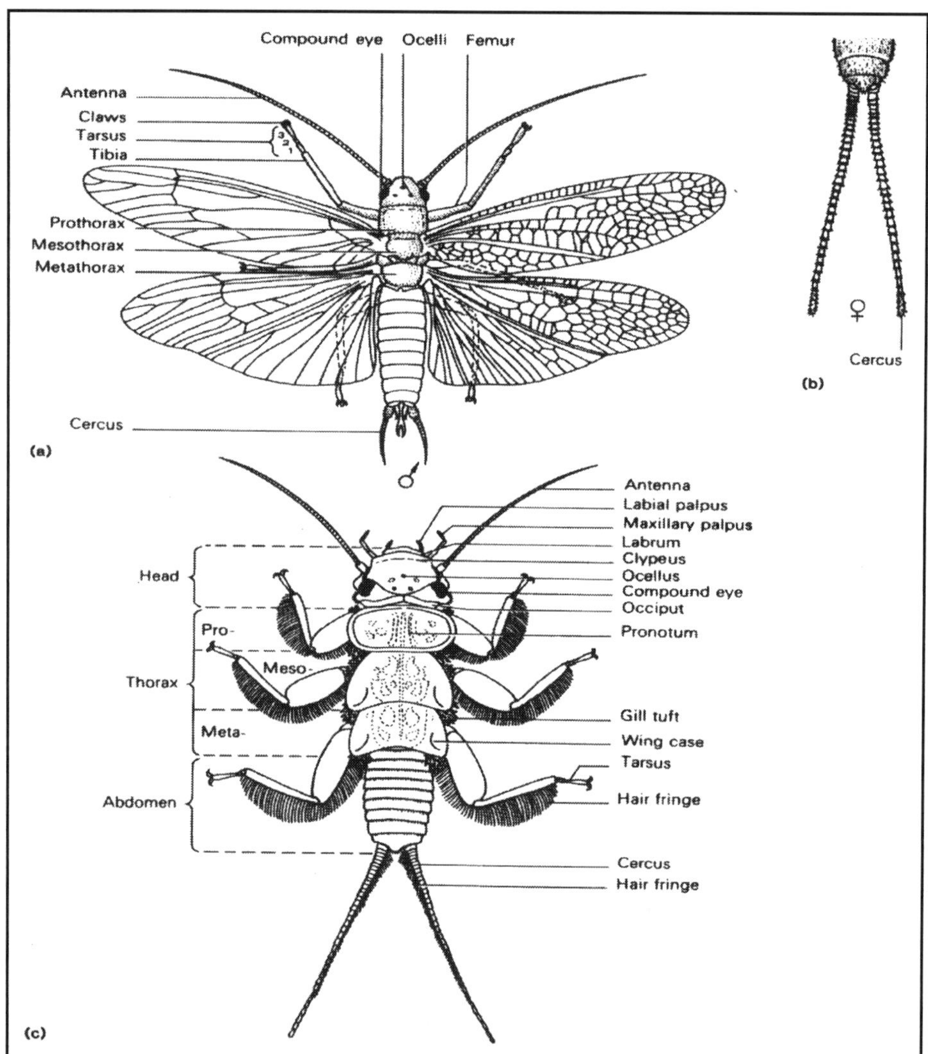

Fig. 193. Plecoptera, stoneflies. (a) Adult male, *Pteronarcys californica.* (b) Terminalia of adult female. (c) Generalized nymph. (a,b) After Essig, 1942; (c) after Claassen, 1931.

Plenary powers: Special powers granted by the International Congress of Zoology (Monaco, 1913; Paris, 1948) to the International Commission on Zoological Nomenclature permitting the suspension of the International Rules of Zoological Nomenclature or decisions as to how they shall apply in specific cases [*see Bul. Zool. Nomencl.*, **4:** 51-56 *et seq.* (1950)].

Pleometrosis: The foundation of a colony of social insects by more than one queen.

Plesiotype: A specimen or specimens upon which subsequent descriptions or figures are based.

Pleura: The membrane covering the lung and lining the inner wall of the thorax. (Gr. rib, side).

Pleural apophysis: Arm like projection of the pleural ridge.

Pleural coxal process: The anterior end of the pleural ridge providing reinforcement for the coxal articulation.

Pleural membrane: The unsclerotized cuticle in the pleural region of the abdomen that separates the terga and sterna.

Pleural ridge: The internal ridge dividing the pterothorax into the anterior **episternum** and posterior **epimeron.**

Pleural suture: A suture on a thoracic pleuron extending from the base of the wing to the base of the coxa, which separates the episternum and epimeron; a vertical or oblique suture marking an internal ridge of the thoracic pleuron, running from the dorsal coxal articulation to the pleural wing process; the externally visible indication of the pleural ridge, running from the leg base to the tergum.

Pleural wing process: The posterior end of the pleural ridge providing reinforcement for the wing articulation; ventral articulation of a wing; a fulcrum for the wing base, formed at the top of the internal ridge formed by the pleural suture.

Pleural: Pertaining to the pleura, or lateral sclerites of the body; lateral.

Pleurite: A lateral or pleural sclerite.

Pleuron (pl. pleura; adj. pleural; dim. pleurite)**:** A lateral plate or sclerite of a thoracic segment; the lateral plate on either side of a somite in arthropods.

Pleurosternal sutures: Sutures dividing the sternum from the laterosternites.

Pleurotergite: A sclerite containing both tergal and pleural elements.

Plexus: A network of interlaced nerves or blood vessels. (L. a plaiting.)

Plumose: Featherlike.

Plumose: Featherlike; plumose antenna, such as found in male mosquitoes; and feathery setae.

Podite: A segment of an arthropod leg, moved by muscles inserted in its base.

Poikilotherm: Organisms with body temperature determined by ambient temperature; cold-blooded.

Poikilothermic: Animals that remain at ambient temperature, that fluctuate with environmental temperatures; "cold-blooded."

Poikilothermous: "Cold-blooded"; animals with little or no control of their body temperature, which usually varies with that of the ambient medium. The opposite is homeiothermous.

Point: A small triangle of stiff paper, used in mounting small insects.

Poison: A chemical or material that can cause injury or death when eaten, absorbed, inhaled, or taken up by plants or animals.

Poison bait: An attractant foodstuff for insects, mollusks, or rodents, mixed with an appropriate toxicant.

Poison glands: Accessory glands that produce poison, as in the stings of Hymenoptera.

Poisonous bait: A food or other substance which is mixed with a poison to attract pests so they will consume it.

Poisson distribution: A function that describes a random distribution, in which the variance is approximately equal to the mean.

Poisson model: A mathematical model that describes a random dispersion of organisms.

Pollen basket: See corbiculum.

Pollen brush: A brush of short stiff hairs used in collecting pollen (bees).

Pollen grain: A microgamete of a seed plant.

Pollen rake: A comblike row of bristles at the apex of the hind tibia of a bee.

Pollination: Transfer of pollen from the male parts of flowers (anthers) to the female receptive organ (stigma).

Pollinium: A mass of pollen grains characteristic of the orchid family; often bearing a stalk with an adhesive disc that facilitates its transfer by insects.

Pollute: To put undesirable chemicals or materials into the environment.

Pollution: It is an undesirable change in physical, chemical or biological characteristics of air, land, and water that may or will harmfully affect human life directly or indirectly.

Poly: Many.

Polychaeta: An order of class Chaetopoda of phylum Annelida, marine *e.g.*, bristle worms, tube worms etc.

Polychloroterpene: A chlorinated hydrocarbon insecticide prepared from camphene, a pine tree derivative (e.g., toxaphene).

Polyculture: The cultivation of several crops intermingled; a mixed stand of crop plants.

Polycyclic: A disease with multiple reproductive cycles per season (cf., monocyclic).

Polyectic: Describes pollinating insects that visit plants of many different taxa; compare with monolectic.

Polyembryony: An egg developing into two or more embryos; mitotic division of an egg resulting in the development of several embryos from a single egg; found among several groups of parasitic Hymenoptera.

Polyethism: Within a social insect **caste,** the division of labour either by specialization throughout the life of an individual or by different ages performing different tasks; the presence of several discrete types of behavior by different groups of individuals in colonies of social insects.

Polygenic resistance (micro-gene resistance): Host-plant resistance conferred by many genes.

Polygenic: A characteristic, such as resistance, conferred by more than one gene.

Polygyny: Social insects that have several queens, either at the same time or sequentially (serial polygyny).

Polylectic: Bees that gather pollen from a wide range of flowers; utilizing a variety of plant species as sources of pollen.

Polymerase chain reaction (PCR): A technique to make copies of DNA fragments.

Polymorphic: Of a species, existing in two or more genetic variants (morphs).

Polymorphism: Having several forms, for example, one or both sexes of a single species may occur in two or more clearly distinct forms; existence of individuals of more then one form in a species. (Gr. *poly* = many + *morphe* = form); the presence of two or more distinct, structurally different types of individuals within the same stage of one species; a form of individual variability; the occurrence together in the same habitat of two or more distinct forms of a species in such proportions that the rarest of them cannot be maintained by recurrent mutation (cf. Dimorphism); the simultaneous occurrence of more than one phenotype in a population.

Polynaviruses (PVs)**:** A group of viruses found in the ovaries of some parasitic wasps, involved in overcoming host immune responses when injected with the wasp eggs.

Polyneuronal innervation: The situation in which insect skeletal muscle units receive more than one motor axon.

Polynominal nomenclature: A system of nomenclature consisting of a scientific designation of a species on the basis of more than three descriptive words; the antecedent of the Linnaean "binomial" system.

Polyp: Mostly fixed form of coelenterates, trunk cylindrical, the distal end fused, proximal and having mouth surrounded by a ring of tentacles e.g. *Hydra.*

Polyphage (adj. polyphagous)**:** An eater of many kinds of food, e.g. many plant species from a range of families; used particularly of phytophages.

Polyphagous: Eating various types of food; insects that feed on plants belonging to several or many different families; Feeding on a broad array of plant or animal species. cf. monophagous, oligophagous.

Polyphenism: Envionmentally induced differences between successive generations, lacking a genetic basis.

Polyphyletic: A group evolutionarily derived from more than one ancestor, recognized by the possession of one or more features evolved convergently; rejected in **cladistics** and **evolutionary systemics;** a term applied to a taxonomic category derived from two or more ancestral sources; not of a single immediate line of descent (cf. Monophyletic). Evolving from more than one taxon.

Polyploidy: A condition in which the nuclear complement of chromosomes is an integral multiple (greater than 2) of the haploid number.

Polypneustic: A respiratory system with at least eight functional spiracles on each side of the body; *see also* **Apneustic; Oligopneustic.**

Polypod larva: A type of larva with jointed legs on the thorax and prolegs on the abdomen (Fig. 194).

Polytopic: Occuring in different places as, for instance, a subspecies composed of widely separated populations.

Polytrophic (ovarioles): Nurse cells within each ovarian follicle.

Polytrophic ovariole: An ovariole in which several nurse cells remain closely attached to each oocytes as it moves down the ovariole.

Polytypic: A category containing two or more immediately subordinate categories, as a genus with several species or a species with several subspecies (cf. Monotypic).

Polyxene (adj. polyxenous): A parasite with a wide range of hosts.

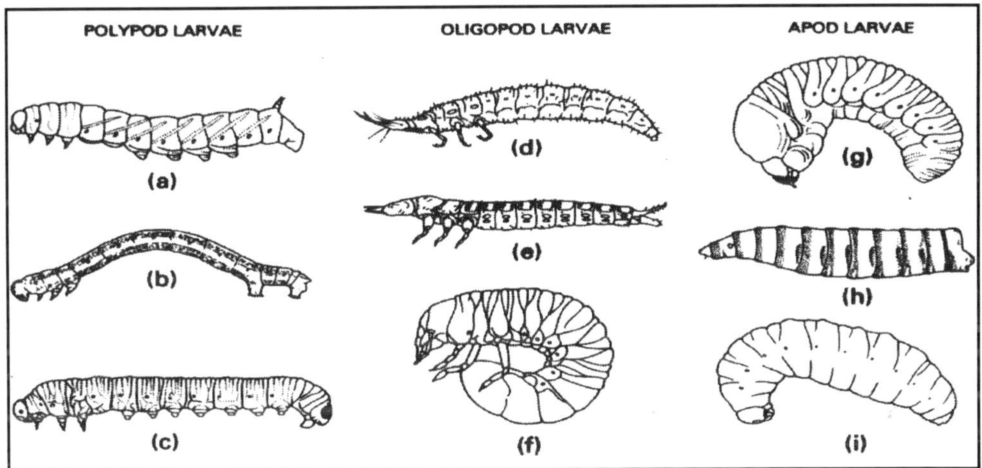

Fig. 194. Examples of larval types. Polypod larvae: (a) Lepidoptera: Sphingidae; (b) Lepidoptera: Geometridae; (c) Hymenoptera: Diprionidae. Oligopod larvae: (d) Neuroptera: Osmylidae; (e) Coleoptera: Carabidae; (f) Coleoptera: Scarabaeidae. Apod larvae: (g) Coleoptera: Scolytidae; (h) Diptera: Calliphoridae; (i) Hymenoptera: Vespidae. After Chu, 1949; afte Borror *et al.*, 1989; after Ferrar, 1987; after CSIRO, 1970.

Population: A group of individuals of the same species living together in a limited and defined space; all the members of a species that occur in a defined area at a specific time; a group of individuals of the same species within given space and time constraints.

Population density: The number of organisms of a species per unit area.

Population dynamics: Changes in population size and structure over time; the interplay between population and environmental forces that influence them; the forces that control population size, and their effects. The study of changes in population of organisms and of the reasons for such changes.

Population growth curve: It is graphic representation of the increases in population numbers over a unit period of time under certain environmental conditions.

Population growth rate curve: It is graphic representation to show the average rate of change in number of organism over a unit period of time.

Population index: An estimate of population size based on insect effects or products.

Population intensity estimate: An absolute estimate of population size expressed as number per unit habitat (e.g., number per leaf).

Population regulation: The maintenance of an approximately constant population size and density, and the forces that control it.

Population resilience: The capacity of a population to adapt to change or to persist in a changing environment.

Population stability: The ability of a population to absorb disturbance and to return to an equilibrium state.

Pore canal (s): A small channel involved in the transport of materials from the epidermis to the body surface; tiny tubes connecting the cellular and cuticular layers of the insect integument.

Pore canals: Fine tubules that run through the cuticle and carry epidermally derived compounds to the **wax canals** and thus to the epicuticular surface.

Pore kettle: The chamber within a chemoreceptor sensillum that has many pores (slits) in the wall (Fig. 195).

Porrect: Extending forward horizontally; porrect antennae. Projecting forward; head or antennae that project anteriorly and horizontally.

Portal vein: One that divides into capillaries before reaching the heart. (L. *porta* = gate).

Postabdomen: The modified posterior segments of the abdomen, which are usually more slender than the anterior segments e.g. Crustacea also the postabdomen in the scorpion.

Postalar bridge: Exoskeletal region of a wing-bearing segement that unites the postnotum with the epimeron.

Postalar callus: A rounded swelling on each side of the mesonotum, between the base of the wing and the scutelum (Diptera).

Post-coxal bridge (s): The pleural area behind the coxa, often fused with the sternum; exoskeletal region of a wing-bearing segment that unites the epimeron with the sternum.

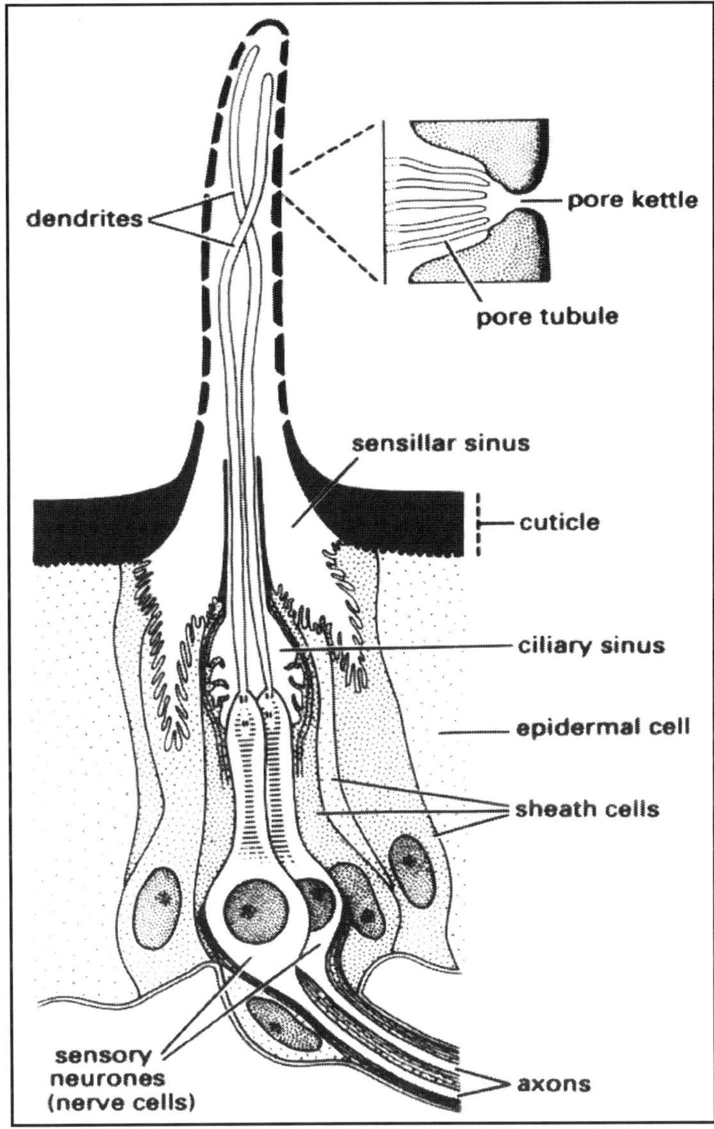

Fig. 195. Longitudinal section through a multiporous olfactory sensillum. After Birch and Haynes, 1982; Zacharuk, 1985.

Postdiapause quiescence: A transitional period of dormancy, following diapause, that is not sensitive to token stimuli.

Postemergence application (POST): Application of pesticides or fertilizers following the plant emergence.

Posterior cell: One of the cells extending to the hind margin of the wing, between the third and sixth longitudinal veins (Diptera).

Posterior cranium: The posterior, often horse-shoe-shaped, area of the head.

Posterior cross vein: A cross vein at the apex of the discal cell (Diptera).

Posterior imaginal ring: Ring of embryonic cells located at the anterior end of the larval hindgut. See also anterior imaginal ring.

Posterior notal process: Articular point for a wing.

Posterior: At or towards the rear.

Posterior: Hind or rear; the hinder part or toward the hinder (tail) end, away from the head, Opposite of anterior. (L. following).

Postgena (pl., postgenae): A sclerite on the posterior lateral surface of the head , posterior to the gena; lateral region of the head above each subgenal suture and between the occipital and postoccipital sutures; the "cheek"; the lateral part of the occipital arch posterior to the post-occipital suture (Fig. 196).

Postgenal bridge: Fusion of lobes of the postgenae found in some insects.

Posthumeral bristle: A bristle on the anterolateral surface of the mesonotum, near the margin of the humeral callus (Diptera).

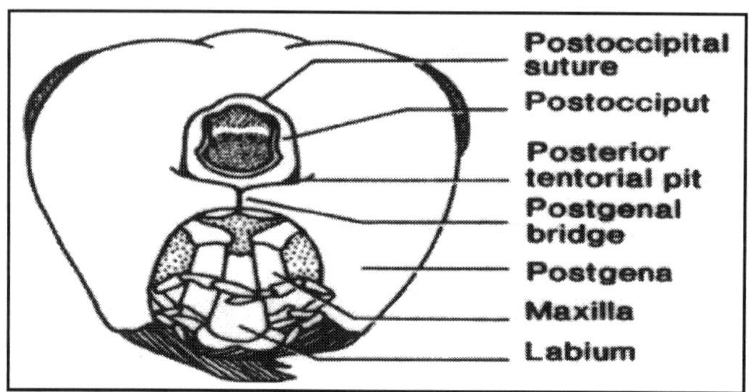

Fig. 196. Modification of the head capsule: Vespid wasp.

Postmarginal vein: The vein along the anterior margin of the front wing, beyond the point where the stigmal vein arises (Chalcidoidea).

Postmentum: The proximal or basal portion of the labium and composed of the proximal submentum and distal mentum (Fig. 197).

Postnodal cross veins: A series of cross veins just behind the costal margin of the wing, distad of the nodus, and extending from the costal margin of the wing to R_1 (Odonata).

Postnotum (pl., postnota): The posterior part of a pterothoracic notum, bearing the phragmata that support longitudinal muscles; an intersegmental, dorsal thoracic skeletal plate; the dorsal sclerite behind the scutellum in adult Diptera.

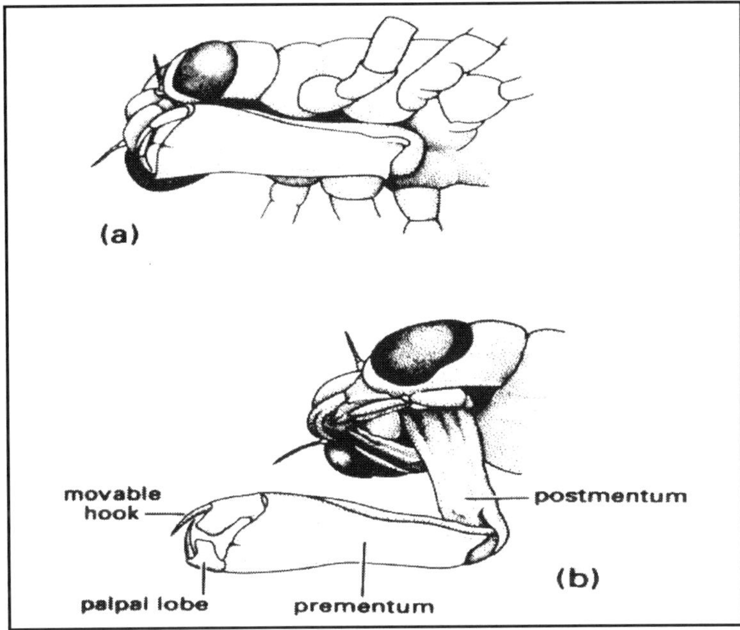

Fig. 197. Ventrolateral view of the head of a dragonfly larva (Odonata: Aeshnidae: *Aeshna*) showing the labial 'mask': (a) In folded position; (b) extended during prey capture with opposing hooks of palpal lobe forming claw-like pincers.

Postoccipital suture: A grove on the head that indicates the original head segmentation, separating the postocciput from the remainder of the head; the transverse suture on the head immediately posterior to the occipital suture; a line surrounding the foramen magnum and indicative of an inflected ridge to which muscles are attached; located posterior to the occipital suture.

Postocciput: Head sclerite posterior to the postoccipital suture; surrounds most of the foramen magnum; the extreme posterior rim of the head, between the postoccipital suture and the foramen magnum.

Postscutellum: A small transverse piece of a thoracic notum immediately behind the scutellum; in Diptera, an area immediately behind or below the mesoscutellum.

Post-tarsus: see **Pretarsus.**

Postvertical bristles: A pair of bristles behind the ocelli, usually situated on the posterior surface of the head (Diptera).

Potential natality: The reproductive rate of individuals in an optimal environment.

Potential pest: A pest which has the potential to be serious but which is usually suppressed by natural regulating factors.

Ppb: Parts per billion. One ten millionth of a percent; equal to micrograms per liter.

PPE: See *personal protective equipment.*

Ppm: Parts per million. One ten thousandth of a percent; equal to milligrams per liter.

Pre-access interval: The interval of time between the last application of pesticide to an area and safe access to the area for domestic livestock, and man.

Preadaptation: Fitness for an environment which the organism does not occupy, or an environmental relationship which it does not maintain, at the time when the adaptation appears; usually applied to a new characteristic, arising by mutation, which permits invasion of a new habitat or the development of a new environmental relationship (cf. Adaptation, Environment).

Preapical: Situated just before the apex.

Precipitin reaction: The formation of a visible precipitate at the interface when an antigen and the corresponding antiserum are brought together (cf. Antigen, Antiserum, Antibody, Quantitative specificity).

Precision farming (site-specific farming): A farm management technology that uses within-field spatial variability to guide application of management actions. Utilizers GPS and GIS systems to allow application as per site-specific requirements.

Precision: A measure of the degree of error in making estimates of a population's size.

Precoccinelline: A defensive allomone produced by lady beetles during auto-hemorrhage.

Precocene: An antijuvenile hormone isolated from the plant *Ageratum houstonianum*; an insect growth regulator produced by certain plants that depresses the source of juvenile hormone.

Precosta: The most anterior wing vein.

Precoxal bridge: Exoskeletal region of a wing-bearing segment that unites the episternum with the sternum.

Precoxal bridge: The pleural area anterior to the coxa, often fused with the sternum.

Predaceous: Feeding as a predator.

Predation: Preying on other organisms.

Predator: A free-living organism that attacks and feeds on other animals, usually amaller or less powerful than itself (prey). The prey is usually killed quickly, and mostly or entirely eaten; many prey individuals are eaten by each predator (L. *praeda* = booty).

Predisposition: Making a plant more susceptible to a pest or disease, usually as a result of genetic, cultural or environmental defects.

Preemergence application (PRE): The application of pesticides or fertilizers after planting, but before the weeds or crop have germinated.

Pre-erythrocytic (also **exo-erythrocytic**) **schizogonous cycle:** In the asexual cycle of a malaria parasite (*Plasmodium* spp.), the schizogonous cycle within the parenchyma cells of the liver, producing a large **schizont**.

Preference: The factor by which certain plants are more or less attractive to insects by virtue of their texture, colour, aroma or taste.

Pregenital: Anterior to the genital segments of the abdomen.

Pregenital segments: The first seven abdominal segments.

Pre-harvest interval: The interval of time between the last application of pesticide and the safe harvesting of edible crops for immediate consumption; This the minimum time that must pass after the last pesticide application before the crop is harvested. Specified on the product label.

Preharvest: The time before the marketable commodity is removed from the field.

Prehensile: Adapted for grasping or holding. (L. *prehando* = seize).

Preimaginal conditioning: Conditioning of an immature insect that persists into the adult stage.

Preirrigation: Irrigation before the crop is planted. Used to apply herbicides or moisten ground to allow for fast emergence of seedlings.

Pre-Linnaean name: A name published prior to Jan. 1, 1758, the starting point of zoological nomenclature and the assumed date of publication for the tenth edition of Linnaeus's *Systema naturae;* such names are unavailable and may not be made available by republication in their original form after Jan. 1, 1758, nor through citation in synonymy (cf. Available name).

Prementum: Apical portion of the labium that bears the labial palpi; the distal part of the labium, distad of the labial suture, on which all the labial mucles have their insertions; the free distal end of the labium, usually bearing **labial palps, glossae** and **paraglossae**.

Preoral cavity: Cavity formed by the mouthparts.

Pre-oviposition period: The period of time between the emergence of an adult female insect and the start of its egg laying; the period between mating and the commencement of egg laying.

Prepatent period: In the course of a disease, the time between infection and first symptoms.

Preplant incorporated (PPI): Pesticide applied and physically mixed into the soil before planting the crop.

Preplant: Decisions or actions made prior to planting a crop.

Prepupa: A quiescent non feeding larval stage occurring just prior to the transformation to the pupal stage in some insects; a quiescent stage between the larval period and the pupa; found in some Diptera and Thysanoptera; a resting stage of the last larval instar, prior to the molt to the pupal stage.

Prescription: Recommendation of a strategy or course of action.

Prescutal suture: Line that delineates the prescutum.

Prescutum: The anterior third of the **alinotum** (either meso-or metanotum), in front of the **scutum**; anterior sclerite divided off from the alinotum of a wing-bearing segment.

Presocial: Behavior that includes one or two of the three characteristics that define social behavior.

Presoldier: In termites, an intervening stage between worker and soldier.

Press: The process on the proximal apex of the tarsus of a bee that pushes pollen into the **corbicula** (Fig. 198).

Presternal suture: The line that separates the presternum from the rest of the sternum.

Presternum: A sclerite on the anterior part of the sternum divided off by the presternal suture; a smaller sclerite of the **eusternum,** lying anterior to the basisternum.

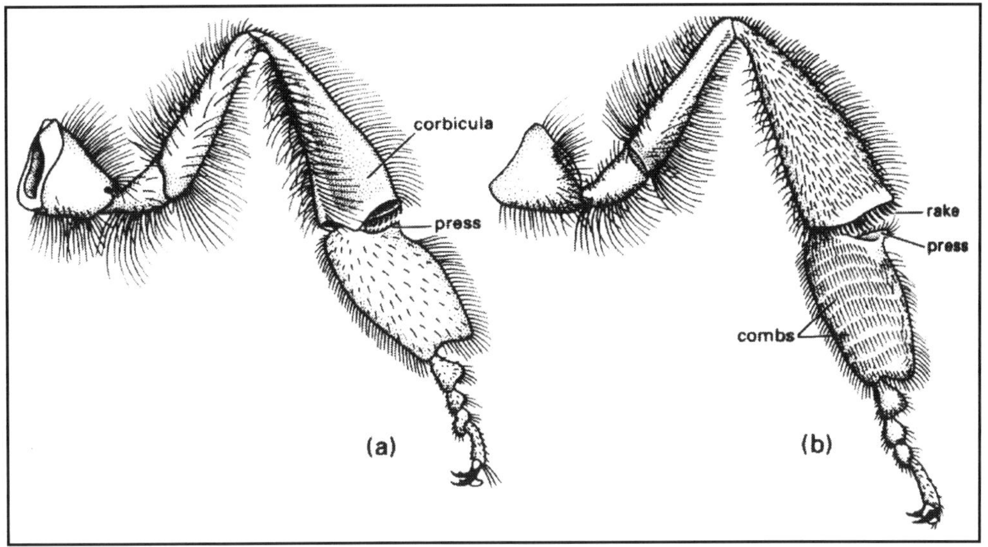

Fig. 198. The hindleg of a worker honey bee, *Apis mellifera* (Hymenoptera: Apidae): **(a)** outer surface showing corbicula, or pollen basket (consisting of a depression fringed by stiff setae), and the press that pushes the pollen into the basket; **(b)** the inner surface with the combs and rakes that manipulate pollen into the press prior to packing.

Prestomal: Refers to the distal end of the proboscis in some adult Diptera; teeth on the labella of some blood-sucking Muscidae.

Pretarsus: The most distal segment of the insect leg, bearing the claws and arolium (Fig. 199).

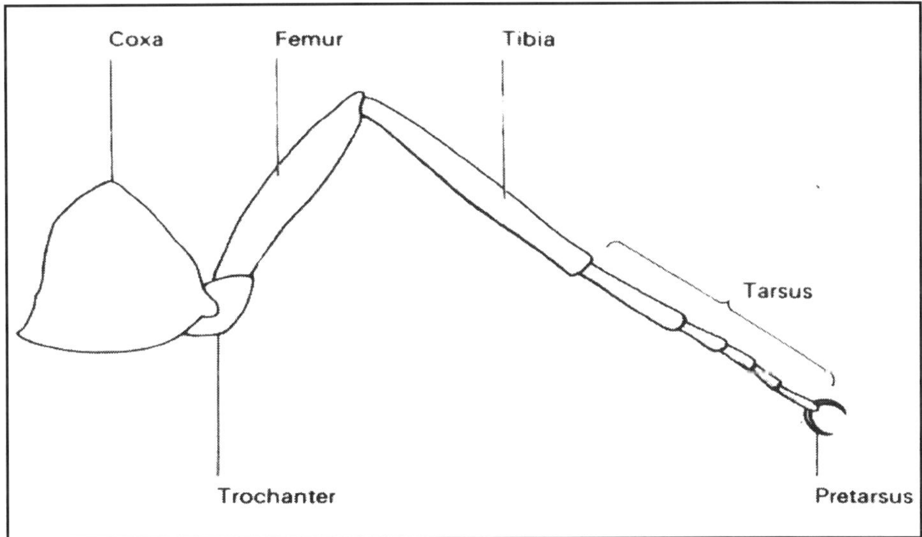

Fig. 199. Generalized insect leg.

Preventative: A measure applied in anticipation of pest attack.

Preventive practice: A pest management approach that attempts to prevent problems.

Prey: An animal eaten for food by a predator. A food item for a predator.

Prgonathous: Anteriorly directed mouthparts (Fig. 200).

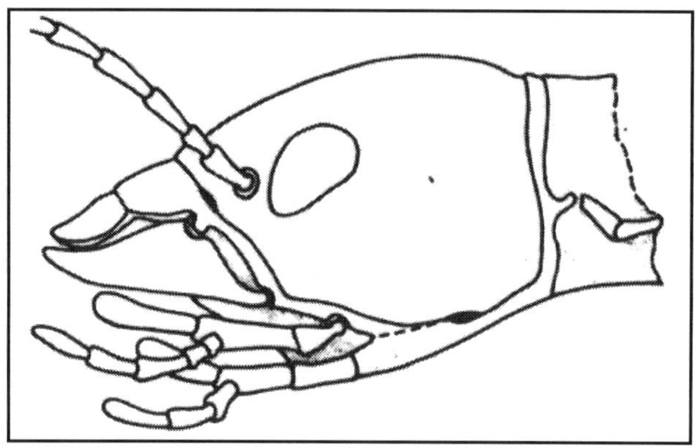

Fig. 200. Position of mouthparts relative to the head capsule: Prognathous.

Primary cycle: In a disease, the cycle that involves the typical host; *see also* **Secondary cycle.**

Primary defense: A defense mechanism that is continuously present, such as crypsis.

Primary homonym: One of two more identical trivial names which, at the time of original publication, were proposed in combination with the same (or an identical) generic name (*e.g., X-us albus* Smith, 1910, and *X-us albus* Jones, 1920); the later of such primary homonyms are to be permanently rejected; also on eof two or more identical names for genera or higher categories (cf. Homonym, Secondary homonym).

Primary host: The main host for a particular disease organism or crop pest.

Primary infection: The initial introduction of a disease into a crop by insect vectors invading from elsewhere.

Primary inoculum: Pathogen propagules that initiate the first infection during a season.

Primary metabolite: A metabolic product necessary for growth and reproduction of a plant.

Primary oocyte: Oocyte that is located proximally in an ovariole and that is ready to undergo vitellogenesis.

Primary orientation: Assumption and maintenance of the body in space.

Primary parasite (as applied to insects): One which is parasitic on a plant feeding, predatory or scavenging species but never on another parasite.

Primary reproductive: In termites, the king and queen founders of a colony.

Primary segmentation: The arrangement of body units such that intersegmental grooves from the attachment points for longitudinal muscles.

Primary sexual characters: The gonads; the ovaries in females, the testes in males (c.f. Accessory sex characters, Secondary sexual characters).

Primary spermatocytes: Diploid cells that will undergo meiotic cell division and eventually form into spermatozoans.

Primer pheromone: A pheromone that acts to modify the physiological state of an animal.

Pro: Before.

Proboscis (alternative term, rostrum)**:** The extended tubular mouthparts (beaklike) of sucking insects through which food is ingested insect; particularly of the piercing / sucking pattern.

Procephalon: In embryology, the anterior head formed by fusion of the primitive anterior three segments.

Proctodael feeding: Ingestion by one individual of material from the anus of another individual.

Proctodaeum: The hindgut of insects; plays important roles in elimination and water and salt resorption.

Proctodeal invagination: Ecotodermal invagination that develops into the hindgut.

Proctodeal trophallaxis: In social and subsocial insects, the transfer of nutritive fluid from the anus to another individual; may be mutual or unidirectional.

Proctodeum (adj. proctodeal): see **Hindgut.**

Procuticle: The inner zone of the insect cuticle, containing chitin and protein, divisible into exocuticle and endocuticle.

Producer: (i) Farmer or person who grows crops. (ii) Ecological term for green plants that form the base of the food chain.

Profundal zone: The bottom and deepwater area beyond the depth of effective light penetration.

Proglottis (pl. progloitides): A series of segments of a tape worm; each segment has its own set of sexual organs. Ripe proglottods full of mature fertilized eggs and capable of reproduction without any assistance from the adult tapeworm.

Prognathous: Having the head horizontal and the mouth parts projecting forward; *see also* **Hypognathous; Opisthognathous.**

Prognosis: Anticipation of pest problems before they occur and prediction of the outcome of these if left unattended.

Progressive provisioning: The supplying of food to the offspring over time, as the offspring grows.

Prokaryote: Microorganisms that lack membranc-bound organelles (e.g., bacteria and mollicutes).

Proleg: A fleshy unsegmented abdominal leg found in caterpillars (Lepidoptera, sawflies), bearing a characteristic arrangement of crochets; a fleshy, unjointed "false leg," occurring ventrally on the abdomen of caterpillars and other larval insects; bilateral abdominal legs characteristic of lepidopteran and many hymenopteran larvae (Fig. 44).

Pronotum (pl. pronota): The dorsal, often shield like sclerite or surface of the prothorax.

Propagative virus: Viruses that increase in numbers inside their insect vectors.

Propagule: A part of an organism that can reproduce the organism, such as spores, sclerotia, seeds, tubers, and root fragments.

Propalis: A resinous substance collected by honeybees from tree buds and used as a cement for nest damage repair.

Propodeum: First abdominal segment of Hymenoptera-Apocrita; it is completely associated with the thorax and anterior to the petiole; in apocritan Hymenoptera, the first abdominal segment is fused with the thirax to form a **mesosoma** (or **alitrunk** in ants).

Propolis: A resinous material or bee- glue of vegetable origin used by bees in addition to wax for the building and repair of their nests.

Proprietory name: Distinguishing name given by the manufacturer to a particular formulated product.

Proprioceptors: Sense organs that responds to the position of body organs; a mechanoreceptor that provides information about an insect's position in its environment and/or the position of the insect's body parts; a sense organ that detects the relative position of parts of an animal's own body.

Prosoma: The foremost body region of arachnids (spiders, mites, ticks) to which the walking legs are attached.

Prostomium: See acron.

Prot-, proto: First; primary.

Protectant: A chemical applied to the target before pest numbers increase.

Protective clothing: Clothing to protect the spray operator from the toxic effects of crop protection chemicals. This may include rubber gloves, boots, apron, respirator, face mask, etc.

Protein epicuticle: The innermost layer of the epicuticle (inner epicuticle).

Protelean parasite: An entomophagous insect that attacks its prey only when the attacking insect is immature, the adult being free living.

Proteolysis: The spiltting of proteins into smaller components by the breaking of peptide bonds.

Prothoracic gland hormone: See ecdysone.

Prothoracic gland: An endocrine organ in the prothorax of immature insects, also called ecdysial gland, that secretes ecdysone (Fig. 201, 202, 203); endocrine organs located in the prothorax of immature insects, secreting molting hormone.

Prothoracicotropic hormone (PTTH): A hormone secreted by neurosecretory cells of the brain and serving to activate the prothotacic glands that controls aspects of moulting and metamorphosis via action on the **corpora cardiaca.**

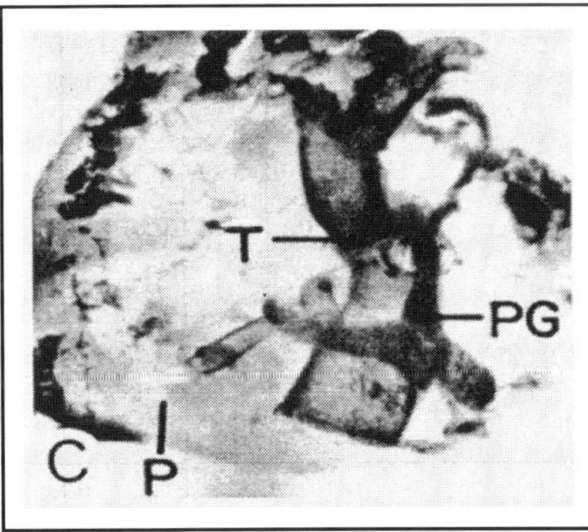

Fig. 201. Endocrine gland: Whole mount dissection of southwestern corn borer, *Diatraea grandiosella,* showing the prothoracic gland (PG), and the associated tracheal trunk (T).

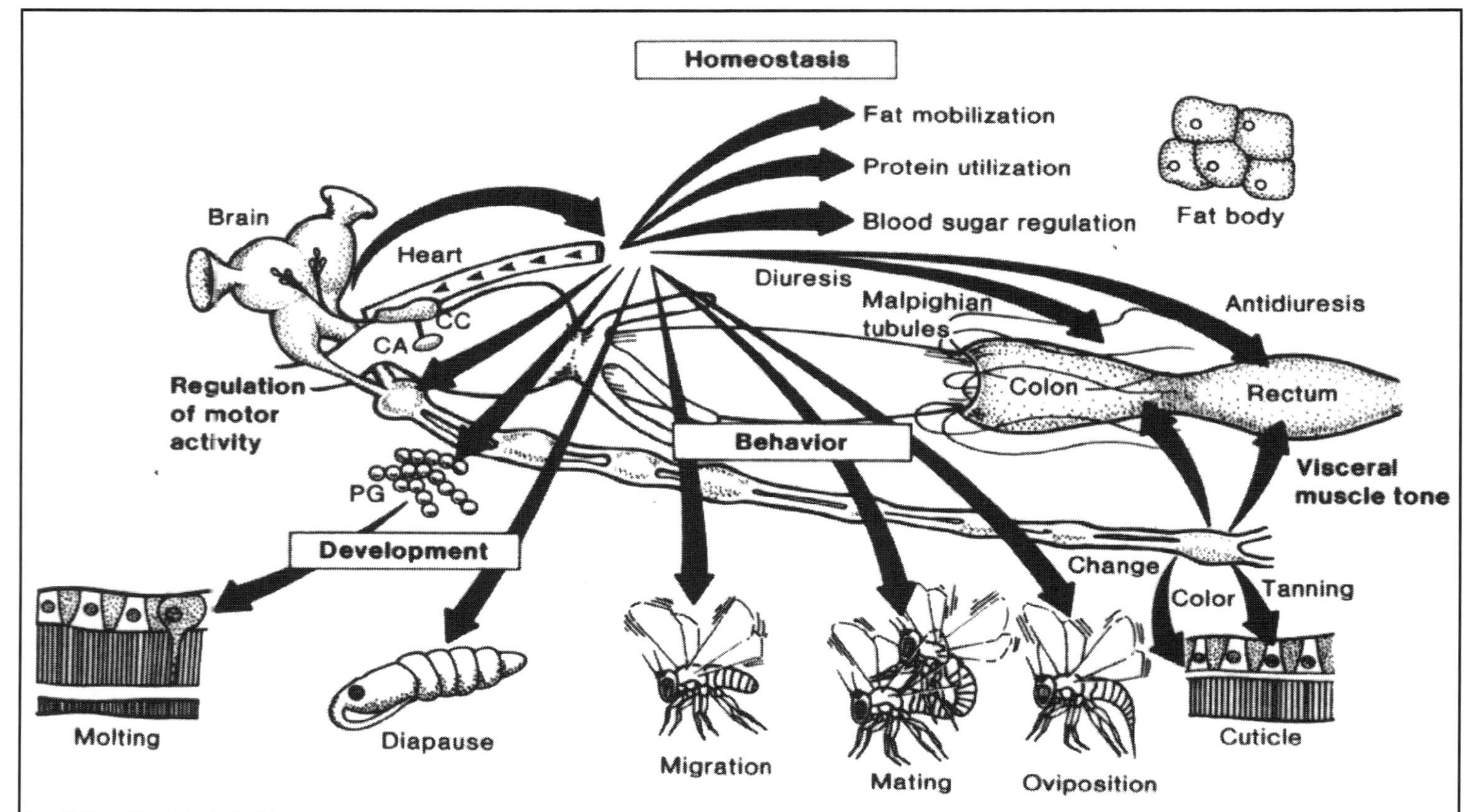

Fig. 202. Major physiological functions regulated by neurohormones in insects. Hormonal distribution is accomplished chiefly by release into the hemolymph, but localized secretion from nerves does occur. CC = corpora cardiaca; CA = corpora allata; PG = prothoracic gland. After Cook and Holman, 1985.

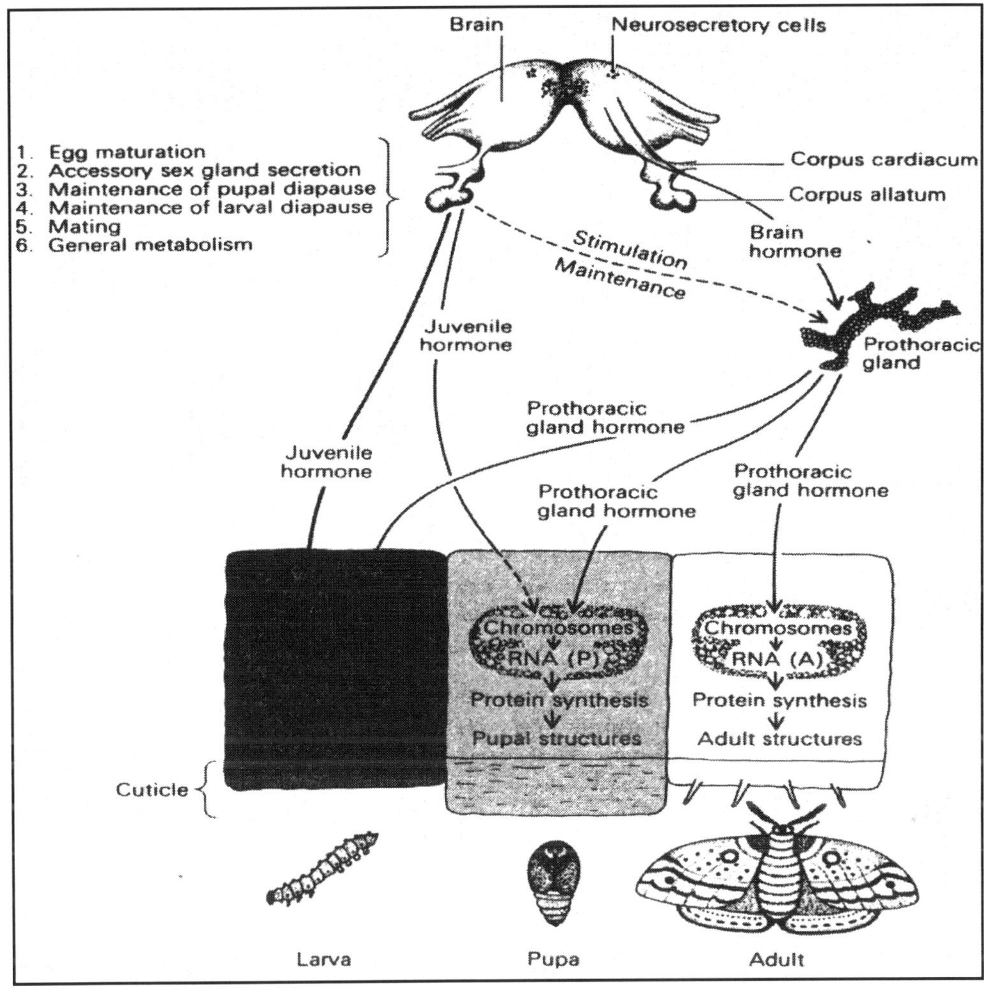

Fig. 203. Principal endocrine tissues in the giant silkworm moth, *Hyalophora ceropia*. After Gilbert, 1964 (Gilbert, personal communication, 1992).

Prothorax: The first, or the most anterior, of the three segments of the thorax.

Prothoracic gland: A gland situated in the fore part of the insect thorax which secretes moulting hormone.

Protoandry: Production of sperms and later of eggs, gonad. (Gr. *proteros* = first + *andros* = main).

Protocephalon: Bilaterally expanded anterior region of the germ band in an insect embryo (Fig. 204).

Protocerebrum: Dorsal lobe of the insect brain, contains optic lobes and other major cell masses and regions of neuropile; the anterior pair of lobes of

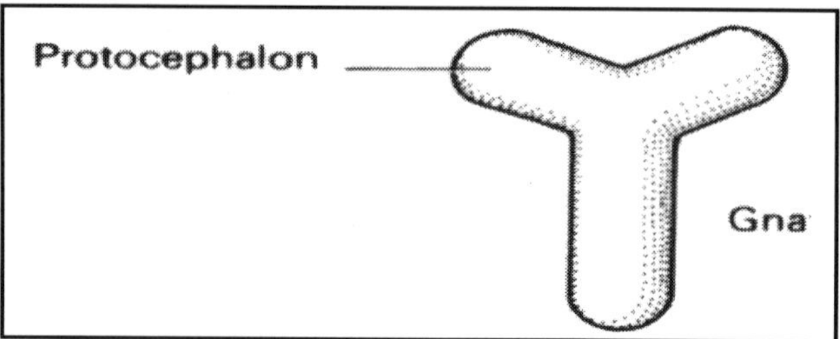

Fig. 204. Segmentation and appendage formation: Prior to segmentation showing Protocephalon.

the brain that innervate the compound eyes and ocelli; the largest and most anterior part of the brain, which includes the optic lobes.

Protolean: A parasite parasitoid; parasitic when immature; free-living as adult.

Protonymph: The second instar in the development of mites and ticks. cf. deutonymph.

Protoplasm: (Gr. *proctos* = first + *plasma* = form). Living substance; the complex colloidal physicochemical system that constitutes living matter and is the viscid, semifluid material of animal and plant cells.

Protoplasmic poisons: Substances that kill by precipitating intercellular proteins.

Protozoa: A phylum of unicellular animals, most free-living, some parasitic.

Protozoology: The science which deals with the study of protozoa (cf. Microbiology).

Protraction: Withdrawal, the converse of extension.

Protrusible (or **exsertile) vesicle:** A small sac or bladder, capable of being extended or protruded.

Protura, order: (*prot,* first; *ura,* tail) The order of insects comprising the telsontails or **proturans** (Fig. 205). It is a primitive order of small wingless insects for which there is no common name.

Proventriculus (also **gizzard):** The grinding organ of the foregut; a structure posterior to the crop that bears sclerotized teeth or spines; the stomodael valve in Diptera; the posteriormost portion of the foregut, just before the midgut, that controls entry of food into the midgut; often lined with sclerotized teeth that grind the food.

Proximal: Adjacent to or nearest the point of attachment or origin; referring to the part of an appendage that is closest to the body; the part of an appendage closer to or at the body (opposite to distal).

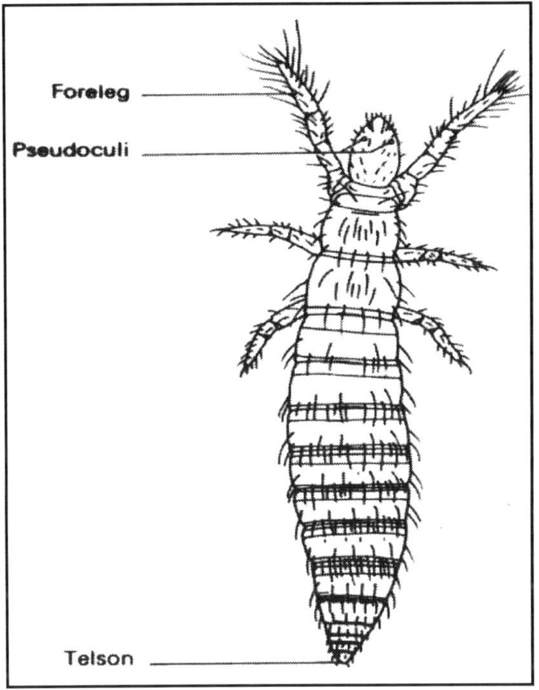

Fig. 205. Proturan, *Acerentulus barberi barberi* Ewing.

Prune: To selectively remove unwanted branches from plants.

Pseudergate: In 'lower' termites the equivalent to the worker caste, comprising immature nymphs or undifferentiated larvae; last instar worker nymphs which retain the potential to moult into different castes.

Pseudo: False; temporary.

Pseudocoel: A body cavity not lined with peritoneum and not part of a blood - vascular system as in nematodes and some other invertebrates. Compare Coelom, Hemocoel. (Gr. *pseudes* = false + *koiles* = hollow).

Pseudocopulation: The attempted copulation of an insect with a flower.

Pseudoplacental viviparity: Viviparity (producing living offspring) in which a **microlethical** egg develops via nourishment from a presumed placenta.

Pseudopod: False leg; pseudopodium; locomotory protuberances on the legless larvae of some Muscidae; false legs on the abdomen of the larvae of butterflies, moths and sawflies (Fig. 44).

Pseudopupa: The larva of certain Coleoptera in a quiescent condition preceding the true pupal stage.

Pseudo-tracheae: Fine tubes which traverse the pad-like foot of the sponging mouthparts of flies; the ridged grooves on the ventral surface of the labella of some adult Diptera; used for absorbing liquid food materials.

Psi: A measure of pressure, pounds per square inch.

Psocoptera, order: (*psoco,* rub small; *ptera,* wings) The order of insects comprising the booklice, barklice and relatives (Fig. 206).

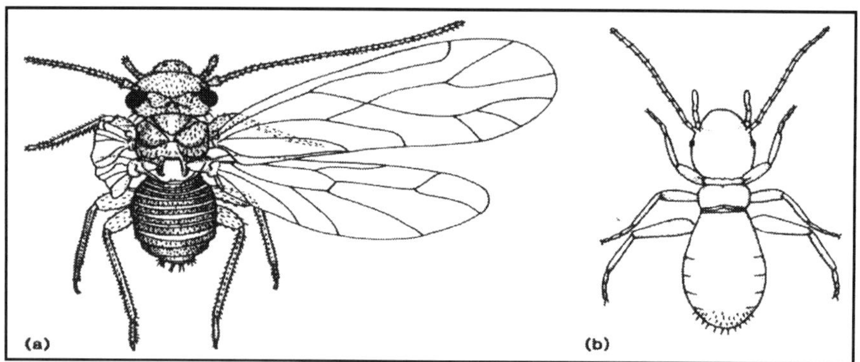

Fig. 206. Psocoptera. (a) Winged species, *Peripsocus californicus.* (b) Wingless species, *Troctes divinatorius.*

Psychophily: Plant pollination by butterflies.

Psyllidophagous: A parasite or predator that feeds on Psyllidae; host-specific to the Psyllidae.

Psyllose: Infestation symptoms of cotton plants after attack by Cotton Psyllids.

Pterins: Red, yellow, and white pigments commonly found as insect eye pigments.

Ptero: Wing.

Pterostigma: A pigmented (and denser) spot near the anterior margin of the fore and sometimes hind wings.

Pterostigma: A thickened opaque or dark spot along the costal margin of the wing, near the tip (e.g. Odonata, Hymenoptera); a pigmented thickened spot near the anterior apex of the wings of Odonata and Mecoptera, and on the forewings of Hymenoptera and Psocoptera.

Pterothorax: Collective term describing the wing-bearing segments of the thorax in pterygotes.

Pterygota (alternative term, Metabola)**:** The major subclass of insects which undergo metamorphosis whether complete or incomplete and which usually possess wings in the adult stage. cf. Apterygota.

Ptilinum: A bladderlike organ that when inflated extends from the front of the head of the adult of some species of flies and ruptures the puparium thereby facilitating emergence; a sac everted from a fissure between the antennae of schizophoran flies (Diptera) that aids puparium fracture at emergence.

Pubescent: Covered with fine hairs. Downy, covered with short fine hairs.

Pulvillus (pl. pulvilli) (also **euplantula**)**:** Ventral pad on some **tarsomeres** of leg; a pad at the base of each tarsal claw; padlike structure on the lower surface of tarsi in some Orthoptera or in association with each pretarsal claw in Diptera; used for gripping the substratum.

Pupa (pl. pupae) (alternative term, pupal stage)**:** The stage in the life cycle between larva and adult in insects which undergo complete metamorphosis, a nonfeeding, usually inactive stage; transitional, metamorphic, inactive non feeding stage of holometabolous insect between the larval and adult stages during which morphological changes occur (Fig. 207) also termed a **chrysalis** in butterflies.

Pupariation: Formation of the puparium by larvae of Diptera.

Puparium (pl.puparia): Specialized form of pupa in which the last larval skin is retained as an extra covering; a protective enclosure of the final instar larva (higher Diptera, Strepsiptera) or last nymphal instar (Aleyrodidae), in which the pupa forms; the exuvium of the third (last) larval instar in higher Diptera within which the pupa develops.

Pupate: Molt from larval to pupal stage; transform to a pupa.

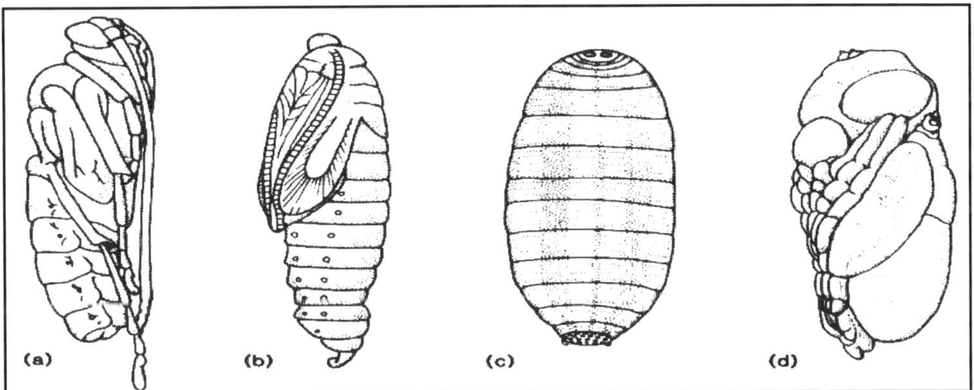

Fig. 207. Pupal types. (a) Exarate, ichneumon wasp. (b) Obtect, moth. (c) Coarctate, house fly. (d) Puparial case removed, exposing exarate house fly pupa. After Boreal Labs key card; after Packard, 1898; after Wigglesworth, 1970.

Pusa bin: A specialized storage structure for grains developed by scientists of Pusa Institute of New Delhi, in which a polythene sheet is inserted in the eathern cell structure.

Pustule: A small elevated blister-like area on a leaf where fungal spores are produced.

Pycnidium: Flask-shaped fungal fruiting structure of some *Fungi Imperfecti* that release spores called conidia when moistened.

Pygidial glands: Exocrine glands that secrete a repugnatorial substance and that are located at the posterior end of the abdomen in some species of beetles.

Pygiduim: The tergum of the last visible segment on the insect abdomen.

Pygopodia: Eversible accessory locomotor appendages that arise from the terminal abdomen segment in some hexapodous larval forms.

Pyloric valve: The valve at the junction of the midgut and hindgut; valvular structure at the anterior extremity of the insect hindgut.

Pylorus: The anterior hindgut where the Malpighian tubules enter, sometimes indicated by a muscular valve.

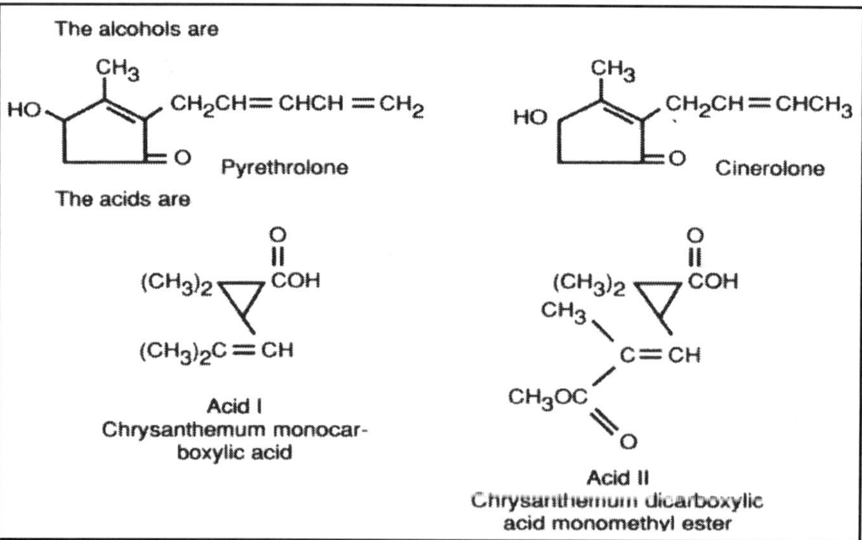

Fig. 208. Pyrethum compounds consist of four esters, which are the combinations of two different alcohols with two different acids. The proportions may vary with the strain of flowers, growing conditions and method of extracting and concentrating these toxins.

Pyrethrin: One of the insecticidal chemicals present in the plant pyrethrum.

Pyrethroid (synthetic pyrethroid): An organic synthetic insecticide with a structure based on that of pyrethrum (e.g., allethrin).

Pyrethroid: Class of insecticides with chemical structure similar to pyrethrum; synthetic chemicals with similarity in structure to pyrethrins.

Pyrethrum: Botanical insecticide extracted from *Chrysanthemum* flowers (Fig. 208).

Q

Quadrangle: A cell immediately beyond the arculus (Odonata, Zygoptera).

Quadrat: A four-sided defined area from which a sample is collected, used in population determination.

Quadrivoltine: Four life cycles per year.

Qualitative defenses of plants: Toxins and small-molecular-weight compounds (such as alkaloids) that are active against the physiological systems of phytophagous insects; complex, digestibility-reducing substances (such as tannins) that reduce the ability of insects to feed on plants.

Quantitative specificity: In serology, the principle that a given kind of antibody will react more strongly, under comparable conditions, with the particular kind of antigen used in its formation than with any other substance (cf. Antibody, Antigen, and Precipitin reaction).

Quarantine: Restrictions to prevent pest spread from infested to non-infested regions; All operations associated with the prevention of importation of unwanted organisms into a territory, or their exportation from it; a legal instrument to prevent or slow the spread of pests from infested into uninfested areas by restricting movement of infested stock.

Quasisocial: Social behavior in which individuals of the same generation co-operate and nest-share without division of labour.

Queen: Female sexually reproductive form of social insects such as termite (Fig. 155); a female belonging to the reproductive caste in eusocial or semisocial insects called a **gyne** in social Hymenoptera.

Queen substance: A pheromone produced by the queen honey bee and serving various functions in the hive as well as during mating and swarming flights; a secretion from the queen honey bee which inhibits sexual development of workers and regulates their building activities.

Quiescence (torpor): Inactivity induced by unfavorable environmental conditions; activity resumes immediately on return of favorable conditions; a slowing down of metabolism and development in response to adverse environmental conditions; *see also* **Diapause.**

Quinone: A benzene derivative in which two hydrogen atoms are replaced by two oxygen atoms. Quinones are utilized in the hardening and darkening of the insect cuticle.

R

Race: A subspecific category that is based on particular attributes, such as behavioral or physiological (other than morphological) differences; similar to a subspecies; often used in describing plant pathogens.

Race: Subspecies (q.v.).

Race: A naturally occurring group within a species usually visually indistinguishable but with some physiological difference from other members of the species, e.g., with a different plant host range. cf. strain.

Radial cell: A cell bordered anteriorly by a branch of the radius; the marginal cell (Hymenoptera).

Radial cross vein: A cross vein connecting R_1 and the branch of the radius immediately behind it.

Radial sector: The posterior of the two main branches of the radius.

Radiation: See Adaptive radiation.

Radius: In wing venation, the fourth longitudinal vein, posterior to the subcosta; with a maximum of six branches.

Radula: The rasp-like tongue of slugs and snails.

Rake: The process on the distal apex of the tibia of a bee that gathers pollen into the press.

Random sample: Sample collection wherein every member of the population has an equal probability of being selected.

Range: The difference between the smallest number and the largest number in a sample.

Rank: The level in a taxonomic hierarchy, e.g. species, genus, family, order.

Raptorial (legs): Forelegs modified for grabbing and holding prey (Fig. 209).

Raptorial: Adapted for capturing prey by grasping; raptorial front legs.

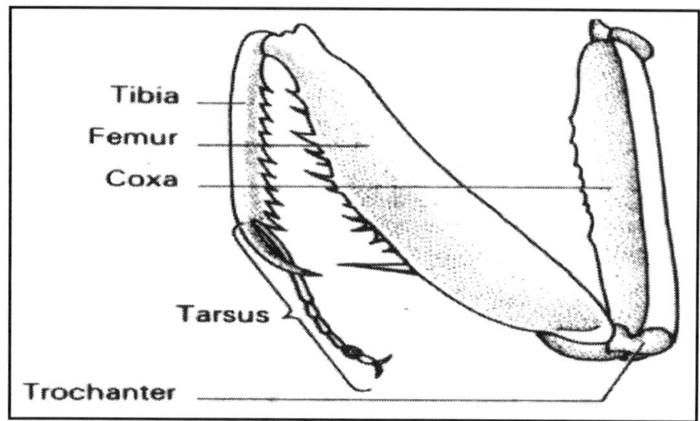

Fig. 209. Raptorial (grasping) forelegs of a praying mantis.

Rassenkreis (Rensch)**:** A polytypic species composed of several subspecies (cf. Subspecies, Polytypic).

Rate: Amount of specific chemical applied per treatment unit, usually area.

Reaction chain: A continuous series of behavioral acts, each of which is dependent on completion of the preceding act.

Recapitulation: The theory that ontogeny recapitulates phylogeny (cf. Ontogeny, Phylogeny).

Receptor apparatus: Portion of an ommatidium of a compound eye in which light is converted to a nervous impulse.

Receptor cells: Portion of a sense organ in which stimuli are converted to a nervous impulse (sense cell).

Receptor fibril: The portion of a neuron that transmits impulses toward the cell body.

Recessive character: A character from one parent that remains undeveloped in offspring when associated with the corresponding dominant character from the other parent. (L. *recessus* = a going back).

Recessive characteristic: A characteristic governed by an allele that does not express itself in the presence of another allele of the same gene. Recessive traits are only expressed in homozygous individuals.

Recessive: A character that is expressed in the phenotype only when the individual is homozygous for the gene producing it (cf. Dominant, Homozygous).

Reclinate: Inclined backward or upward.

Recommendation: A suggestion from, or device given by, a farm advisor, extension specialist, pest control advisor, or other agricultural authority that relates to the growing of a crop and the control of pests.

Recruitment: The addition of new individuals to a population, usually by either birth or immigration.

Rectal pad: Thickened sections of the epithelium of the rectum containing enlarged cells, responsible for active water and ion uptake from the faecal contents of the rectum.

Rectal papillae: Pads projecting into the lumen of the rectum that play an important role in the excretory system.

Rectum (adj. rectal)**:** Muscular, enlarged, posteriormost region of the hindgut; the termninal enlarged portion of the digestive tract of insects (and other animals) (L. *reetus* = straight).

Recurrent nerve: Neural connection between the frontal ganglion and the hypocerebral ganglion.

Recurrent vein: One of two transverse veins immediately posterior to the cubital vein (Hymenoptera); a vein at the base of the wing between the costa and subcosta, extending obliquely from the subcosta to the costa (Neuroptera).

Recurved: Curved upward or backward.

Redistribution: Movement of pesticide subsequent to the initial application to other parts of the plant, usually by rain.

Reduced rate: A lowered rate of pesticide application to permit survival of beneficials.

Reduction division: That division of the maturing germ cells by which the somatic or diploid number of chromosomes is reduced to the haploid number.

Reduviid: A member of the Reduviidae (hemiptera, Heteroptera).

Reentry interval: The time required after a pesticide application so that workers can go back into the area and be safe without wearing personal protective clothing or equipment.

Reflex action: Action resulting from an afferent sensory impulse on a nerve center and its reflection as an efferent motor impulse independent of higher nerve centers or the brain. An automatic response to a stimulus. (L. *re* = back + *flecto* = bend).

Reflex bleeding: Release of hemolymph in response to a threat.

Reflex: A functional unit of the nervous system involving the entire pathway from a receptor cell to an effector.

Reflex: A simple response to a simple stimulus.

Reflex: An involuntary movement or other response elicited by a stimulus applied peripherally, conducted through the central nervous system, and conducted back peripherally.

Refractory period: The time interval during which a nerve will not initiate another impulse; in reproduction, the period during which a mated female will not re-mate.

Refuge: A safe place.

Regeneration: Replacement of parts lost through mutilation or otherwise.

Regenerative cells: Replacement cells at the bases of the midgut epithelial cells.

Registered nursery stock: Plant propagation materials registered by a government agency as free from particular pests. See *certified stock.*

Registered pesticide: A pesticide approved by the U.S. Environmental Protection Agency (EPA) for use as specified by the label.

Regles Internationales de la Nomenclature Zoologique: The International Rules of Zoological Nomenclature adopted by the Fifth International Congress of Zoology at Berlin (1901) and subsequently amended at succeeding congresses. The French text is the official version.

Regression: The amount of change in one variable associated with a unit change in another variable.

Regulated pest: An organism designated by a state or fedral agency as a pet requiring regulatory restriction to protect the host, the public, and the environment.

Regulation (population regulation): The natural phenomenon of population numbers fluctuating within certain bounds for long periods of time.

Regulatory control: The use of enforceable regulations to prevent the spread of a pest or to suppress or eradicate it.

Rejected name: An otherwise available name which has been permanently rejected by the International Commission on Zoological Nomenclature under the plenary powers (cf. Official Index of Rejected and Invalid Specific Trivial Names in Zoology, Official Index of Rejected and Invalid Generic Names in Zoology).

Relapsing fever: A disease caused by *Borrelia* species of bacteria and transmitted by body lice.

Relative population estimate: An estimate of population size based on the kind of sampling technique used (e.g., number per trap).

Relative variation (RV): Percentage standard error of a mean; used as a measure of sampling precision.

Release: Introduction of biocontrol agents, particularly insects, onto a particular site; specifically, to stimulate a particular behavior.

Releaser pheromone: A pheromone that acts via the central nervous system to produce a quick behavioral response.

Releaser: A particular stimulus whose signal stimulates a specific behavior; an environmental or communicative stimulus that triggers a fixed action pattern.

Releasing mechanism: An innate capacity to respond in a particular way to a specific stimulus.

Remigium: The anterior part of the wing, usually more rigid than the posterior **clavus** and with more veins.

Renal: Pertaining to a kidney. (L. *renes* = kidneys).

Reniform: Kidney-shaped.

Repellant: A chemical which has the property of including avoidance by a particular pest; or material that deters pests from moving to plants or animals; or a chemical that causes insects to orient their movements away from a source; a chemical or physical source which induces insects to move away from it. cf. attractant.

Replacement (replacement pest; secondary pest outbreak): When a major pest is suppressed and continues to be suppressed by a tactic but is replaced in importance by another pest, previously with minor status; or a minor pest species becomes a key pest after a control tactic targeting a different major pest species allows the minor pest to increase in number and cause damage.

Replacement cells: See regenerative cells.

Replete: An individual ant which is distended by liquid food.

Replicate: Specifically of disease organisms, to increase in numbers.

Reproduction: The maintenance of a species from generation to generation.

Reproductive isolation: A condition in which interbreeding between two or more populations is prevented by intrinsic factors (cf. Geographic isolation, Ecologic isolation, isolating mechanism).

Reproductive potential (alternative term, biotic potential)**:** The maximum possible rate of increase of an organism in the absence of any limiting factors.

Reservoir host: See *alternative host.*

Reservoir: A site where organisms can survive or increase in number; of diseases, the natural host and geographical range.

Resident population: A pest population in a defined area such as a field or an orchard.

Residual pesticide: A pesticide that remains in the environment for a relatively long time. The pesticide may continue to be effective for weeks, months, or years.

Residual properties: The ability of a pesticide to remain effective over time.

Residue: (i) The amount of toxicant remaining at the target site after application. (ii) Plant debris that remains in the field after harvest. (iii) Amount of pesticide remaining in or on plant tissues (or in soil) after a given time, or in produce especially at harvest time.

Resilin: A cuticular protein with rubber like or elastic properties in some insect; sometimes found in pure form in skeletal articulations.

Resistance (of insects to insecticides): The developed ability of a strain within an insect species to withstand an insecticide to which it was formerly susceptible.

Resistance (of plants to insects): See plant resistance.

Resistance: (i) Genetically determined capacity of a plant to inhibit pest attack, pest reproduction, or injury due to a pest. (ii) The ability of a significant portion of a pest population to survive a pesticide at rates that once killed most individuals of that population; or the ability of a strain to tolerate or avoid factors that would prove lethal or reproductively degrading to the majority of strains in a normal population; or the natural or induced capacity to avoid or repel attack by pests (or parasites). Also the ability to withstand the toxic effects of a pesticides or a group of pesticides, often by metabolic detoxification.

Resonating muscles: Muscles capable of undergoing several successive by a single nerve impulse; usually associated with insect flight and sound production; also called asynchronous muscles.

Resource tracking: A relationship, for example between parasite and host or plant and pollinator, in which the evolution of the association is based on ecology rather than phylogency; *see also* **Coevolution; Phyletic tracking.**

Respiration: A metabolic process in which substrates (food) are oxidized using molecular oxygen; also used inappropriately to mean breathing, as through spiracles or gas exchange across thin cuticle; intracellular oxidative breakdown of carbohydrate to carbon dioxide and water; the release of energy by the oxidation of fuel molecules within a cell; obtaining oxygen from the surrounding medium and giving of carbon dioxide. (L. *re* = *back* + *spiro* = breathe).

Respirator: A face mask that filters poisonous gas and particles from the air, enabling a person to breathe and work safely despite the presence of a toxicant. Used to protect the nose, mouth, and lungs from pesticide poisoning.

Respiratory exposure: Pesticide inhalation.

Resting potential: The slight charge that can be measured in an unstimulated nerve cell.

Restricted use pesticide (restricted material): In the United States, a pesticide that may only be applied by applicators certified by the state in which they work, because the EPA has determined that the pesticide is harmful to the environment or hazardous to workers; or a pesticide that may be applied only by applicators certified by the state in which they work.

Resurgence (pest resurgence): A situation in which a population, after having been suppressed following application of an insecticide, rebounds to numbers higher than before suppression occurred.

Reticulate evolution: Evolution "dependent on repeated intercrossing between a number of lines, and thus both convergent and divergent at once" (Huxley).

Reticulate: Net-like; meshed; a network of lines.

Reticulo-punctate: A descriptive term for cuticle sculpturing; a combination of a network of lines with a series of regular punctures.

Retina: Portion of a dorsal ocellus in which light is converted to a nervous impulse (retinular cells); or the cell layer of an eye containing the receptors of light impulses (L. *rete* = net).

Retina: The receptive apparatus of an eye.

Retinaculum: Particularly used of the specialized hooks or scales on the base of the forewing that lock with the **frenulum** of lepidopteran hind wings in flight; also the retaining hook of the springtail **furcula** (spring).

Retinula cell: A sensory cell of the light receptor organs (**ommatidium, stemma** or **ocellus**) comprising a rhabdom of rhabdomeres and connected by nerve axons to the optic lobe; one of a group of cells comprising the receptor unit of an ommatidium; or a monopolar sensory neuron within an ommatidium of the compound eye.

Retinula: The lower light sensitive region of an ommatidium in the insect compound eye.

Retinular cells: Cells in photosensitive organs in which light is converted to a nervous impulse.

Retractile: Capable of being pushed out and drawn back in.

Reversion: The reappearance of ancestral character traits or that have been in abeyance for one or more generations. (L. *re* = back + *verto* = turn).

Review: Synopsis (q.v.).

Revision: In taxonomy, the presentation of new material or new interpretations integrated with previous knowledge through summary and reevaluation (cf. Synopsis, Monograph).

Rhabdom: A rodlike structure formed by the inner surfaces of adjacent retinula cells in an ommatidium ; or centrally located rod structures formed by contributions from each retinular cell; or the central, rod like element in an ommatidium, consisting of several rhabdomeres, one from each retinula cell.

Rhabdom: The central zone of the retinula consisting of microvilli filled with visual pigment; comprising rhabdomeres belonging to several different retinula cells.

Rhabdom: The dense rod-like structure lying below the crystalline cone in each ommatidium of the insect compound eye.

Rhabdomere: One of typically eight units comprising a rhabdom; portion of each retinular cell contributed to the formation of the rhabdom.

Rheophilic: Liking running water.

Rheotaxis: Directed response in relation to water current.

Rhizome: A horizontal underground stem that forms roots and buds to produce new plants.

Rhizomorph: A thick strand of fungal hyphae that can grow between roots and infect adjacent trees.

Rhizosphere: A zone in the soil surrounding the roots of plants, usually richer in fungi and bacteria than elsewhere in the soil.

Rhodopsins: Light-absorbing pigments involved with photosynthesis.

Rhythm: Cyclic happening of any event.

Rickettsia: An obligate intracellular bacteria like parasite of arthropods, many types of which are pathogenic for man and other animals. They are intermediate between bacteria and viruses.

Riker mount: A thin, glass-topped exhibition case filled with cotton.

Ring gland: Small tissue mass found in cyclorrhaphous Diptera that is thought to be homologous with the corpora allata, the corporacardiaca, and the thoracic glands.

Riparian: Riverside.

Risk: The probability that a pesticide will cause harm.

RNA (ribonucleic acid): A polynucleotide that is involved in protein synthesis in cells, it exists in several forms that have specific functions. It serves directly as the primary genetic information in some viruses.

Rocky Mountain spotted fever: A human disease caused by *Rickettsia rickettsi* and transmitted by ticks and bed bugs.

Rodent: Any animal of the order Rodentia, e.g., mice, rats, squirrels, gophers, and woodchucks.

Rodenticide: Poison used to kill rodents; or a toxicant effective against rodents.

Rogue: To remove undersirable plants from a crop.

Rootstock: The root system of a plant to which the scion is grafted or budded.

Rostellum: In tapeworms on the broader end of the head is a rounded prominence, the rostellum round the base of which lie the hiiks. Rostellum can be protrude or retracted to some extent.

Rostrum (alternative term, Beak or snout or proboscis)**:** A facial extension that bears the mouthparts at the end (Fig. 210); The extended forepart of the head of an insect consisting of or bearing the mouthparts. The term is applied to insects such as weevils in which biting mouthparts are at the tip of a snout as well as to insects with piercing/sucking mouthparts.

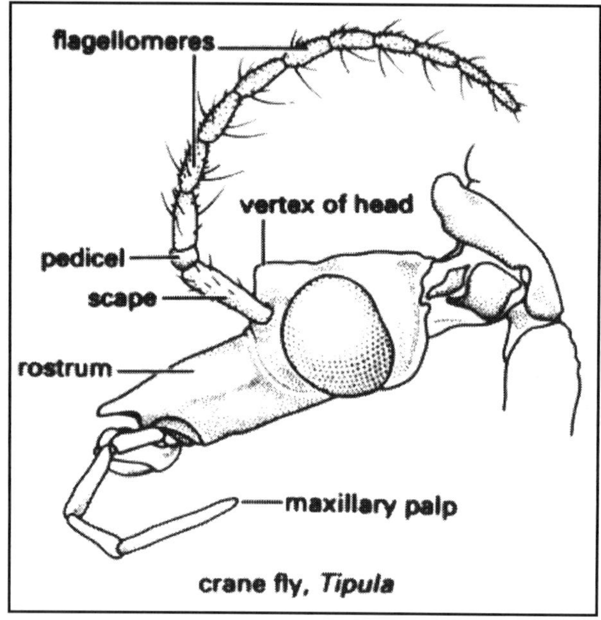

Fig. 210. Crane fly, *Tipula*: Rostrum.

Rostrum: Proboscis in the Hemiptera; the snout-like anterior extension of the head in the Curculionoidae.

Rot: Physical decay of plants and plant products caused by fungi and bacteria.

Rotation: See *crop rotation.*

Rotenone: A particular legume-derived botanical chemical with insecticidal and other toxic properties extracted from the roots of *Derris* and *Lonchocarpus* species of plants.

Rouging: The removal of unhealthy or unwanted plants from a crop.

Round dance: A communication dance of honey bees (Fig. 211).

Round dance: A form of recruitment in the honey bee, used when a food source close to the hive is communicated to other bees.

Royal jelly: A secretion from the hypopharyngeal glands of worker honeybees on which larvae are fed; or a nutritive substance produced by glands in the heads of worker honey bees and fed to the larvae. Larvae fed this diet throughout development produce queens.

R-strategist (r-strategist): A species characterized by having rapid development, high motility, and a high reproductive rate relative to a K-strategist; or a species characterized by high reproductive rate and

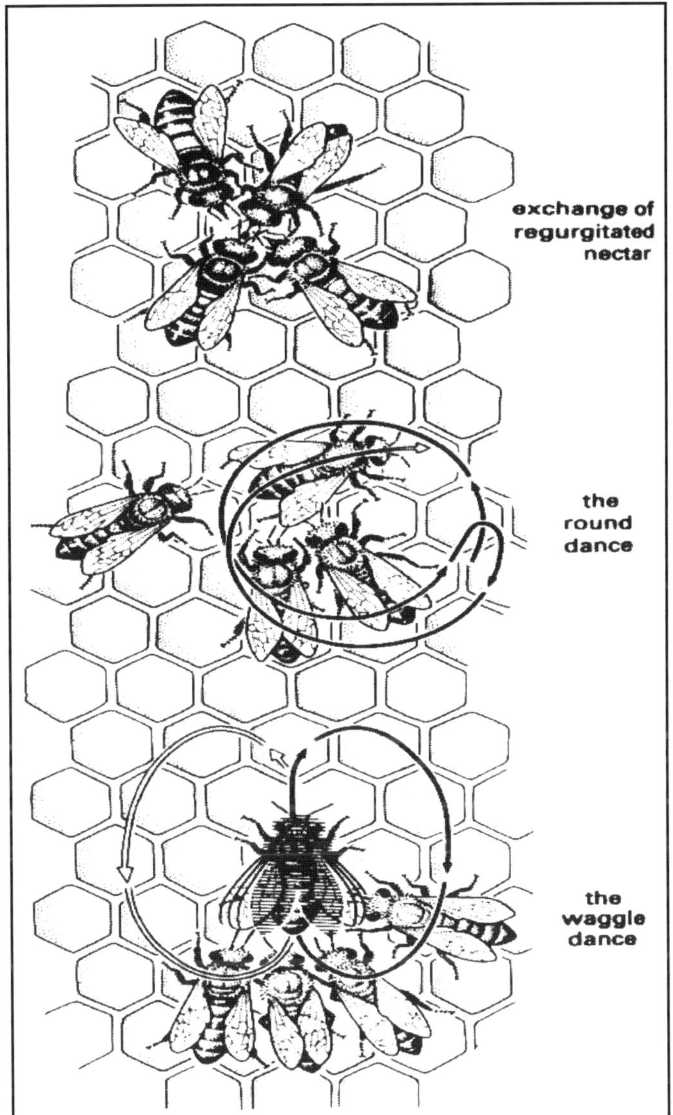

Fig. 211. The dance language of bees: Round dance.

low survival rate; an ecological term applied to organisms which produce large numbers of offspring but with poor survival. cf. **K** –strategist.

Ruderal: Plants growing in artificial habitats created by people (waste ground, roadsides, and others); in some sense considered weeds; secondary pioneers.

Rudimentary: Reduced in size, poorly developed, incompletely developed or having no function. Compare vestigial (L. *rudus* = unwronght).

Rufous: Reddish colored.

Rugose: Wrinkled.

Runner: See *stolon.*

Runoff: (i) The liquid that does not stay on the target being sprayed. (ii) Water draining off a field; or the process of spray shedding from a plant surface during and immediately after application, when droplets coalesce to form a continuous film and surplus liquid drops from the surface.

Rust: A fungal disease that cause symptoms of reddish brown or black pustules. Caused by fungi in the order Uredinales.

Ryania: A botanical alkaloid insecticide extracted from the stem and roots of the shrub *Ryania speciosa.*

S

Sabadilla: A botanical insecticide extracted from seeds of *Schoenocaulon officinale*

Safener: A chemical additive that reduces the phytotoxic properties of a pesticide.

Salivarium or **salivary reservoir:** The cavity into which the salivary gland opens, between the hypopharynx and the labium.

Salivary gland: The exocrine gland that produces saliva and typically associated with the labial segment; highly variable in size, structure, and function.

Saltation: Discontinuous variation produced at a single step by mutation (cf. Mutation).

Saltatorial (legs): Legs adapted for jumping or springing as a means of locomotion (Fig. 212).

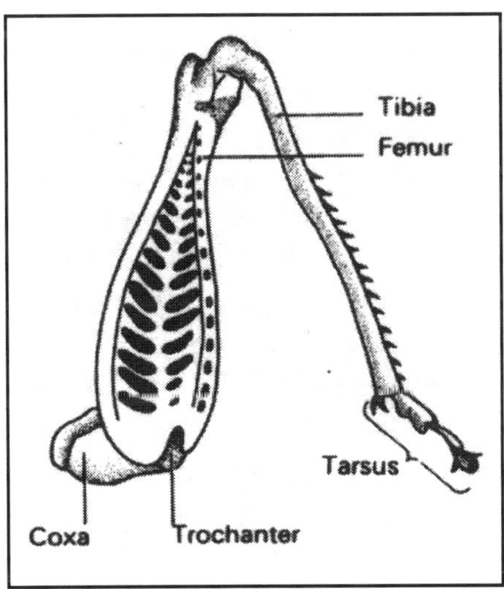

Fig. 212. Modification of insect leg: Salatorial (jumping) hindleg of a grasshopper.

Sample: That portion of a true population which is actually available to the taxonomist.

Sanitation: Destruction of crop and weed residues and cleaning of equipment in order to reduce pest inoculum.

Sap: The term for the liquid in the vascular tissue of plants; water and dissolved materials moving through the conductive tissues of plants.

Saprophage (adj. saprophagous): An eater of decaying organisms.

Saprophagous: Feeding on dead or decaying organic matter or on fungi.

Saprophyte: An organism that feeds on dead or decaying organic matter.

Saprozoic: Animal living on or in dead or decaying matter.

Sarcolemma: The outer sheath of a striated muscles fibre (Fig. 213).

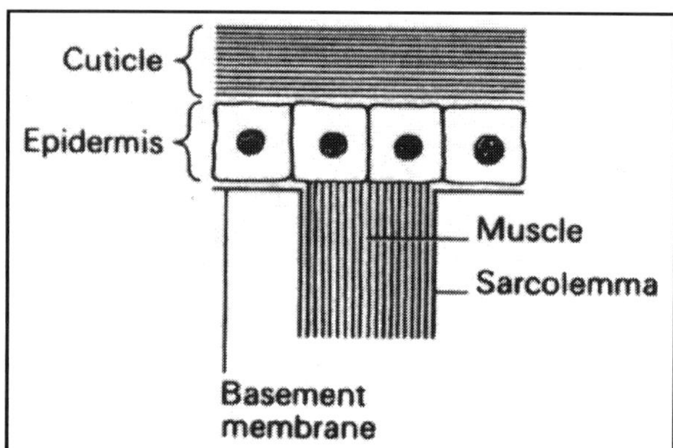

Fig. 213. Skeletal muscle attachments: Muscle attached directly to epidermal cells.

Sarcoplasm: The cytoplasm of striated muscle fibers.

Scab: Crust-like disease lision on the surface of a plant.

Scabies: A contagious skin disorder caused by burrowing of the itch mite *Sarcoptes scabiei.*

Scald: (i) An injury to the bark of a plant caused by frost or excessive drying by sun or wind. (ii) A fungal disease of cereals that causes leaf browning and plant death.

Scale: Flattened setae (unicellular outgrowths of the cuticle).

Scape: The first or basal segment of the insect antenna.

Scapula (also called **parapsis**) (pl., **scapulae**): One of two sclerites of the mesonotum immediately lateral of the parapsidal sutures (Hymenoptera).

Scarabaeiform larva: A grub-like larva, with a thickened cylindrical body, well-developed head and thoracic legs, without abdominal prolegs, and sluggish in behaviour. Oligopod larva; a white grub of the Scarabaeidae with a fat, fleshy C-shaped body and sluggish behaviour.

Scarification: To cut or abraid the seed coat to stimulate seed germination; scarring; appearing scratched; often applied to leaf surface after feeding by thrips or spider mites.

Scavenger: An animal that feeds on dead and decaying animals or plants or on animal wastes or faeces.

Scent gland: A gland producing an odorous substance.

Schizonl: A stage in the life cycle of malerial parasitic, following trophozoite stage of sporozoan paraite and reproduce in the host by multiple fission.

Schizont: The second stage in the asexual cycle of a malaria parasite (*Plasmodium* spp.), produced by the **pre-erythrocytic schizogonous cycle** or by division of a **trophozoite**.

Schwann cell: A cell surrounding the axon of a scolopidium.

Scientific name: A Latinized name, internationally recognized, of a species or subspecies. The scientific name of a species consists of the generic and specific name and the name of the describer of the species, and that of a subspecies consists of generic, specific, and subspecific names and the name of the describer of the subspecies. Scientific names (excluding author's names) are always printed in italics, the binominal or trinominal designation of an animal; the formal nomenclatural designation of a taxonomic category (cf. Vernacular name).

Scion: The part of the plant above the graft union and rootstock in vegetatively propagated crops; the designated variety.

Sclerite: A hardened plate forming part of the exoskeleton of insects; any piece of skeletal cuticle bounded by either sutures or membranes.

Sclerotin (s): A protein that contributes to the hardness of the cuticle; or cuticular protein that has been hardened and darkened through cross-link-age of the molecules or stabilized proteins responsible for the hard, horny character of cuticle.

Sclerotium (pl. sclerotia): A vegetative resting body of a fungus composed of a dense mass of thick-walled mycelium, round or irregularly shaped, and able to remain dormant and resistant to unfavorable conditions for long periods.

Sclerotization: Stiffening of the cuticle by cross-linkage of protein chains.

Sclerotization: The hardening of the body wall.

Sclerotize: Hardening of the body cuticle.

Sclerotized: Hardened by the presence of sclerotin.

Scolex: Head of a tapeworm *e.g., Taenia* sp. (Gr. *skelax* = worm).

Scoli: Tubercules in the form of spiny projections on the body of some insect larvae (some Lepidoptera and Coccinellidae).

Scolopale cell: In a chordotonal organ, the sheath cell that envelops the **dendrite**.

Scolopidia: In a chordotonal organ, the combination of three cells, the **cap, scolopale** and **dendrite**.

Scolopidium: A sensillum located beneath the cuticle and modified for the reception of vibrations.

Scolopophorous organ: Typically, a bundle of sensilla consisting of one or more specialized, bipolar neurons (scolopidia; also called scolopophores) stretched between two internal integumental surfaces; often has a proprioceptive function; also called chordotonal organ.

Scolytoid larva: A fleshy larva resembling the larva of a scolytid beetle.

Scopa (pl., scopae): A pollen- collecting apparatus, either a corbicula (which see) or a brush of hairs on the ventral side of the abdomen (bees); a mass of bristles on the hind leg (tibia) of adult bees; abdominal fringe of long scales on some male Lepidoptera.

Scopula (pl., scopulae): A small, dense tuft of hair.

Scotopic (eye): See superposition (eye).

Scouting: See *surveillance.*

Scramble competition: Competition in which many individuals "scramble" for a limited resource, such as food.

Scraper: The ridged surface drawn over a **file** to produce stridulatory sounds; the sharpened anal angle of the front wing (tegmen) of a cricket or long-horned grasshopper, a part of the stridulating mechanism.

Scrobe: A groove or furrow (antennal scrobe).

Scutellum: A shield-shaped sclerite of the thoracic notum; appears as a triangular plate between the base of the wings in Hemiptera or posterior plate of the alinotum.

Scutoscutellar suture: Line dividing the alinotum into the anterior scutum and posterior scutellum.

Scutum: Relatively large plate on the alinotum anterior to the scutoscutellar suture; the shield of ticks; a shield-shaped dorsal plate on the thorax, occupying most of the dorsum of adult Diptera.

Seasonal cycle: The sequence of life cycles of a species that occurs over a one-year period.

Seasonal dimorphism: If two or more generations are produced in a year, there may be differences in size, shape or colour between the off springs of different seasons.

Sebaceous glands: Glands secreting fatty or oily material.

Secondary compound: See allelochemics and secondary plant substances.

Secondary cycle: In a disease, the cycle that involves an atypical host; *see also* **Primary cycle.**

Secondary defense: A defensive mechanism that is brought into play only in the presence of a threat-for example, an aggressive display.

Secondary homonym: One of two or more identical trivial names which, at the time of original publication, were proposed in combination with different generic names but which, through subsequent transference, reclassification, or combination of genera have come to bear the same (or an identical) combination of a generic and trivial name [for nomenclatural status *see Bul. Zool. Nomencl.* **4:** 97-105 (1950)] (cf. Homonym, Primary homonym).

Secondary host: A host species attacked less commonly than the principal or primary host.

Secondary infection: Microorganisms that enter the host through an injury initially caused by another pest; the infection of healthy plants within a crop after the disease has entered the crop from outside (primary infection).

Secondary metabolite: A plant-produced chemical that is nonessential in primary metabolism but may play a role in defense against herbivory.

Secondary orientation: Positioning of the insect body relative to various external or internal stimuli; superimposed on primary orientation.

Secondary parasite: See hyperparasites.

Secondary pest outbreak: Previously harmless insects becoming pests following insecticide treatment for a primary pest.

Secondary pest: A pest that is normally not a problem until control tactics applied for another pest or pests enable it to increase and cause damage; pest species that are usually present at low levels and are held in check by the action of natural enemies; can assume full pest status when natural enemies are destroyed by a pest management tactic (e.g., insecticide application); species whose numbers are usually controlled by biotic and abiotic factors which sometimes break down, allowing the pest to increase in numbers; or an insect that does not normally attain pest status except when insecticides destroy its natural enemies.

Secondary plant compounds: Plants chemicals assumed to be produced for defensive purposes.

Secondary plant substance (s): A substance produced by a plant that plays no role in the basic metabolism of the plant; or chemicals, found in plants, that have no apparent metabolic role and that often influence the behavior and /or physiology of herbivorous insects.

Secondary segmentation: Any segmentation that fails to match the embryonic segmentation; more specifically, the insect external skeleton in which each apparent segment includes the posterior (intersegnebtal) parts of the primary segment preceding it.

Secondary segmentation: Characterized by membranous areas between adjacent segments slightly anterior to the attachment points of the longitudinal muscles; typical of adult and many larval insects.

Secondary sexual characters: The characters which distinguish the two sexes of the same species but which do not function directly in reproduction (cf. Primary sexual characters, sexually dimorphic).

Secondary spermatocytes: Developing male gametes that form from the first meiotic divisions of the primary spermatocytes; they form into spermatids (Fig. 214) with second meiotic division.

Secondary spread: The spread of a pathogen within a field after the initial or primary infection.

Secretion: A useful substance produced in the body by a cell or multicellular gland; also, the process of its production and passage. Compare Excretion. (L> *secretus* = separated)

Section 18: Section in FIFRA (Federal Insecticide, Fungicide, and Rodenticide Act (USA)) that provides for emergency exemption from federal registration; provides the state with the authority to approve new uses of federally labeled pesticides where an emergency is justified and documented.

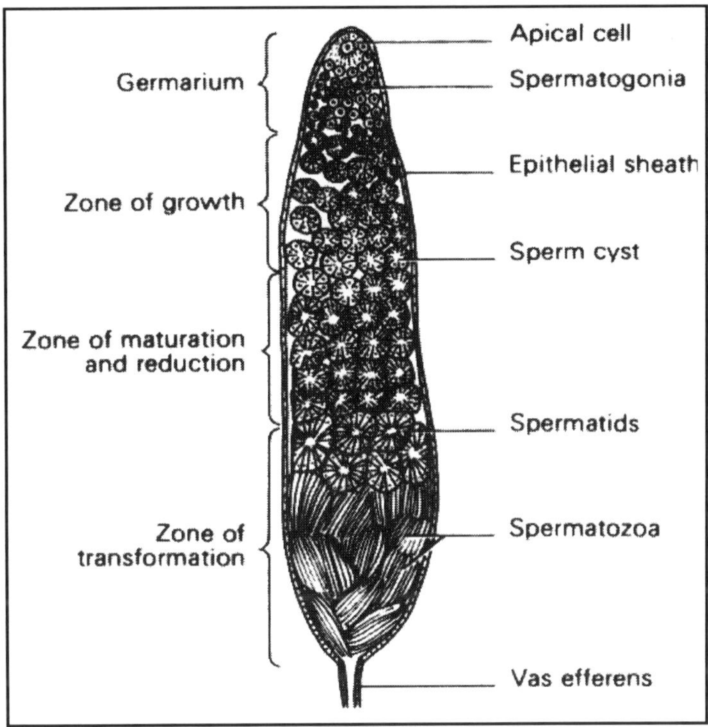

Fig. 214. Section of a testicular follicle.

Section 24 (c), special local need (SLN) registration: Issued for five years in an area with specific, verified pest problems either to be chemical formulator (first-party applicant) or to third-party applicants who supply the necessary data.

Section 3 label: Section within FIFRA that provides full registration of pesticides with a new active ingredient or a subregistration of a material already registered.

Section: A neutral term usually employed with reference to a subdivision of a taxonomic unit or a series of related elements in one portion of a higher taxonomic category (cf. Higher category, Neutral term, Group).

Sector: A major wing-vein branch and all of its subdivisions.

Sectorial cross vein: A cross vein connecting two branches of the radial sector.

Sedentary endoparastic nematode: Nematodes that enter the host, moves to a feeding site, and then become immobile.

Sedentary: Remaining in one location after becoming established; immobile; in size, shape or colour between the off springs of different seasons.

Sedimentary rocks: Rocks which form from the wastage of pre-existing rocks; include the fragmentary rocks deposited as sheets of sediment on the floor of seas, lakes, rivers and on land.

Seed borne: Disease caused by pathogens that survive in, or are carried on, seeds.

Seed dressing: A coating (either dry or wet) of protectant pesticide applied to seeds before planting. Dry seed dressings are often physically stuck to the testa of the seed by a sticker such as methyl cellulose.

Seed treatment: Substances coated on seeds as slurries, solutions, or as dry mixtures prior to planting. Usually insecticides and/or fungicides are used for control of soil insects, and soil-borne pathogens.

Seedling: A small plant after it has germinated from a seed.

Segment (s): Repeating body units of insects and of many other animals; a subdivision of the body or of an appendage, between joints or articulations.

Segmental muscles: The principal longitudinal muscles running between the intersegmental constrictions associated with primary segmentation.

Segmentation: Repetition of a pattern of elements belonging to each of the main organ systems of the body along the antero-posterior axis; distinct in Arthropoda and Annelida.

Selection: see Natural selection.

Selective insecticide: One which is effective against only a narrow range of insects.

Selective pesticide: (i) A pesticide, usually insecticide, that kills a specific target pest but is generally not harmful to most other organisms. (ii) A herbicide that kills weeds but does not harm crop plants.

Selective pressure: The main force that induces a genetic change through natural selection.

Self Pollination: Transfer of pollen from anthers to stigma within the same flower.

Sematectonic communication: Stimulation of behavior or physiological change in one animal in response to stimuli associated with the result of evidence of work by another animal.

Semelparous: A type of life history in which the animal reproduces only once during its lifetime.

Semen: Fluid discharged at ejaculation by the male, consisting of spermatozoa and secretions of glands associated with the gential tract.

Semiaqatic: Living in wet places, or partially in water.

Semi-aquatic: Living in saturated soils, but not immersed in free water.

Semiaquatic: Living in wet places, or partially in water.

Semi-endoparasitic nematode: A parasitic nematode that becomes only partially embedded in host tissue, with the posterior of the body remaining outside of the plant tissue.

Semi-looper: Caterpillar from the subfamily Plusiinae (Noctuidae) with two or three pairs of prolegs, which locomotes in a somewhat looping manner.

Seminal vesicle (*vesicular seminalis*)**:** A saclike storage structure that holds seminal fluid of the male prior to discharge; dilated region of the vas deferens in which spermatozoa are stored; a sac-like structure for the storage of fluid, containing spermatozoa *e.g.,* in earthworm; an expansion of the vas deferens of the male in which sperm are stored.

Semiochemical (s): Any chemical involved in communications among organisms; or chemicals that deliver behavioral messages; any chemical used in intra and interspecific communication, such as pheromones, allomones, and kairomones.

Semisocial: Social behaviour in which individuals of the same generation co-operate and nest-share with some division of reproductive labour.

Semispecies: The species of which a superspecies is composed (cf. Superspecies); semispecies are a special kind of species, not a category different from the species.

Senior homonym: The earliest published of two or more identical names for the same or different taxonomic categories (cf. Homonym, Junior homonym).

Senior synonym: The earliest published of two or more available synonyms for the same taxonomic unit (cf. Synonym, Junior synonym).

Sense cone or **sense peg:** A minute cone or peg, sensory in function.

Sense organ: An organ containing a part sensitive to the stimulus of a particular kind of sense *e.g.* tongue for taste, eyes for sight.

Sensescence: Change in structure and function that decrease an individual's capacity for survival and lead to death.

Sensilla sing Sensillum: Specialized structures that collect information from the external and internal environments and transmit their information to the central nervous system (Fig. 215).

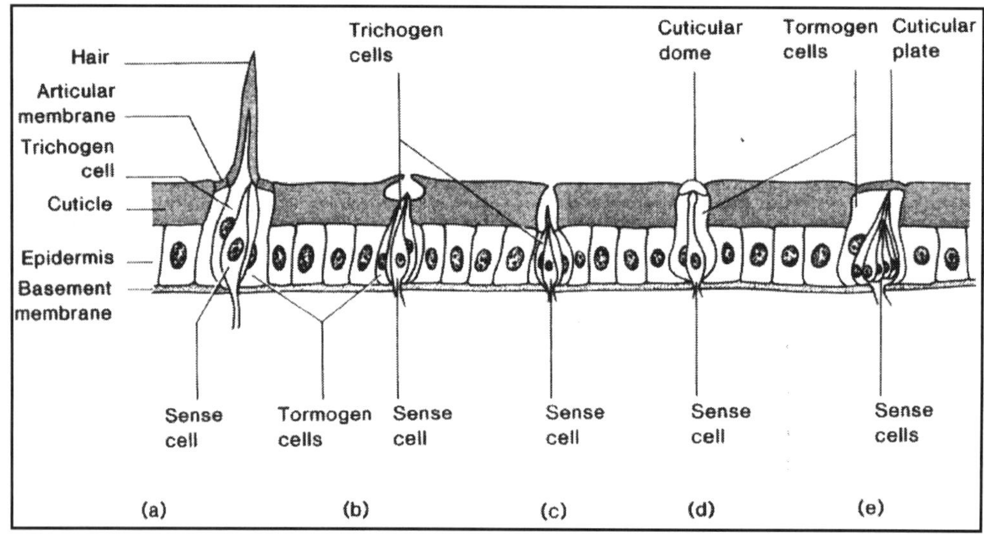

Fig. 215. Types of sensilla related to the trichoid type. (a) Trichoid. (b) Coeloconic. (c) Ampullaceous. (d) Campaniform. (e) Placoid.

Sensitive area: Locations where pesticide applications could be particularly harmful (e.g., stream, residence, park).

Sensitive: Low tolerance to pesticide effects.

Sensory filtering: The process of receiving only certain specific stimuli impinging on the body.

Sensory neuron: A neuron leading from a receptor cell to the central nervous system; usually bipolar neuron with cell body located peripherally in the insect and dendrites associated with sensory structures; also called afferent neuron; a nerve cells that receives and transmits stimuli from the environment.

Septum (pl. septa): A dividing wall or partition between two cavities or structures. (L. *sepes* = fence).

Sequential sampling: Sampling wherein the number of samples taken is variable and depends on whether the results obtained as sampling progresses yield a definite answer about the abundance of a pest.

Seres: The various communities appearing and disappearing between the establishment of pioneer and climax communities.

Serial homology: The occurrence of identically derived features on different segments of the same insect.

Sericulture: The propagation of silkworms for silk production; rearing and breeding of the silk moth to obtain the raw silk fibre; the commercial rearing of silkworms and silk production.

Series: In taxonomy, the sample which the collector takes in the field or the sample available for taxonomic study (cf. Material, Hypodigm0; also neutral term employed especially with reference to a sequence of taxonomic categories or forms (cf. Neutral term, Taxonomic categories, Form).

Serology: The study of the nature and interactions of antigens and antibodies (cf. Antigen, Antibody).

Serosa: Outer membrane around the yolk, amnion, and embryo.

Serous: Secreting watery colourless serum, as by a gland or serous membrane (L.).

Serrate: Toothed along the edge like a saw; serrate antenna.

Serum globulin: The blood fraction in which antibodies, if present, are to be found (cf. Antibody).

Serum: The plasma of blood that segarates from a clot and contains no cells or fibrin (L.).

Sesamin: A mixed-function oxidase inhibitor produced by the *Chrysanthemum* plant.

Sessile: Attached or fastened, incapable of moving from place to place; attached directly, without a stem or petiole.

Sessile: (i) Refers to immobile stages of some insects, in particular, scales and whiteflies that lack the ability to move. (ii) Without stalk or stem.

Seta (pl. setae)**:** A cuticular extension, a **trichoid sensillum;** also called a hair **or** macrotrichium; a bristle or hair like process; a movable hair of the integument, typically forming a sensillum; or epidermal bristle of invertebrates *e.g.*, setae of earthworm and mites; macrotrichia; unicellular, cuticular processes (Fig. 216).

Seta (pl. setae): Slender body hair of filament.

Setaceous: Bristlelike; setaceous antenna.

Setae: Bristles; macrotrichia; sclerotized hairlike projections of skeletal cuticle. Note: in strict usage never called "hairs".

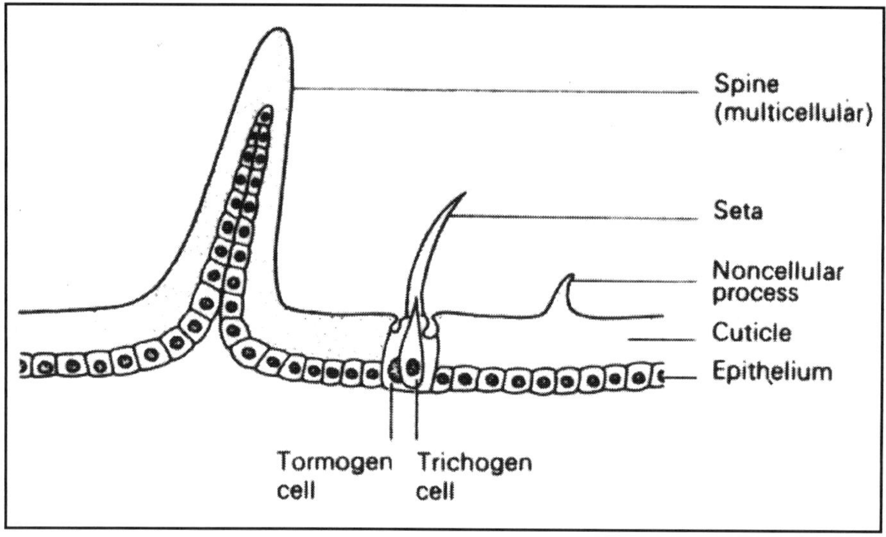

Fig. 216. External integumentary processes.

Setate: Provided with bristles.

Setulose: Bearing short, blunt bristles.

Seventy-five per cent rule: The rule that population *A* can be considered subspecifically distinct from population *B* if 75 per cent of the individuals of *A* are different from "all" the individuals of population *B* (cf. Coefficient of difference).

Severe pest: A pest with a general equilibrium position above the economic-injury level, which makes the pest a constant problem.

"Sewall Wright effect": See Genetic drift.

Sex attractant: (related term, sex pheromone) A volatile chemical substance produced by one sex of an insect (nearly always the female) to attract the opposite sex.

Sex chromosome: A special chromosome, not occurring in identical number or structure in the two sexes and usually concerned with sex determination; the X chromosome or Y chromosome (cf. Chromosome, Autosome).

Sex pheromone (related term sex attractant)**:** A volatile chemical substance produced by one sex of an insect which produces some specific reaction in the opposite sex. Most sex pheromones are powerful attractants but male pheromones may act as aphrodisiacs, A chemical signal from one sex that attracts members of the opposite sex; or an intraspecific chemical messenger produced for the purpose of bringing the two sexes together.

Sex ratio: Within a population, the proportion of males compared with that of females.

Sex-limited character: A character belonging to one sex only (cf. Secondary sexual character, Sex-linked character).

Sex-linked character: A character for which the determiner is located in the sex chromosome (cf. Sex chromosome, Sex-limited character).

Sexual dimorphism: Differences between the sexes.

Sexual: Any act pertaining to sex.

Sexuales: Sexually reproductive male and females of Aphidoidea and Coccoidea.

Sexually dimorphic: With pronounced difference in the morphological expression (form) of the two sexes of a single species (cf. Dimorphism).

Sexuparae: In Aphidoidea the alate parthenogenetic females that produce males and sexual females.

Sheath cell: Any cell enveloping another.

Shell: Hard outer calcareous/siliceous/bony/horny/chitinous covering of animals *e.g.*, in snails.

Shellac: Commercial lac produced as a red resinous substance from the scales of female lac insects.

Shoot: The growth of a plant in the form of stem and its leaves.

Shothole: (i) A plant disease symptom in which small, round pieces drop out of leaves. (ii) The visible holes left by wood-boring beetles of the family Scolytidae, as they leave their host, shothole borers.

Sibling: Full brother or sister.

Sibling species: Pairs or groups of closely related species which are reproductively isolated but morphologically identical or nearly so (cf. Species); closely related species that are difficult to distinguish by ordinary means.

Sieve plates: Porous covers associated with spiracles that are thought to retard water loss and prevent entry of airbone particles.

Sigmoid: S-shaped; the S-shaped growth curve representing development of either an individual organism or a population plotted graphically.

Sign (of disease): Characteristic visible indicators of disease visible on a host, for example mildew mycelium on the surface of a leaf (cf., symptoms).

Sign stimuli: See token stimuli.

Sign stimulus: A stimulus for which an animal has evolved a specific response pattern.

Signal word: A word required on every pesticide label denoting the relative toxicity of the pesticide. The signal words are "Danger Poison" with a skull and crossbones logo for highly toxic, "Warning" for moderately toxic, and "Caution" for low-order toxicity materials.

Siliqua: The fruiting body of Crucifera; a long pod divided in two by a false septum.

Silvicide: Chemical to kill woody shrubs or trees.

Silviculture: The order of insects comprising the fleas.

Simple: Unmodified, not complicated; not forked, toothed, branched, or divided.

Single cycle disease: A disease involving one species of host, one parasite and one insect vector.

Single-feeding bait: Rodent bait containing acute toxins such as strychnine or zinc phosphide and requiring only toxicity materials.

Siphonaptera (alternative term, Aphaniptera)**:** (*siphon,* a tube; *aptera,* wingless) The order of insects comprising the fleas (Fig. 217).

Siphunculata (alternative term, Anoplura)**:** The order of insects comprising the sucking lice.

Siphunculi: The paired protruding organs near the terminal end of the abdomen of Aphidoidea, also called cornicles, through which a waxy secretion is extruded.

Siphunculus (pl. siphunculi): see **Cornicle.**

Sister group: The closest related group of the same rank as the group under study.

Skeletal muscles: Muscles attached at both ends to the integument.

Skeleton, endo: Skeleton lying inside the body *e.g.,* the bony skeleton of vertebtrates.

Skeleton, exo: Skeleton lying outside – the body characteristic of athropods in general. It provides a much larger area for attachment of muscles, protects the muscles from mechanical injury and affords an excellent mechanism to protect insects against desiccation and keeps the form of body of the insect in proper condition.

Skeleton: Supportive shell; or a system of hard parts (External or internal) of the body to serve for protection and support of softer organs and often for the attachment of muscles.

SLN: See *Section 24(c).*

Slug: A relative of snails but having the shell rudimentary or entirely wanting.

Slurry: A thick pesticide suspension made from wettable powder and water; Paste-like liquid used as a seed coating.

Smell: The olfactory sense, the detection of air-borne chemicals.

Smoke: Aerial dispersal of minute soild particles of pesticides through the use of combustible mixtures.

Smut: A plant disease caused by fungi in the Basidiomycetes, a sign of which is the presence of sooty spore masses.

Snail: A member of the phulum Mollusca having a single, usually coiled shell and a broad, flat foot.

Social: Living together in complex colonies; e.g., ants bees.

Social hormone: A pheromone produced by the royal pair of termites and transferred throughout the colony by faecal trophollaxis. This presumably inhibits the production of secondary reproductives from pseudergates.

Social insects: Insects that live in a common nest in groups or colonies generally with differentiated castes, division of labour, effective means of communication, care of young ones and high degree of coordination. The most highly organized social colonies in insects exist among bees, wasps, ants and termites.

Social parasite: An insect that invades or lays its eggs in the nest of another insect and develops on food in the nest.

Sociality: The condition of living in an organized community.

Society: When aggregating group cooperates in altering the environment in its favour, it attains the status of a society.

Sociotomy: Method of nest foundation among the termites in which immatures and secondary reproductives separate from the parent colony or a migrating colony divides into two daughter colonies.

Soft rot: Disease characterized by slimy softening and decay of the affected plant tissues.

Soil application: Pesticide placation to the soil surface rather than to foliage.

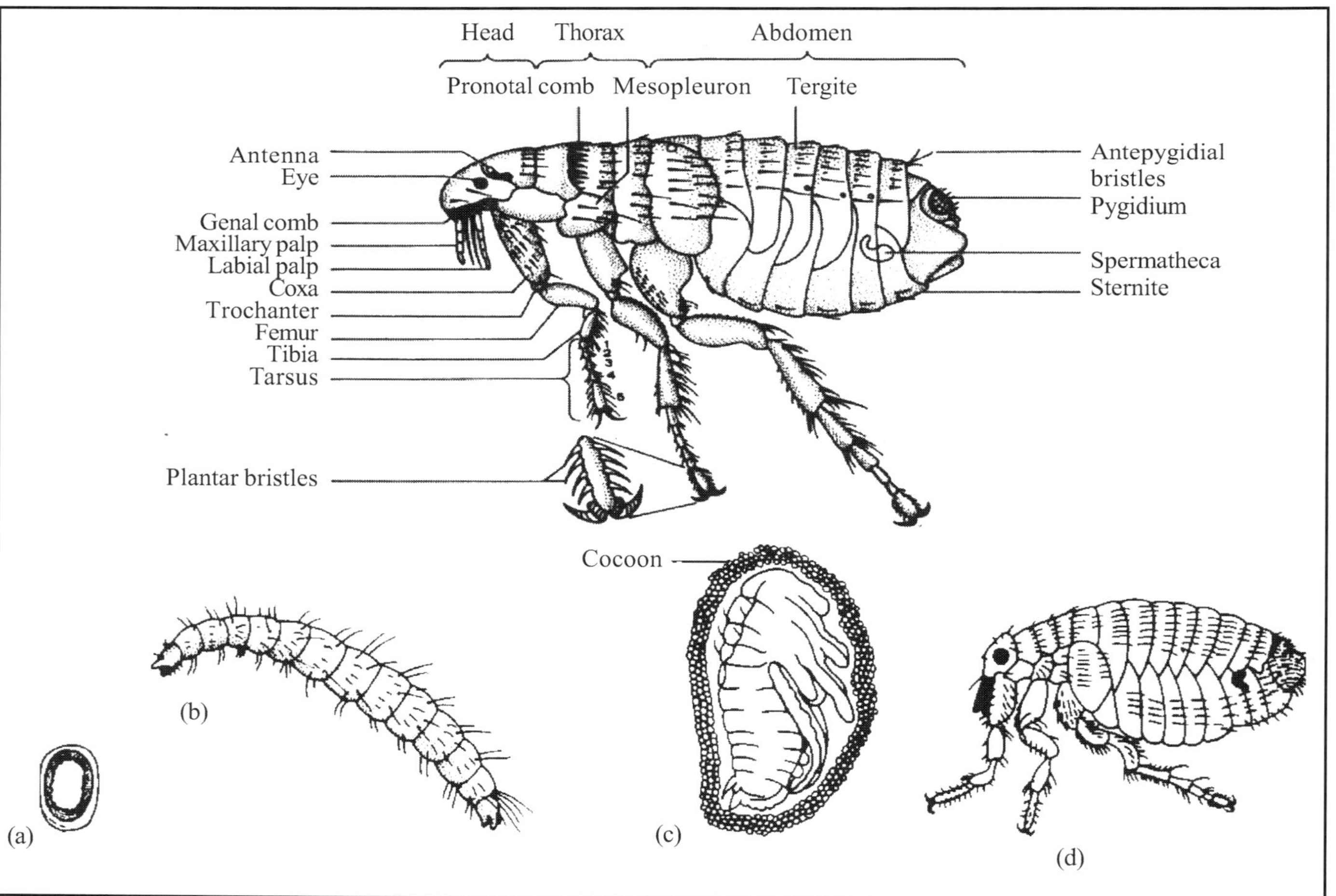

Fig. 217. Siphonaptera, a flea. Life stage of a flea. (a) Egg. (b) Larva. (c) Pupa. (d) Adult. After U.S. Public Health Service.

Soil borne: Disease caused by organisms that live in, or are found in, the soil.

Soil compaction: Reduced volume of cultivated soils caused by pressure applied to the particles; results in decreased permeability to air and water, thus hindering plant growth.

Soil drainage: An assessment of the ability of a soil to hold water.

Soil fumigant: A pesticide that moves through soil as a gas or a vapor to control pests.

Soil incorporation: Mechanical mixing of a pesticide into the soil.

Soil injection: Pesticide placement beneath the soil surface without mixing or stirring the soil.

Soil persistence: The length of time that a pesticide application on or in soil remains effective.

Soil solarization: Placement of clear plastic tarp over wet, fallow soil to trap the radiant heat of the sun and raise the temperature of soil beneath the tarp to control soil-borne pests.

Soil sterilant: Toxicant added to, or injected into, soil for the purpose of killing pests and pathogens; or a pesticide with long residual properties that kills all organisms in the soil; only a few chemicals are sterilants.

Soldier: In social insects, an individual worker belonging to a subcaste involved in colony defence; a caste present in some social insects (especially termites and ants), whose task is to defend the colony.

Solid cone: Jet with air-core reduced to give a cone of spray droplets.

Solitary: Non-colonial, occurring singly or as mating pairs.

Soluble powder (SP): A dry pesticide formulation that forms a true solution when added to water; does not need further agitation; or a pesticide formulation consisting of a finely ground solid material that dissolves in water or some other liquid.

Soluble powder: An insecticide formulation which dissolves in water to form a solution.

Solute: A substance that will dissolve or go into solution, as salt in water. (L. *se* = a part + *lus* = set free).

Solution (s): Dissolved water-soluble pesticide that can be mixed directly with water; usually contains spreaders and stickers; or a concentrated liquid pesticide formulation that may be used directly or require diluting.

Solvent: A fluid capable of dissolving substances; or chemical used to dissolve organic pesticides in water; carrier solution in which the pesticide (technical product) is dissolved to form the concentrate.

-soma, somat-, -some: Body.

Soma, somatic: Pertaining to the body or body cells as contrasted with germ cells. (Gr. *soma* = body).

Soma: See perikaryon.

Somite (pl. somites): A serial segment or homologous part of the body metamere. (Gr. *soma* = Body).

Sooty mold: A fungus with dark mycelium that grows on insect honeydew.

Soporific: A chemical that, once ingested, causes birds to enter a state of stupor and allows capture and removal.

Spanandrous: A shortage of males.

Spatulate: Spoon-shaped; broad apically and narrowed basally, and flattened.

Speciation: The splitting of a phyletic line; the process of the multiplication of species; the origin of discontinuities between populations due to the development of reproductive isolating mechanisms (cf. Allopatric speciation, Sympatric speciation).

Species (adj. specific): A group of all individuals that can interbreed, mating within the group (sharing a gene pool) and producing fertile offspring, usually similar in appearance and behaviour – but *see also* **Polymorphic** – and sharing a common evolutionary history.

Species complex: See Complex.

Species group: See Group.

Species (pl. also species): The basic unit of classification of animals or plants. A group of individuals or populations that are similar in structure and physiology and are capable of interbreeding and producing fertile offspring; the largest group of individuals within which effective gene flow occurs or could occur.

Specific (or **pair-wise) coevolution:** Concerted evolutionary change that takes place between two species, in which the evolution of a trait in one leads to reciprocal development of a trait in a second organism in a feature that evolved initially in response to a trait of the first species; *see also* **Guild coevolution.**

Specific epithet: Second part of a species name, indicating a particular species.

Specific name: "The binominal combination of a generic name and a specific trivial name which constitutes the scientific designation of a species" (International Commission, 1948) (cf. Species, Scientific name, Binominal nomenclature); also used by many workers and in the original Rules in place of *trivial* name.

Specific trivial name: The second term of the binominal designation of a species (cf. Species, Binominal nomenclature).

Spectrum of activity: The range of insect species against which an insecticide is active.

Speed sprayer: See *air-blast sprayer.*

Sperm competition: In multi-mated females, the syndrome by which sperm from one mating complete with other sperm to fertilize the eggs.

Sperm precedence: The preferential use by the female of the sperm of one mating over others; in multiple mating, the tendency for sperm from the most recent mating to fertilize the eggs.

Sperm: The male cell which fertilizes the female cell ovum. (Gr. *sperma* seed).

Spermaphagy: Seed-eating.

Spermatheca (pl. spermathecae): A saclike structure in which sperm from the male is received and stored in the female; or outpocketing of the vagina in which spermatozoa are stored prior to fertilization; or a small sac associated with the median oviduct of the female, in which sperm are stored following copulation.

Spermathecal gland: A tubular gland off the spermatheca, producing nourishment for sperm stored in the spermatheca.

Spermatids: Developing male gametes that form from division (second meiotic) of the secondary spermatocytes.

Spermatocyte: A cell that divides to form the spermatozoa of male animals; or diploid cells that undergo meiotic division and become spermatozoa.

Spermatodesms: Bundles of spermatozoa.

Spermatogenesis: Development of male reproductive cells (spermatozoa) in the testicular follicles.

Spermatogonia: Male germ cells that proliferate mitotically and that are capable of developing into spermatocytes.

Spermatophore: A packet containing sperm produced by some male insects; or spermatozoa encased in a gelatinous packet that is introduced into the female reproductive tract or deposited on the substrate and taken up by the female (Fig. 218); or a sac produced by accessory glands of male insects and transferred to the female reproductive tract, containing sperm and often proteinaceous material. An encapsulated package of spermatozoa (Fig. 220).

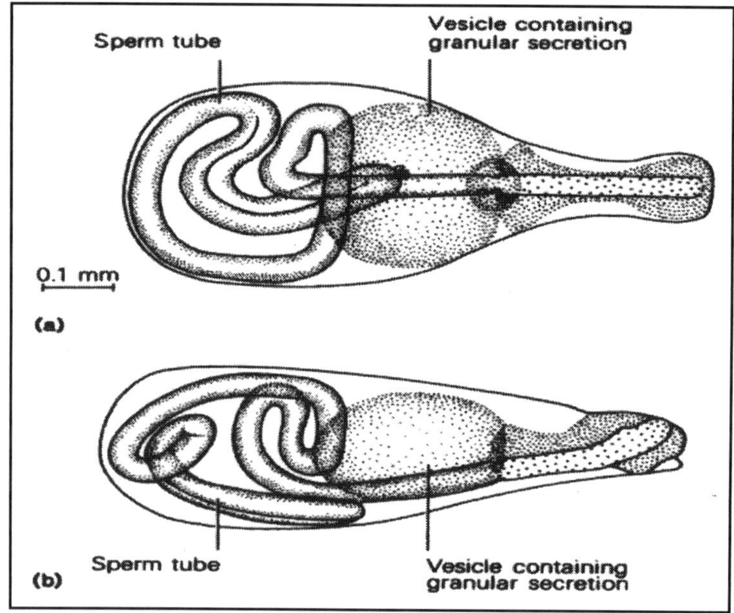

Fig. 218. Spermatophore from a male silverfish, *Lepisma saccharina*.

Spermatophylax: In katydids, a proteinaceous part of the spermatophore eaten by the female after mating (Fig. 219).

Spermatozoa: Male reproductive cells (Figs. 219 & 220); or the matured and functional male sex cells or male gametes. (Gr. *sperma* = seed + *zoon* = animal).

Sperozoite: Spares liberated through the dissolving of membrane of sporacyet, a phase in the life history of sporozoan *e.g.* in *Plasmodium*. (*sporas* = seed + *zoon* = animal).

Sphecophily: Plant pollination by wasps.

Spicule: A small needlelike spine.

Spider: Any member of the arthropod order Araneae having eight legs and feeding exclusively on animal tissues.

Spina: An internal apodeme of the intersegmental sternal plate called the spinasternum.

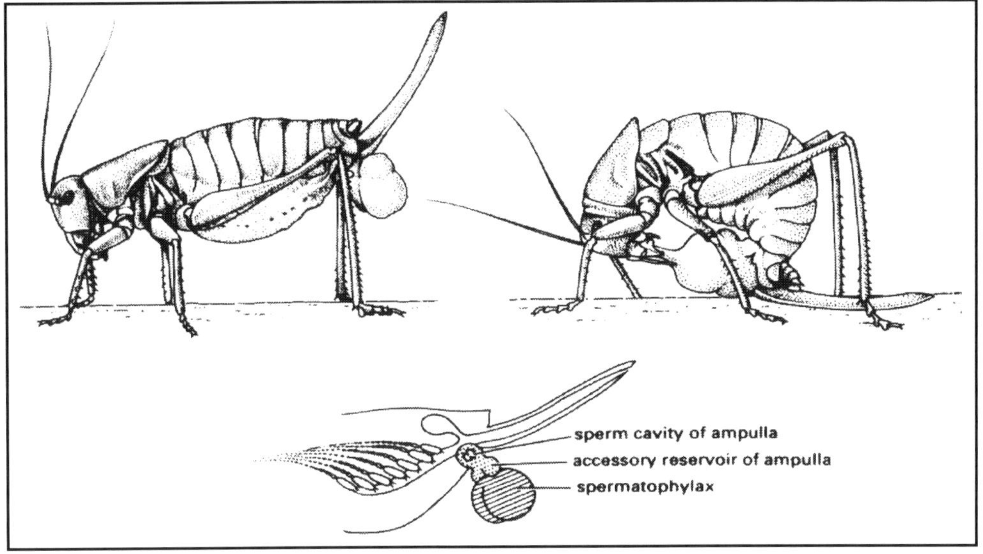

Fig. 219. Nuptial feeding: Spermatophylax.

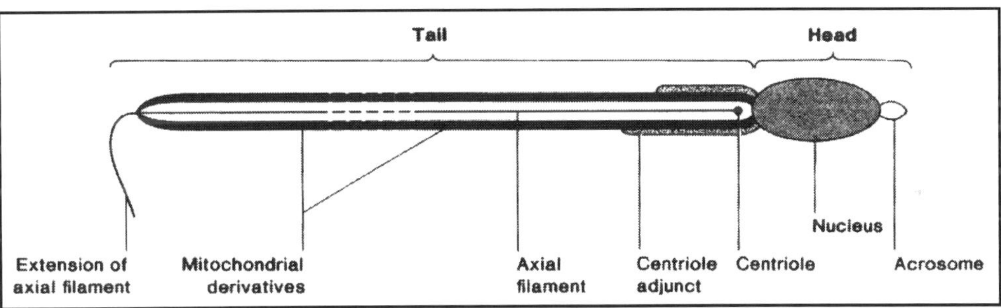

Fig. 220. Generalized insect spermatozoon.

Spina: Internally inflected median process on the spinasternum of the thorax.

Spinal pit: The externally evident indication of the spina on the spinasternum.

Spinasternum: An **intersternite** bearing a spina, sometimes fused with the eusterna of the prothorax and mesothorax, but never the metathorax; an intersegmental sclerite of the thoracic venter which bears a median apodeme or spina, associated with or united with the sternal sclerite immediately anterior to it; also called the intersternite.

Spindle-shaped: Elongate and cylindrical, thickened in the middle and tapering at the ends.

Spine: A multicellular unjointed cuticular extension, often thorn-like outgrowth of the cuticle.

Spinner: The fully mature adult form of mayflies after moulting of the dun (sub-imago) has taken place.

Spinneret: The central silk spinning bristle present in the mouthparts of caterpillars (larvae of Lepidoptera) (Fig. 221).

Spinose: Beset with spines (a spinose costa in Diptera).

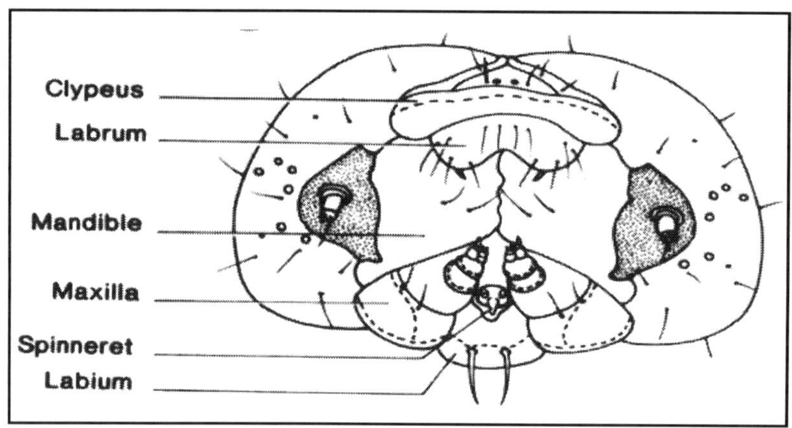

Fig. 221. Example of mouthparts from immature forms: Moth larva (caterpillar).

Spiracle: An external opening of the tracheal system; a breathing pore that are the respiratory system in insects (Fig. 222); or in insects, an external small opening to the tracheal or respiratory system that communicate with the external environment; in cartilaginous fishes, the modified first gill slit. (L. *spiraculnm* = air hole).

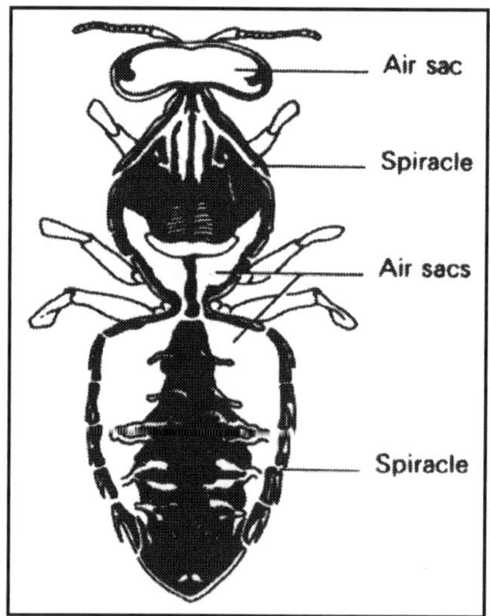

Fig. 222. Tracheal structure: Air sacs in the honey bee showing Spiracle.

Spiracular bristle: A bristle very close to a spiracle (Diptera).

Spiracular gills: In aquatic insects with closed ventilatory systems, filamentous outgrowth of thin cuticle that open directly into the tracheae; also called cuticular gills (Fig. 223).

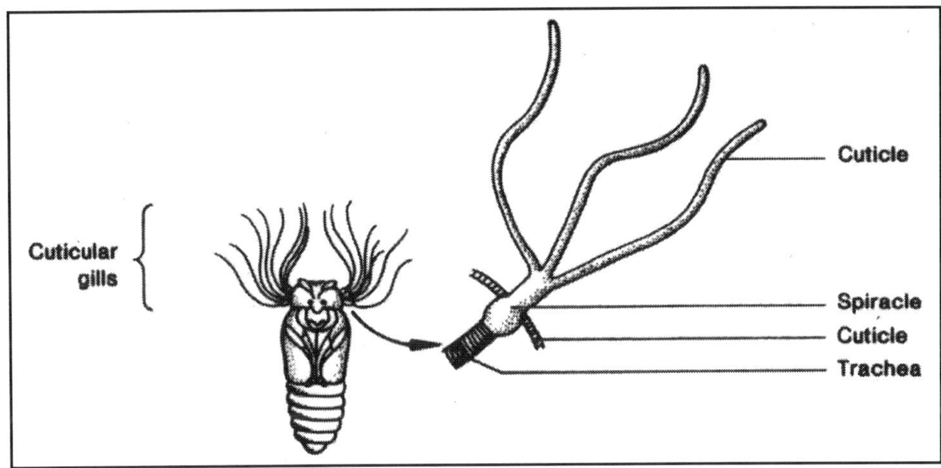

Fig. 223. Ventilatory structure in aquatic insects: Cuticular gills on the thorax of a black fly pupa.

Spiracular plate: A platelike sclerite next to or surrounding the spiracle.

Spiroplasmas: Motile mollicute with helical shape, associated with plant diseases such as citrus stubborn.

Splitter: In taxonomy, one who divides his material more finely than the average; one whose criteria for determining the level to be assigned to a given taxonomic category are such that the effect of his work is to push existing classification upward (as subfamilies to families, subspecies to species); also one who attempts to express even small differences nomenclaturally (cf. Lumper).

Sporangium (pl. sporangia): The structure in which asexual spores are formed.

Spore: A small reproductive structure produced by certain fungi and other microorganisms, capable of growing into a new individual under proper conditions; analogous to seeds of green plants; or a cell in a resident covering capable of developing independently into a new individual. (Gr. *spora* = seed).

Sporogony: The asexual cycle of a malaria parasite (*Plasmodium* spp.) in the mosquito, in which an **oocyst** undergoes nuclear division to produce **sporozoites.**

Sporozoties: In the asexual cycle of a malaria parasite (*Plasmodium* spp.) in the mosquito, the product of **sporogony;** stored in the salivary gland and passed to the vertebrate host during mosquito feeding.

Sporulate: The production and liberation of spores.

Spot treatment: Pesticide treatment of small area within a larger locale.

Spray angle: Angle between the sides of a jet leaving the orifice.

Spray drift: See *drift.*

Spray operator: Person operating a sprayer, and applying a spray.

Spray: 1. Air-carried: spray propelled to target in a stream of air. 2. Coarse: dispersion of droplets of mass median diameter over 200 μm. 3. Concentrate: undiluted commercial pesticide preparation. 4. Fine: dispersion of droplets of mass median diameter from 50-150 μm. 5. Floor: spray applied to the litter on the ground surface. 6. High-volume: over 1200 1/ha on bushes and trees; over 700 1/ha on ground crops (or over 400 1/ha according to definition). 7. Low-volume: spray of 250-600 1/ha on bushes and trees; 50-250 1/ha on ground crops (or 5-4001/ha). 8. Median-volume: 600-1200 1/ha on bushes and trees; 250-700 1/ha on ground crops. 9. Mist: dispersion of droplets of 50-100 μm in diameter. 10. Ultra-low-volume: less than 50 1/ha on ground crops; less than 250 1/ha on trees and bushes (or less than 51 1/ha according to definition).

Sprayer: Apparatus for applying pesticide sprays; not to be confused with 'Spray operator'.

Spread: Uniformity and completeness with which a spray deposit covers a continuous surface, such as a leaf or a seed.

Spreader: Agent added to a pesticide formulation to enhance its ability to increase area of target covered. See *wetting agent;* or material added to a spray to lower the surface tension and to improve spread over a given area (=wetter).

Spur: An articulated movable spine, when on a leg segment usually located at the apex of the segment or on tibia. A serrulate tibial spur is characteristic of the Delphacidae (Homoptera).

Spurious claw: A false claw; a stout bristle that looks like a claw (spiders).

Spurious vein: A veinlike thickening of the wing membrane between two true veins; an adventitious longitudinal vein between the radius and the media, crossing the r-m cross vein (Diptera).

Squama (pl., squamae): A scalelike structure; a calypter; the palpiger (Odonata); a membranous lobe at the extreme base of the wing of Diptera; also called calypter.

Stability: The ability of a pesticide formulation to resist chemical degradation over a period of time.

Stabilizing agent: Chemicals that retard pesticide decomposition during storage.

Stable age distribution: Age distribution approached by a population of stable age schedule of birth and death rate, when growing in unlimited space.

Stadium (pl. stadia): The period between molts in a developing insect; or interval of time passed in an instar; or the instar duration or intermoult period.

Stage: The insect's developmental status (e.g., egg stage).

Stalked: With a stalk or stem; with a narrow stemlike base; fused together to form a single vein.

Stand decline: The gradual decrease in the number or vigor of plants in an area.

Stand establishment: Seed germination and the early phases of seedling growth.

Stand: A group of plants growing together in one area, usually the density of plants per unit area in a particular field or crop.

Standard (or label) rate: The recommended dosage rate of a pesticide.

Standard deviation, S.D.: The square root of the sum (?) of the squared deviations (d) from the mean, divided by *N*:

$$C.D.=\frac{S.D.}{\sqrt{N}}$$

Standard deviation: The square root of the arithmetic averages of the squares of the deviations from the mean; a measure of variation in set of samples.

Standard error (of the mean): Standard deviation divided by the square root of the sample size (*N*):

$$S.E._{M}=\frac{\sqrt{\sum d^2}}{N}$$

Standard error: The measure of standard deviations of a population of means.

Startle display: A display made by some cryptic insects upon discovery, involving exposure of a startling colour or pattern, such as eye spots.

Statary: The sedentary, stationary phase of army ants.

Stem mother: A wingless, parthenogenic, viviparous female aphid that arises from an overwintering aphid egg.

Stemma (pl. stemmata)**:** The simple eye of many holometabolous larval insects, sometimes aggregated into a more complex visual organ; simple eyes found on the sides of the insect head when compound eyes are lacking (Fig. 224); also called lateral ocelli; e.g. many insect larvae.

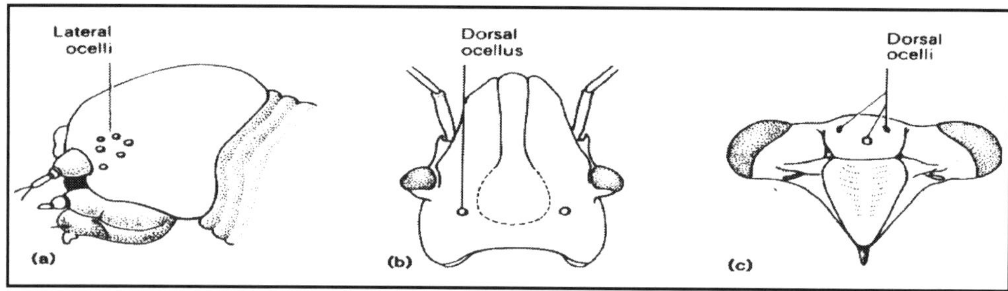

Fig. 224. Ocelli. (a) Stemmata (lateral ocelli) of a caterpillar. (b) Two dorsal ocelli, stink bug. (c) Three dorsal ocelli, cicada.

Stenogastrous: Having a shortened or narrow abdomen.

Stenohygric: Adapted for life in a small range of relative humidity. The opposite is euryhygric.

Stenotopic: Restricted to living in a single type of habitat. The opposite is eurytopic.

Stereomicroscope: A microscope with dual eye piece and objective lens systems which provides stereoscopic vision.

Sterile insect release method (SIR): The introduction of sterile insects into a population to limit reproduction.

Sterile male technique: A means of controlling insects by swamping populations with large numbers of artificially sterilized males; a method of pest control which involves the release of large numbers of sterile male insects.

Sterilization: The elimination of living organisms from containers or material by means of heat or chemicals.

Sternacosta: Apodeme associated with the sternacostal suture.

Sternacostal suture: A suture of the thoracic sternum, the external mark of the sternal apophysis or furca, separating the basisternum from the sternellum; line on the sternum that connects the furcal pits and that is associated with the sternacosta.

Sternal apophyses: Internal projections arising fron the eusternum.

Sternauli (pl., sternaulices): A longitudinal furrow on the lower part of the mesopleuron (Hymenoptera).

Sternellum: One of the three parts of the eusternum in many insects; the part of the eusternum posterior to the sternacostal suture; the small sclerite of the eusternum lying posterior to the basisternum.

Sternite: A sclerite on the ventral side of the body; the ventral sclerite of an abdominal segment.

Sternopleural bristles: Bristles on the sternopleuron (Diptera).

Sternopleurite: One of the three theoretical subcoxal sclerites that have come to form the insect pleuron.

Sternopleuron (pl., sternopleura): A sclerite in the lateral wall of the thorax, just above the base of the middle leg (Diptera).

Sternum (pl. sterna; adj. sternal; dim. Sternite)**:** The ventral surface or sclerite of the insect thorax or abdomen.

Sticker: Additive that increase the amount of spray adhering to a treated surface; or a material of high viscosity used to stick powered seed dressings on to seeds; two commonly used stickers are paraffin and methyl cellulose. A solution of methyl cellulose can be added to a spray to increase retention on plant foliage; or an ingredient added to a pesticide formulation to improve its adherence to a surface (e.g., casein).

Sticky trap: Traps containing a sticky substance that holds insects to be counted.

Stigma (pl., stigmata): Pigmented cell immediately posterior to the costal vein and near the apex of the wings of dragonflies and damselflies; thickened portion of the wing membrane near the apex of the costal edge of the wing (Fig. 225).

Stigma: A dark mark near the fore-margin of the wings of many insects; a spiracle. Also (botanical), the female receptive part of a flower.

Stigmal vein: A short vein extending posteriorly from the costal margin of the wing, usually a little beyond the middle of the wing (Chalcidoidea).

Stimulant: A chemical or physical factor which increases some specific insect activity (particularly feeding or oviposition). cf. deterrent.

Stimulus filtering: The process by which the multitude of stimuli that impinge on the insect nervous system are reduced to those which are pertinent to the behavior of the insect. *See* Sensory filtering.

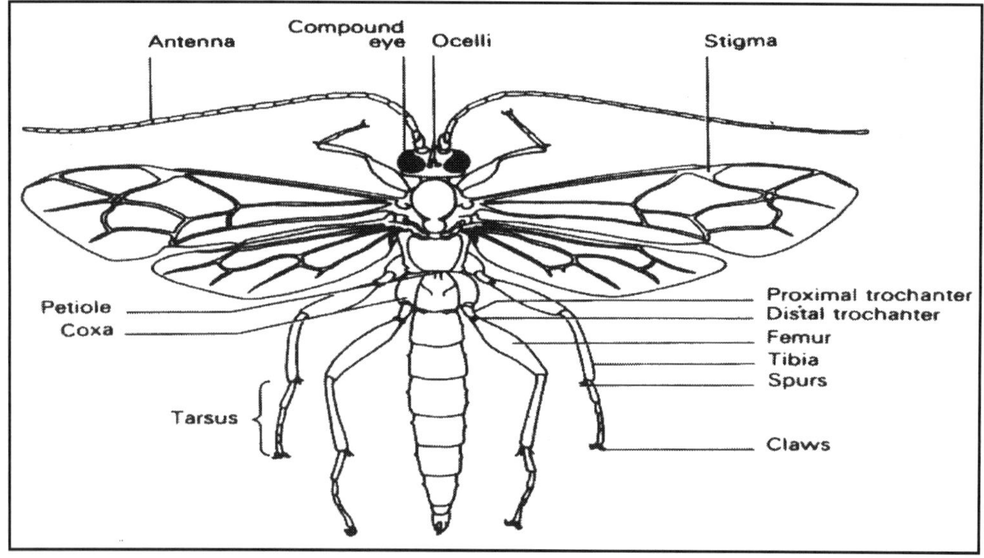

Fig. 225. External features of Hymenoptera. An ichneumon wasp (Ichneumonidae).

Stimulus (pl. stimuli): Environmental change of an organism sufficiently intense to being about change in the activities of the living material without itself providing energy for the new activities.

Stipes (pl., stipites): The second segment or division of a maxilla, which bears the palp, the galea, and the lacinia; lateral lobes of the millipede gnathochilarium.

Stolon: A modified horizontal stem above ground, creeping and rooting at the nodes, or curved and rooting at the tip.

Stoma (pl. stomata): Small aperture in the leaf epidermis through which water vapor is released during transpiration and gases are exchanged between the atmosphere and the plant.

Stomach poison: Pesticide that acts through the stomach after ingestion; or toxicants that must be ingested to have a lethal effect; or a toxicant (poison) which operates by absorption through the intestine after having been injested by the insect, usually on plant material; or an insecticide that is fatal only after being eaten, entering the insect body through the gut (e.g., boric acid).

Stomatogastric nervous system: The nerves associated with the fore- and midgut.

Stomatogastric system: The part of the visceral nervous system associated with the brain, aorta, and foregut; also called the stomodael system.

Stomodaeal nervous system: The main component of the visceral nervous system; controls activities of the anterior gut and dorsal vessel; or a set of small ganglia and their connections, laying on the surface of the foregut.

Stomodael invagination: The embryonic ectodermal invagination that forms into the foregut.

Stomodael valve: A structure formed by an intussusception of foregut and midgut tissues.

Stomodaeum (adj. stomodeal): The foregut of insects, beginning with the mouth and ending with the proventriculus.

Stomodeal trophallaxis: see **Oral trophallaxis.**

Strain: The progeny of a single isolation maintained in pure culture. Sometimes refers to a group of isolates with similar physiological characters or host relationships; a group within a species that is in some way physiologically different from other members of the species, e.g. resistant to an insecticide. cf. race.

Strepsiptera, order: (*strepsi,* turning or twisting; *ptera,* wing) The order of insects comprising the twisted-winged parasites stylopids (Fig. 226); a small order of insects parasitic on other insects and commonly known as stylopids.

Stress: Any one or combination of factors that decrease the vigor of an organism.

Stretch receptors: Mechanoreceptors composed of multipolar neurons associated with muscles, the alimentary canal, and other viscera; these receptors produce a nervous impulse when associated tissue undergoes a change in length.

Stria (pl., striae): A groove or depressed line.

Striate: With grooves or depressed lines.

Striated muscle: Muscles in which myosin and actin filaments overlap to give a triated effect.

Strickland Code: A code of nomenclature prepared by a committee of the British Association of the Advancement of Science under the secretaryship of H.E. Strickland and first published in 1842.

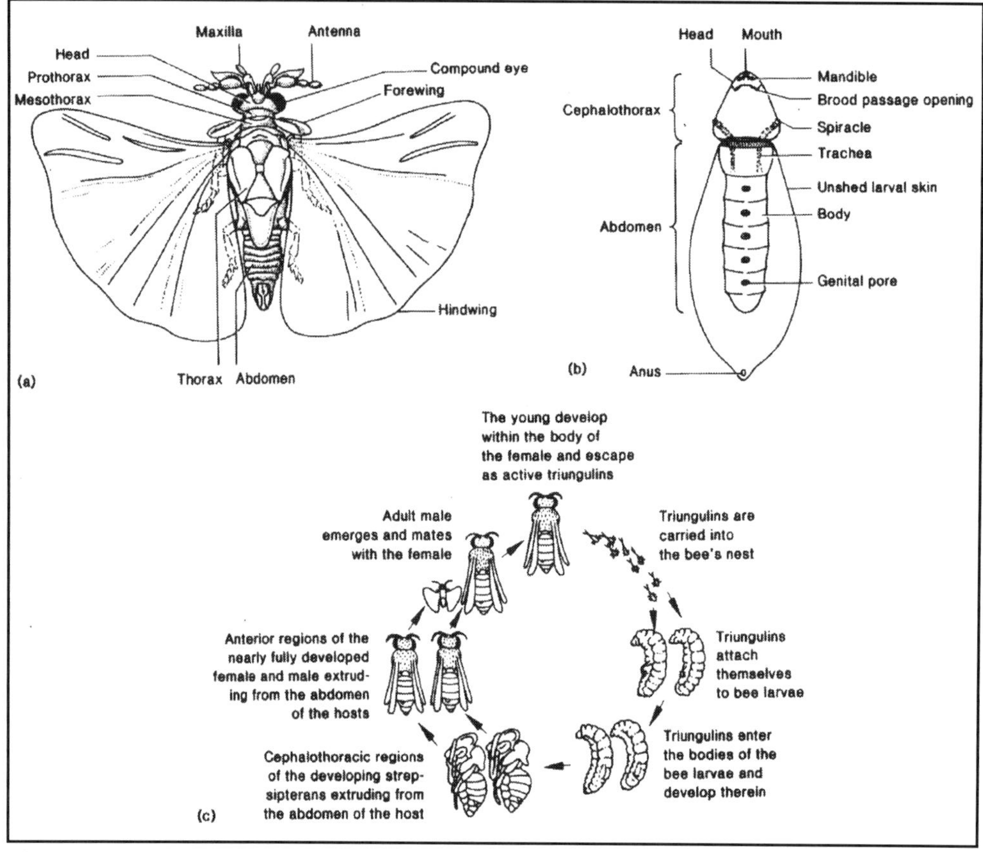

Fig. 226. Strepsitera, twisted-wing parasite. (a) Male, *Stylops medionitans.* (b) Female. (c) Life cycle. After Essig, 1942.

Stridulate: To make a noise (chirp or creak) by rubbing two rough structures or surfaces together.

Stridulation: Production of sound by means of a frictional mechanism; or sound production accomplished by rubbing two rough or ridged surfaces of body parts together.

Strip harvesting: Harvesting of different areas of a single crop at different times.

Stripe: A longitudinal color marking.

Strobilus, pl. strobilic: The part of tapeworm beyond the neck region also known as body consisting of proglottids which are broader than the segments of the neck region.

Stubble: Plant stems and stalks left in a field after harvest. See *residue.*

Stunting: A retarding effect on growth and development.

Stylate: With a style; stylelike; stylate antenna (Fig. 227).

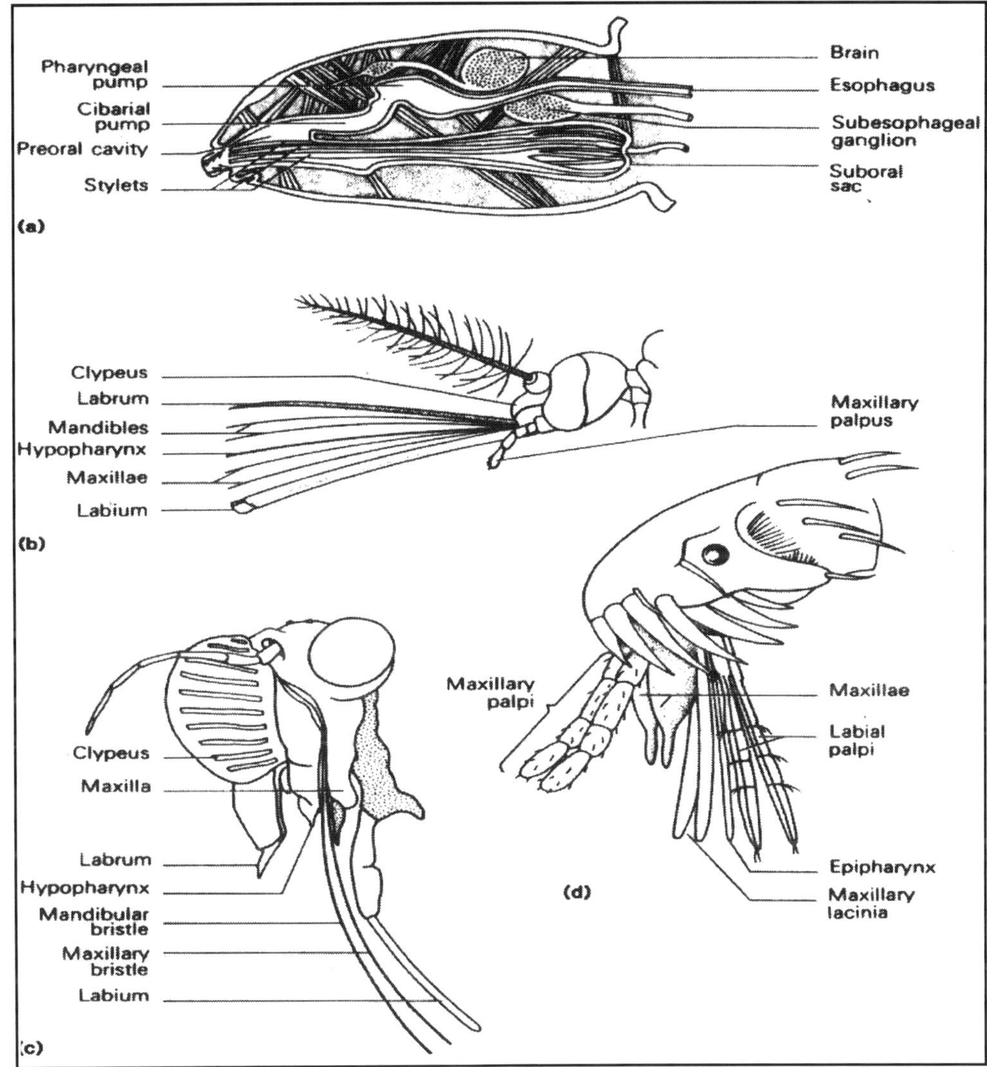

Fig. 227. Examples of stylate haustellate mouthparts. (a) Sagittal section of head of a sucking louse. (b-d) Mouthparts spread out to show details of structure: (b) mosquito; (c) cicada; (d) flea. After Vogel, 1921; after Snodgrass, 1959; after James and Harwood, 1969.

Style: A bristlelike process at the apex of the antenna; a short, slender, fingerlike process.

Style: In apterygote insects, small appendages on abdominal segments, homologous to abdominal legs.

Style: Small, pointed, fingerlike process, non-articulated; many different references in insect anatomy, often to parts of the male genitalia.

Stylet: A needle-like part of the mouthparts of insects and mites that feed with a piercing/sucking action; often refers to part of the mouthparts in Diptera and other orders, used for piercing or sucking. The protrusible oral spear that is the feeding structure of all plant-parasitic nematodes, and also of some other nematodes. It is one of the piercing structures in piercing-sucking mouthparts; or a piercing organ adopted for offensive action against plant cell walls or animal integument in case of nematodes; or a thin, sclerotized lance formed of modified mouthparts, capable of piercing a plant or animal.

Stylet-borne virus: A nonpersistent virus carried on the mouth parts of its vector.

Stylet-borne: Plant pathogens, especially viruses that are transferred between hosts on the stylets (mouthparts) of their vectors.

Styli: Simple, bilateral appendages borne on the abdominal segments (Fig. 228).

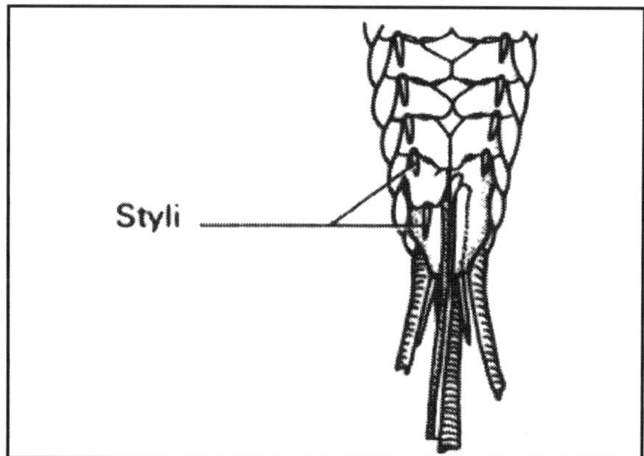

Fig. 228. Nongenital abdominal appendages: Venter of posterior portion of the abdomen of a silverfish.

Stylophore: The part of the head (capitulum) of mites that bears the stylets.

Stylopized: A condition in which an insect, such as a bee, is parasitized by a Stresipteran.

Stylus (pl., styli): A ventral, unsegmented appendage on the abdomen of bristletails; a short, slender, fingerlike process (Thysanura).

Sub -imago (alternative term, dun)**:** Winged but sexually immature mayfly adult which has to undergo another moult.

Subacute toxicity: The poisonous or injurious action of a toxin to an organism after repeated sublethal doses.

Subalar sclerite: Small sclerite posterior to the pleural wing process.

Subalare (pl. subalaria): A small sclerite in the upper part of the pleuron that articulates with the axillary sclerites; a small sclerite, one of the epipleurites that lies posterior to the pleural wing process, an attachment for the direct flight muscles.

Subantennal suture: A suture on the face extending ventrally from the base of the antenna.

Subapical: Located just proximad of the apex.

Subbasal: Located just distad of the base.

Subclass: A major subdivision of a class, containing a group of related orders.

Subcorticolous: Beneath the bark of trees.

Subcosta: In wing venation, the third longitudinal vein, between the costa and the radius.

Subcuticle: The innermost part of the endocuticle, representing newly secreted cuticle in which the microfibers are not yet oriented.

Subcuticular space: The narrow space between the endocuticle and the epidermal cells, formed during molting.

Subdiscal (or subdiscoidal) vein: The vein forming the posterior margin of the third discoidal cell (Hymenoptera).

Subeconomic pest (noneconomic pest): A pest with a general equilibrium position far below the economic-injury level; highest pest populations do not reach the economic-injury level.

Subequal: Approximately, or almost, equal in size or length.

Subesophageal ganglion: A knotlike swelling of nerve tissue at the anterior end of the ventral nerve cord, below the oesophagus (Fig. 229); or the ganglion, ventral to the foregut, that innervates sense organs and muscles associated with mouthparts, salivary glands, and the neck region; or a composite ganglion that innervates the mouthparts, located in the head below the digestive tract.

Subfamily name: The scientific designation of a subfamily, recognized by the termination *inae*, which termination, by action of the International Commission on Zoological Nomenclature, may not be used for names of other categories [for minor exceptions in trivial names, *see Bul. Zool.*

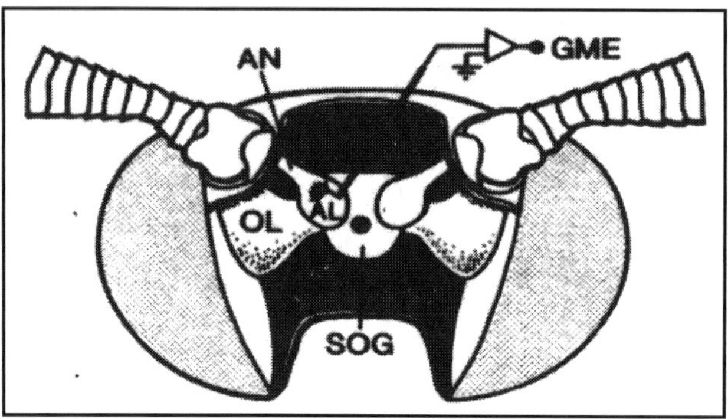

Fig. 229. An enlargement showing how one can fill an individual neuron with a dye (hexamine cobaltic chloride) and also record from the identified neuron using a glass microelectrode (GME). AN = antennal nerves; OL = optic lobe; SOG = subesophageal ganglion; AL = antennal lobe.

*Nomencl.***4:** 262 (1950)] (cf. Family name). A major subdivision of a family, containing a group of related tribes or genera. Subfamily names end in-*inae, e.g.* **Melolonthine.**

Subgenal plate: A platelike sternite that underlies the genitalia.

Subgenal suture: The horizontal suture below the gena, just above the bases of the mandibles and maxillae, a lateral extension of the epistomal suture.

Subgeneric name: The name of an optional category between the genus and the species, enclosed in parentheses when cited in connection with a binominal or trinominal combination and therefore excluded from consideration when determining the number of words of which a specific, subspecific, or infrasubspecific name is composed [*e.g., X-us (Y-us)albus rufus* is a trinominal – *see Bul. Zool. Nomencl.,* **4:** 96-97 (1950)] (cf. Nominotypical subgenus).

Subgenual organ: A chordotonal organ on the proximal tibia that detects substrate vibration. Group pf chordotonal sensilla located in the basal portion of the tibial leg segment; or an organ on the tibia of many insects, consisting of a group of scolopidia and sensitive to vibrations transmitted through the legs.

Subgenual sutures: Lines on the insect head located immediately above the bases of the mandibles and maxillae.

Subgenus (pl., subgenera): A major subdibvision of a genus, containing a group of related species. In scientific names, subgeneric names are capitalized and placed in parentheses following the genus name.

Subimago (also **subimaginal instar**): In Ephemeroptera, the winged penultimate instar; subadult. The first of the two winged instars of a mayfly after it emerges from the water; winged "subadult" that undergoes one molt and becomes the adult; unique to mayflies (Ephemeroptera).

Subjective synonym: A conditional or taxonomic synonym dependent upon an author's opinion (subject to subsequent revision) that two or more available names proposed for nominally different categories actually represent a single taxonomic category (cf. Synonym, Objective synonym).

Sublittoral: Shore zone below the littoral; shallow water zone at sea or lake edge.

Submarginal cell: One or more cells lying immediately behind the marginal cell (Hymenoptera.

Submarginal vein: A vein immediately behind and paralleling the costal of the wing (Chalcidoidea).

Submedian cell: The cell behind the median cell, in the basal posterior portion of the wing (Hymenoptera).

Submentum: Proximal portion of the postlabium; the basal portion of the labium in insect mouthparts.

Subocular suture: A suture extending ventrally from the compound eye.

Subocular sutures: Vertical lines that may run beneath the compound eyes.

Suboesophageal ganglion: The fused ganglia of the mandibular, maxillary and labial segments, forming a ganglionic centre beneath the oesophagus; the ganglion (swelling) of nervous tissue below the oesophagus in insects.

Suborder: A major subdivision of an order, containing a group of related families; a taxonomic category below the level of order, e.g. Sub-order Homoptera within the order Hemiptera.

Subphylum (pl., subphyla): A major subdivision of a phylum, containing a group of related classes.

Subquadrangle: A cell immediately behind the quadrangle (Odonata, Zygoptera).

Subsocial: A social system in which adults look after immature stages for a certain period.

Subspecies: A geographically defined aggregate of local populations which differs taxonomically from other such subdivisions of the species (cf.

Polytypic, Cline); a subdivision of a species, usually a geographic race. The different subspecies of a species are ordinarily not sharply differentiated, intergrade with one another, and are capable of interbreeding; The trinominal combination of a generic name, a specific trivial name and a subspecific trivial name which constitutes the scientific designation of a subspecies (International Commission) (cf. Subspecies, Nominotypical subspecies); the third term of the trinominal designation of a subspecies [for nomenclatural status *see Bul. Zool. Nomencl.*, **4:** 89-96 (1950)] (cf. Infrasubspecific name, Subspecific name, Subspecies).

Substitute name: A name proposed to replace a preoccupied name and automatically taking the same type and type locality (cf. New name).

Substitutional ovipositor: An ovipositor formed from extensible posterior abdominal segments; *see also* **Appendicular ovipositor.**

Subtriangle: A cell or group of cells proximal to the triangle (Odonata, Anisoptera;).

Succession: The natural sequence by which one kind of ecological community is gradually replaced by another; the progressive changes in vegetation and animal life that occur over time; groups of species that successively occupy a given habitat as the conditions of that habitat change.

Sucker: An organ to assist in locomotion and for attachment, *e.g.*, in *Taenia* sp.

Suctorial: Sucking; adapted for sucking.

Sulcate: With a groove or furrow.

Sulcus (pl., sulci): An alternate name for a suture. A suture formed by an infolding of the body wall.

Summer annual: Plants that germinate in the spring, grow in the summer, produces flowers or seeds, and dies in the fall.

Summer oil: A highly refined petroleum oil that can be applied to trees in full foliage without having phytotoxic effects.

Sunscald: The destruction of plant tissue caused by exposure to the sun.

Supercooling: Resistance to freezing by lowering the temperature at which freezing of body fluids begins.

Superfamily: The taxonomic category immediately above the family and below the order (cf. Family). A group of closely related families. Superfamily names end in *-oidea.*

Superficial cleavage: A type of embryonic development in which the cleavage nuclei migrate to the surface of the egg; cell membranes then form about each nucleus.

Superficial layer: The outermost of the three layers of epicuticle often bearing a lipid or wax and / or cement layer, with outer and inner epicuticle beneath it; *see also* **Inner epicuticle; Outer epicuticle.**

Superior appendage: One of the two upper appendages at the end of the abdomen, a circus (Odonata).

Superlingua (pl. superlinguae): A lateral lobe of the hypopharynx, the remnant of a leg appendage of the third head segment.

Supernormal stimulus: A stimulus that exceeds normal with respect to size or other properties and elicits an exaggerated response.

Superparasitism: The occurrence of more parasitoids within a host than can complete their development within the host.

Superparasitoid: A parasitoid that produces several offspring per individual host.

Superposition (eye): Compound eye type in which a superposition image is formed.

Superposition eye: A type of compound eye occurring in nocturnal insects, in which the ommatidia are not surrounded by a shield of pigment.

Superposition image: Image formed when the knight reaching the rhabdom of an individual eye unit, ommatidium, enters via the diopteric apparatuses of more than one adjacent eye unit.

Superspecies: A monophyletic group of entirely or largely allopatric species (cf. *Artenkreis,* Allopatric, Semispecies).

Supplement (spray): Adjuvant.

Supplement: An adventitious vein formed by a number of cross veins being lined up to form a continuous vein, located behind and more or less parallel to one of the main longitudinal veins (Odonata).

Supplementary reproductive: In termites, a potential replacement reproductive within its natal nest, which does not become alate; also called a **neotenic** or **ergatoid.**

Suppressive soil: Soil in which antagonistic microorganisms prevent plant pathogens from causing disease.

Supraalar bristles: A longitudinal row of bristles on the lateral portion of the mesonotum, immediately above the wing base (Diptera).

Supraesophageal ganglion: A nerve mass that lies in the head above the esophagus; the insect brain.

Supraspecific: A term applied to a taxonomic category or to taxonomic and evolutionary phenomena at a level higher than the species (cf. higher category).

Surfactant: An additive to liquid mixtures of chemicals to improve emulsifying, dispersing, spreading, or wetting characteristics of the material applied or a pesticide formulation.

Surveillance (scouting): The watch kept on a pest for detection of the species presence and determination of population density, dispersion, and dynamics.

Survivorship curve: A graphic representation of deaths over time for a given population.

Susceptible: Unable to inhibit or deter pest feeding and reproduction (cf., resistance); or capable of being easily infested or infected; not resistant.

Suspension: A liquid pesticide suspension of active ingredient particles. Constant agitation is required to maintain uniform distribution of the suspension.

Suture (s): A seam or line indicating the division of distinct parts of the body wall; or a line of indentation in the cuticle, usually forming an internal strengthening ridge; or external grooves in the insect skeleton that may show the fusion of two plates; also called sulci.

Swarm: A concerted departure or association of a large group of insects from a nest or breeding habitat.

Swath: (i) The width of ground path treated by a single passage of a sprayer of a pesticide application device other application equipment. (ii) Row of harvested crop placed in the field to dry.

Sweep: (i) Type of knife used to cultivate. (ii) A net used to collect insects off plants.

Swimmeret: An abdominal appendage that functions as a swimming organ in Crustacea.

Syconium: A receptacular fruit; for example, fig, the fruit of *Ficus* spp.

Sym-, syn: Together.

Symbiont: An organism living in symbiosis with another organism; or an organism living in intimate association with another organism.

Symbiosis: Two or more organisms living in close association and interacting with one another that benefits one or both of the organisms; The general term for two organisms that live together in a partnership, sometimes beneficial; includes commensalisms, inquilinism, mutualism, and parasitism; or interrelation between 2 organisms of different species (Gr. *syn* = together + *bios* = life); The intimate, dependent relationship between two dissimilar organisms to their mutual advantage.

Symmetry: A definite pattern of body organization; bilateral symmetry, a type of body organization in which the various parts are arranged more or less symmetrically on either side of a median vertical plane; that is, shere the right and left sides of the body are essentially similar.

Sympathetic nervous system: see **Visceral nervous system.**

Sympatric: A term applied to two or more population which occupy identical or broadly overlapping geographical areas (cf. Allopatric).

Sympatric: Overlapping geographic distributions of organisms or taxa; *see also* **Allopatric.**

Sympatric hybridization: The occasional production of hybrid individuals between two otherwise well-defined sympatric species.

Sympatric speciation: Species formation in the absence of geographic isolation (cf. Allopatric speciation). Division of a species into two or more descendant species within the same area.

Sympatric species: Species which occupy the same geographical range but isolated from each other. In other words, individuals living close enough to each other in the same area allowing the possibility of interbreeding are sympatric.

Symphilid: A member of the order Symphyla within the class Myriapoda.

Symphyla: An order of class Myriapoda comprising the symphilids.

Symphypleona: A sub-order of springtails (Collembola) which are globular in form and without visible segmentation. cf. Arthropleona.

Symptom: (i) The visible manifestation of disease in an organism. (ii) Plant injury caused by a herbicide.

Symptomless carrier: An organism infected by a pathogen without showing symptoms.

Synanthropic: Associated with humans or their dwellings.

Synanthropy: A close association with man.

Synapse (plural, **synapses**): Region of close association between terminal arborizations and dendrites; the junction between two neurons or between a neuron and another cell; the site of approximation of two nerve cells at which they may communicate; the point of interaction between adjacent neurons or between a neuron and a muscle or gland, involving a chemical neurotransmitter.

Synaptic cleft: Space between the terminal arborizations of one neuron and the dendrites of the next.

Synaptic poison: A chemical that interrupts normal transmission of impulses across synapes.

Synchronic species: Species which occur at the same time level (cf. Allochronic species).

Synchronous muscle: A muscle that contracts once per nerve impulse.

Synecology: The study of groups of interacting species a particular community and their environment.

Synergism: (i) The enhancement of the effects of two substances that is greater than the sum of their individual effects. (ii) Response to a pesticide mixture greater than the combined response of each pesticide applied alone. (iii) Damage caused by two pathogens infecting simultaneously that is greater than the sum of the damage caused by each pathogen alone; or the action of two or more factors together achieving a greater effect than the sum of the individual effects of the factors.

Synergist: A nontoxic or relatively safe chemical that, when added to a pesticide, increases the toxicity of the pesticide (e.g., piperonyl butoxide); or a substance that enhances the effectiveness of a second substance.

Synomone: A communication chemical that benefits both the receiver and the producer; *see also* **Allomone; Kairomone.**

Synonym: In nomenclature, one of two or more different names for the same taxonomic unit (cf. Senior synonym, Junior synonym, Objective synonym, Subjective synonym); two or more names for the same thing.

Synonymy: A chronological list of the scientific names which have been applied correctly or incorrectly to a given taxonomic unit, including the dates of publication and the authors applying the names; in its most abbreviated form designed for nomenclatural purpose only (cf. Full bibliographical synonymy).

Synopsis: In taxonomy, a brief summary of current knowledge of a group; the inclusion of new material or new interpretations is not necessarily implied (cf. Review, Revision, Monograph).

Synthetic organic pesticide: Manmade pesticide which contains carbon, hydrogen, and other elements.

Syntype: "One of a number of specimens of equal nomenclatural rank which formed all or part of the material before the original author in those cases where the author did not designated or indicate a holotype" (I.C.Z.N.) [isee Bul. Zool. Nomencl., **4:** 186 (1950)] (cf. Cotype, Lectotype).

System: An interrelated set of elements that can be identified and around which a boundary can be drawn.

Systematic serology: The application of serology to taxonomic problems; comparative serology (cf. Serology).

Systematics (alternative term, taxonomy)**:** The study of the naming and classification of living organisms; the science of biological classification; the classification of animal and plant species into their higher taxa; sometimes regarded as synonymous with taxonomy; or the study of the diversity and classification of organisms.

Systemic: (i) A chemical that when applied to an organism spreads throughout the organism through the vascular system. (ii) A pathogen infection that spreads throughout the host.

Systemic insecticides: Having the property of being absorbed into and translocated by the sap stream of a plant or the bloodstream of an animal; an insecticide taken into the body of a host (plant or animal) that kills insects feeding on the host; a synthetic organic insecticide that, when absorbed by plants or ingested and absorbed by animals, makes them toxic to insects (e.g., aldicarb).

Systems technology: The assembly, organization, and employment of an identifiable set of interrelated elements that constitute a system.

Systole: Contraction phase of the heartbeat cycle.

T

Tactile: Referring to the sense of touch.

Taenidium (pl. taenidia): The spiral thickening of the tracheal wall that prevents collapse; helical fold in the cuticular lining of tracheae and tracheoles (Fig. 230); cuticular ridges that support the walls of tracheae; a spiral or circular thickening of the inner surface of the tracheae.

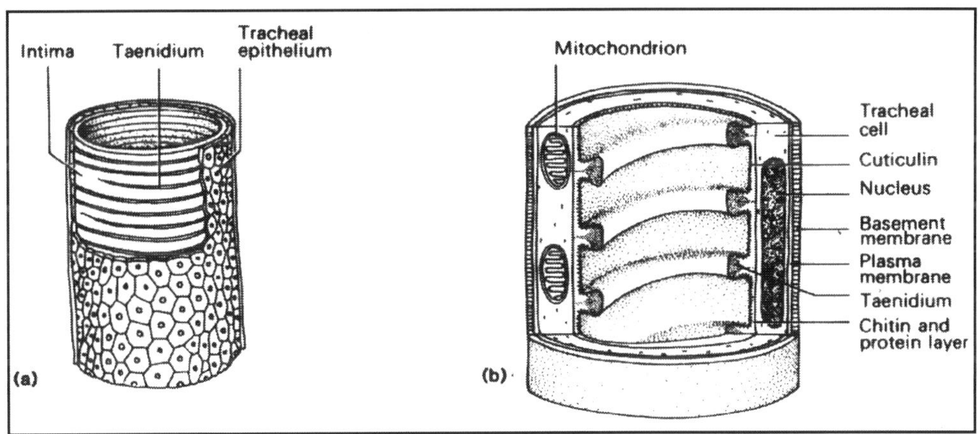

Fig. 230. Tracheal structure. (a) Portion of trachea. (b) Fine structure of trachea.

Tagma (pl. tagmata): The group of segments that form a major body unit (head, thorax, abdomen).

Tagmata: The three major regions of the insect body; head, thorax, and abdomen.

Tagmosis: The differentiation of the body into distinct functional regions, or tagmata; the organization of the body into major units (head, thorax, abdomen).

Taint: Unwanted flavour in fresh or processed food from a pesticide used on the growing crop.

Tandem running: A form of recruitment behaviour in which one individual follows another, maintaining antennal contract.

Tandem: One behind the other, the two connected or attached together.

Tanglefoot: A sticky substance used in traps to immobilize insects. Sometimes applied in a band around tree trunks to prevent insects from climbing the trunk.

Tank mix: Two or more formulated pesticides combined in a spray tank for single application.

Tanning hormone: See bursicon.

Tapetum: A reflective layer at the back of the eye formed from small tracheae; a basal layer of tracheae in an ommatidium that reflects light.

Taproot: The large primary root that grows vertically downward, from which lateral roots arise.

Target cover: The degree to which the crucial portion (target) of a plant (or other sprayed object), receives a deposit of insecticide.

Target surface: The surface intended to receive a spray or dust application.

Target: Object onto which a pesticide is to be deposited by spraying or other means; the subject towards which a control tactic, such as pesticide application, is directed.

Target-site resistance: Increased tolerance by an insect to an insecticide through reduced sensitivity at the target site.

Tarsal claw: A claw at the apex of the tarsus.

Tarsal formula: The number of tarsal segments on the front, middle, and hind tarsi, respectively.

Tarsomere: A subdivision, or "segment", of the tarsus.

Tarsus (pl. tarsi): The last (terminal) segment of the leg, immediately distal to the tibia; consists of one or more subdivisions or tarsomeres and bearing the pretarsus with its claws.

Taste: Chemoreception of chemicals in a liquid dissolved form.

Taxes: Directed response relative to a stimulus source.

Taxis (pl., taxes): A directed (orientated) response involving the movement of an animal toward or away from a source of stimulation.

Taxon (pl. taxa): Any taxonomic unit or category (pl., taxa) (species, genus, family, phylum, etc.).

Taxonomic category: One of a hierarchy of levels into which natural populations are classified, such as subspecies, species, genus, and family.

Taxonomic character: Any attribute of an organism or of a group of organisms by which it differs from an organism belonging to a different taxonomic category or resembles an organism belonging to the same category (cf. Biological character).

Taxonomy (alternative term, systematics)**:** The study of the naming and classification of living organisms.

Taxonomy (adj. taxonomic)**:** The science of classification into categories of varying rank, based on similarities and differences, and the describing and naming of these categories (cf. Classification, Systematics); the theory and principles of classifying organisms; or the laws of classification as applied to natural history; identification of plant and animal species.

Technical material: Undiluted, nonformulated pesticide in its pure form as produced by the manufacturer.

Technical product: The usual form in which a pesticide is prepared and handled prior to formulation; usually at a high level of purity (95-98%) but not completely pure.

Technical-grade material: The relatively pure form of a pesticidal compound that comprises the active ingredient in the final pesticide mixture.

Tegmen (pl. tegmina): A leathery, hardened fore wing; serving as a protective covering of the hindwing, as in in the order Orthptera and Dictyoptera e.g. grasshoppers, locusts, crickets, katydids, and related forms.

Tegula (pl. tegulae): A small scalelike structure overlying the base of the front wing; one of the articular sclerites of the neopteran wing, lying at the base of the costa; *see also* **Axillary sclerites; Humeral plate**.

Tegula (plural, tegulae): A small, sclerotized flap overlying the base of the forewing of certain insects.

Teleology: Ascribing purposiveness to natural phenomena.

Telispore: A thick-walled dark spore of rust and smut fungi that is able to survive adverse conditions.

Telotaxis: Orientation and movement directly towards a visual stimulus.

Telotrophic (or **acrotrophic**) **ovariole:** An ovariole in which the nurse cells are only within the germarium; the nurse cells remain connected to the oocytes by long filaments as the oocytes move down the ovariole; *see also* **Panoistic ovariole; Polytrophic ovariole**.

Telson: The last abdominal segment of crustaceans. See periproct.

Temporal memory: The ability to compensate for the passage of time during locomotory behavior such as foraging from the nest.

Tenacity index: Ratio of the quantity of residue per unit area at the end of a given period of weathering to that present at the beginning.

Tenacity: The property of a pesticide deposit or residue to resist removal by weathering.

Tenaculum: The clasplike structure associated with the abdominal structure called furcula found on the venter of Collembola; serves as a "catch" for the furcula prior to "springing" (Fig. 231).

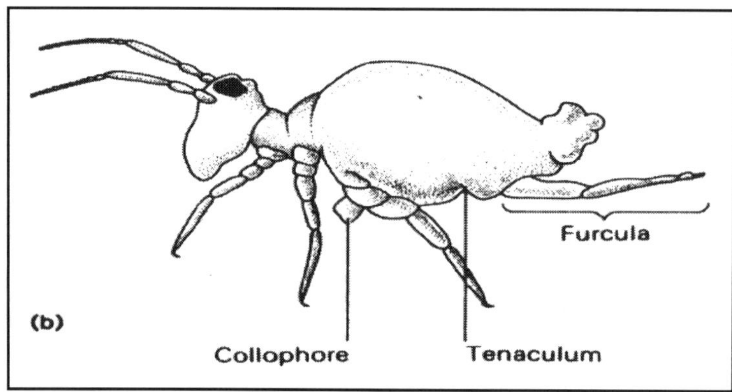

Fig. 231. Springtail: Tenaculam.

Tendinous epidermal cells: Specialized cells that produce the microtubules that make up the tonofibrillae.

Tenent hairs: Tiny hairs on the pulvilli and tarsal pads that allow some insects to cling to smooth surfaces (Fig. 232).

Teneral: Newly emerged, pale, soft-bodied adult insect individuals prior to the completion of melanization and sclerotization.

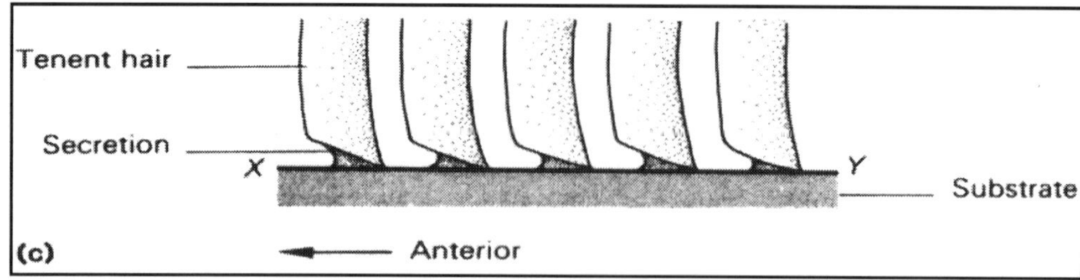

Fig. 232. Tenent hairs that compose the climbing organ *Rhodnius prolixus* (Hemiptera; Reduviidae).

Tentacle: An elongate flexible appendage usually near the mouth. (L. *tento* = from + *tenin* = hold).

Tentiform: Shaped like a tent.

Tentorial bridge: See corporotentorium.

Tentorial pits: Pitlike depressions on the surface of the head that mark the points of union of the arms of the tentorium with the outer wall of the head. There are usually two tentorial pits in the epistomal suture and one at the lower end of each postoccipital suture.

Tentorium: An internal framework in the insect head that provides rigidity and areas for muscles attachment; the endoskeleton of the head, formed from invaginations of the body wall; a set of apodemes that form the internal bracing of the head.

Teratogen: Something or a material capable of causing birth defects.

Teratogenesis: The malformation of a fetus, with structural abnormalities present at birth or becoming evident after birth.

Teratogenic: Including birth deformities.

Teratology: The study of structural abnormalities, especially monstrosities and malformations.

Tergite (the *g* has the soft or *j* sound)**:** A sclerite of the tergum; the dorsal surface of an abdominal segment.

Tergosternals: Abdominal muscles in a dorsoventral orientation.

Tergum (also called **tergite**) (pl. terga; adj. tergal; dim. tergite)**:** Dorsal region of the thorax and abdomen; usually called notum in reference to the thorax.

Terminal arborizations: Branching processes found at the end of an axon.

Terminal filament: A thread made up of a contribution from the apex of each ovariole, which anchors the cephalad end of each ovary to the dorsal diaphragm.

Terminal: At the end; at the posterior end (of the abdomen); the last of a series.

Terminalia: The terminal abdominal segments involved in the formation of the genitalia.

Termitaria: The nests build by termites under or above the ground.

Termiticide: An insecticide that kills termites.

Termitophile: A symbiont living in the nest of termites.

Termitophilous: Living in the nest of termites; a nest inquiline.

Terpene: A naturally-occuring compound in which the basic structural component is isoprene, $CH_2C\ (CH_3)\ CH_2CH_2$.

Terrestrial: Belonging to or living on the dry land or ground or earth. (L. *terra* = earth).

Territoriality: Occupation of a site that is defended from other individuals of the same sex and species.

Tertiary parasite (as applied to insects)**:** One which is parasitic on a secondary parasite.

Testes: In the male reproductive system, the bilateral, paired structures housing the germ cells.

Testicular follicles: Tubules in which the sperm are formed. It is one to several sperm tubes composing the testes.

Testis (pl. testes): One of (usually) a pair of male gonads; the male sex organ that produces sperm; or the male sex cells; in vertebrates sex hormones are also produced in this organ. *e.g. Hydra,* materials and others.

Tetrapoda: A vertebrate typically with 4 limbs – the amphibians, reptiles, birds, and mammals. (Gr. *tetra* = four + *pous* = foot).

Thanatosis: Feigning death.

Thecogen cell: A cell that surrounds the glial cell and the outer section of a sensory neuron.

Thelytoky (also **thelytokous parthenogenesis**): Parthenogenesis producing only female offspring.

Therapeutic practice: A pest management approach that attempts to cure existing problems.

Thermal constant: The number of degree-days required for an event (e.g., pupation) to occur.

Thermocline or metalimnion: The region of freshwater media between epilimnion and hypolimnion which shows the maximum change in temperatures.

Thermoreception: The reception of stimuli as heat.

Thermoregulation: Adjustment of body temperature.

Thigmotactic: Contact-loving; insects that inhabit crevices; also refers to insects that live very close together-touching.

Thigmotaxis: A taxis based on lateral contact.

Thinning: The removal of some plants or plant parts from row crops or orchard trees so that the remaining plants have appropriate room to grow and develop.

Thiocyanate: An organic insecticide that contains the SCN group; has a creosotelike odor; interferes with cellular respiration and metabolism.

Thoracic ganglia: Three ganglia located posterior to the subesophageal ganglion and innervating the three thoracic segments.

Thoracic glands: Masses of tissue usually closely associated with tracheae, that secrete the molting hormone, ecdysone; also called prothoracic glands or ecdysial glands.

Thorax, cephalo: Head and thorax coleased into one *e.g.* in Arachnida.

Thorax: The body region between the head and abdomen which bears the legs and wings.

Thorax: The central tagma of an insect's body between the head and the abdomen comprising pro-, meso- and metathorax that bears the legs and the wings.

Threshold: The minimum level of stimulus required the initiate (release) a response.

Thysanoptera, order: (*thysano,* a fringe; *ptera,* wing) The order of insects comprising the thrips (Fig. 233).

Thysanura, order: (*thysan,* fringe; *ura,* tail) The order of insects that includes three-pronged bristle tails comprising the silverfish, firebrats, and relatives (Fig. 234).

Tibia (pl. tibiae): The fourth segment of the leg, between the femur and the tarsus.

Tibial spur: A large spine on the tibia, usually located at the distal end of the tibia.

Tick: A member of the order Acari within the class Arachnida; ectoparasitic on higher animals.

Till or tillage: To cultivate and prepare the soil for use in agriculture, to kill weeds and make it more favorable for crop growth.

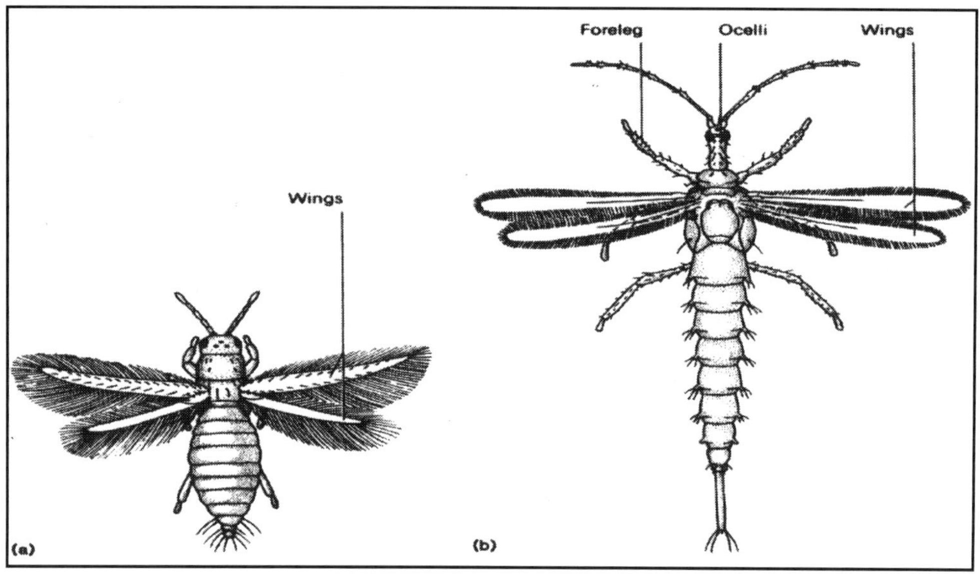

Fig. 233. Thysanoptera, thrips. (a) Member of Terebrantia. (b) Member of Tubulifera, female.

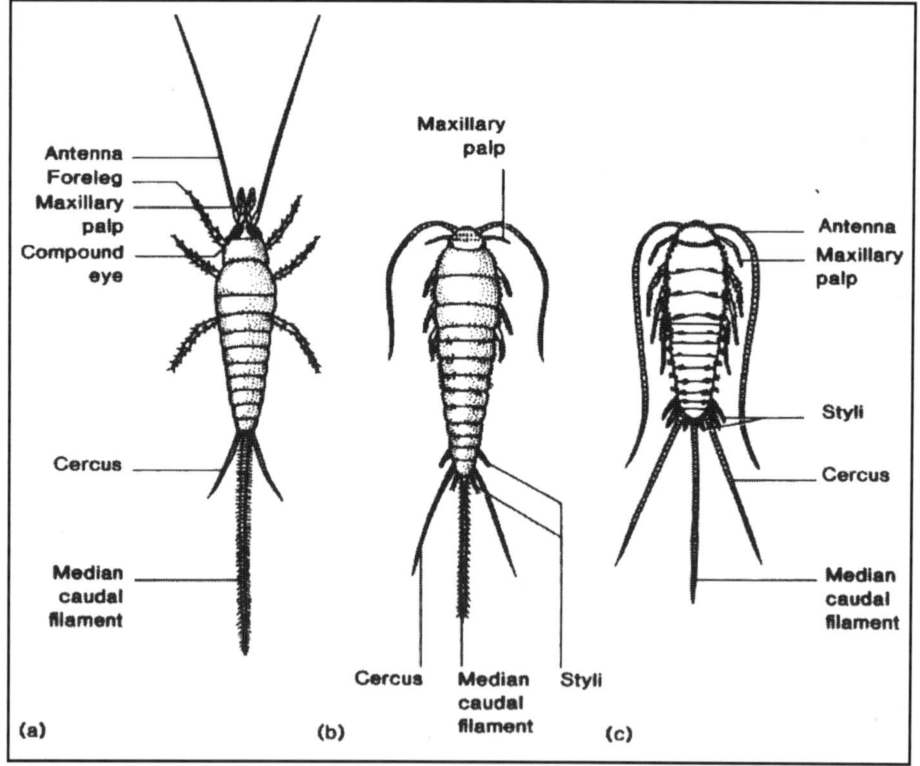

Fig. 234. (a) Archeognatha, jumping bristletail, *Machilis* sp. (b) Thysanura, common silverfish, *Lepisma saccharina.* (c) Thysanura, firebrat, *Thermobia domestica.*

Tiller: The young vegetative basal shoot in grasses.

Tillering: The process when a grass or cereal plant develops secondary shoots.

Time interval: The time that is required between the final pesticide application and harvesting; the time interval ensures that the legal residue tolerance will not be exceeded.

Tissue: A group of cells structurally and functionally similar. (L. *texo* = weave).

Token stimuli: Specific environmental cues that induce diapause or elicit a specific behavioral response only at certain times; or stimuli that by themselves may not have relevance to survival (e.g., substances that lack nutritive value but that act as olfactory attractants to a food source).

Tolerance (as applied to pesticide): Legally defined maximum permitted level of a pesticide on or in produce at harvest.

Tolerance (as applied to plant resistance to pests): A mechanism of plant resistance to pests where by the plant is able to grow and produce despite pest injury; the ability of a plant to withstand insect attack and recover.

Tolerance, permitted: Maximum amount of toxicant allowed in foodstuff for human consumption.

Tolerance: (i) The ability of an organism to withstand unfavorable conditions such as pest attack, extreme weather, or pesticides. (ii) The amount of residue (in ppm) of pesticide that may legally remain in or on any food crops at the time of sale. (iii) Ability to endure infestation (or infection) by a particular pest (or pathogen) without showing severe symptoms of distress. (iv) The ability of a host to withstand injury by pests; involves a plant response only.

Tolerant: (i) Pests that are not controlled by a particular pesticide treatment. (ii) Plants that can withstand pest attack or pesticide application with no appreciable reduction in yield.

Tonic: Characterized by continuous potency of a material.

Tonofibrillae: Fibrillar structures (bundles of microtubles) of skeletal muscle that anchor muscle to the integument (Fig. 3.18b, 3.19), microtubules.

Tonofibrillae: The fine connective fibrils extending from the ends of the skeletal muscles into the cutcle.

Top sample: A pest management monitoring sample taken from the top of the tree.

Topotype: A specimen collected at the type locality.

Tormogen cell: An epidermal cell that forms the socket and membrane around a seta; or an epidermal cell secreting a ring of cuticle that connects a cuticular hair to the cuticle.

Tortricid: A member of the family Tortricidae (Lepidoptera).

Toxic: Poisonous. Able to cause injury by contact or systemic action to plants, animals, and people.

Toxicant: Poison. An agent capable of being toxic; or chemical exhibiting toxicity.

Toxicity: The degree to which a chemical is poisonous to an organism; the ability to injure; or ability to poison, or to interfere adversely with vital processes of the organism by physico-chemical means.

Toxicogenic: Producing disease symptoms as a result of an introduced toxin.

Toxicognath: A poison jaw (centipedes) (a modified leg).

Toxicology: The science or study of poisons.

Toxin: A poison.

Tracer: Additive to facilitate location of a deposit, by radioactive or fluorescent means.

Trachea (pl. tracheae): A cuticle-lined air-conducting tube of the respiratory system of the insect body that begins with the spiracle and ends with the tracheoles; an air tube; the windpipe of land vertebrates from the glottis to the lungs; part of the respiratory system in insects and other arthropods (Gr. *trachys* = rough); a system of branching tubules through which a gaseous exchange is accomplished in insects a tube of the respiratory system, lined with taenidia, ending externally at a spiracle, and terminating internally in the tracheoles (Fig. 235).

Tracheal end cell: End of a trachea; typically branches into several tracheoles.

Tracheal gill (s): A heavily tracheated extension of the body, permitting extraction of oxygen from water; or integumental evaginations, covered by thin cuticle and supplied with tracheae and tracheoles; found in many aquatic (Fig. 235) and some endoparasitic insects.

Tracheal system: The insect gas exchange system, comprising trachea and tracheoles.

Tracheole (s): A delicate tubule extending from a trachea, the site of gas exchange between the tracheal system and the tissues; the minute terminal branches of the tracheal system.

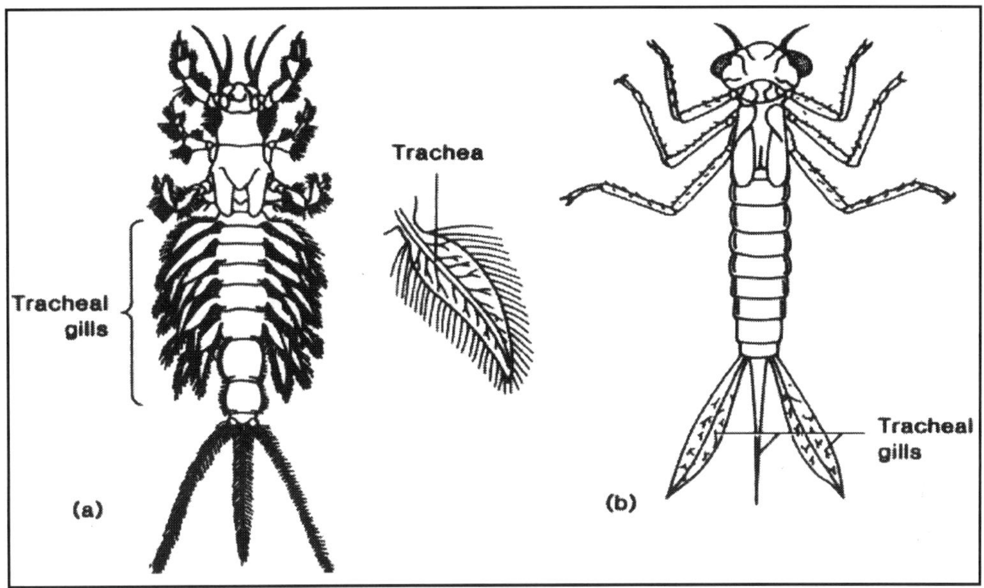

Fig. 235. Ventilatory structures in aquatic insects. (a) Lateral abdominal tracheal gills in a mayfly nymph. (b) Terminal abdominal tracheal gills in a damselfly nymph.

Trade name (proprietary name, brand name): Name given to a pesticide by the manufacturer or formulator (e.g., Sevin).

Trail-marking pheromone: An intraspecific chemical messenger produced by foraging ants and termites to indicate sources of requisites to other colony members.

Transgenic: Plants into which genes have been inserted from another organism using genetic engineering technology. Often inaccurately described by the term GMO.

Translaminar: A pesticide which passes through from one surface of a leaf to the other (from lamina to lamina) through the leaf tissue.

Translocation: Movement of water, nutrients, chemicals, pathogens, or photosynthates within a plant; the uptake of a pesticide into part of a plant body and its subsequent dispersal to other parts of the plant body.

Translucent: Allowing light to pass through, but not necessarily transparent.

Transovarial transmission: The transmission of micro-organisms between generations via the eggs.

Transovarial: Transmission via the egg from mother to offspring, for example, transovarial transmission of symbiotic microbes from one generation to the next.

Transpiration: The evaporation of water from plant leaves through stomata; or loss of water through the integument or tracheae by evaporation.

Transverse costal vein: A cross vein in the costal cell (Hymenoptera).

Transverse cubital vein: A transverse vein connecting the marginal and cubital veins (Hymenoptera).

Transverse marginal vein: A cross vein in the marginal cell (Hymenoptera).

Transverse median vein: A cross vein between the median or discoidal vein and the anal vein (Hymenoptera).

Transverse orientation: Body oriented at a fixed angle relative to stimulus direction.

Transverse radial vein: Same as transverse marginal vein.

Transverse suture: A suture across the mesonotum (Diptera).

Transverse tracheal commissures: Connections between longitudinal tracheal trunks.

Transverse: Across, at right angles to the longitudinal axis.

Transverse: At right angles to the longitudinal axis.

Trap crop: Plants attractive to certain pests, planted to divert these pests from the main crop; or a small area of a crop used to divert pests from a larger area of the same or another crop; or crop of plants (sometimes wild plants) grown especially to attract insect pests, and when infested either sprayed or collected and destroyed. Trap plants usually grown between the rows of the crop plants or else peripherally; or a crop planted because of its attractiveness to certain pests and then destroyed or treated so as to destroy the insects.

Trapping-out: The removal of individuals from a pest population, in significantly large numbers, by means of trapping (often using u.v. light traps).

Traumatic insemination: Ejaculation of semen directly into the hemocoel; characteristic of bed bugs and close relatives; also called hemocoelic insemination.

Trawling: A method adopted in deep-sea fishing, a wide-mouthed bag-net dragging along the bottom (Fr. *Trauler* = troller + to go hither and thither).

Treated area: Location where a pesticide has been applied.

Treatment level: See *economic threshold.*

Trench fever: A nonfatal disease of humans caused by infection with the rickettsia *Rochalimaea quintana* and transmitted via the faeces or crushed bodies of the louse *Pediculus humanus humanus.*

Triad (adj. triadic): Triplets of long wing veins (paired main veins and an intercalated longitudinal vein).

Triangle: A small triangular cell or group of cells near the base of the wing (Odonata, Anisoptera).

Tribe: A unit of taxonomic category intermediate between the genus and the subfamily into which groups of related genera are placed. Insect tribe names always end in-ini, e.g. Scarabaeini.

Trichobothria: Minute sensory hairs on the tarsi (spiders).

Trichogen cell: An epidermal cell that secretes a cuticular process, such as a hair; or cell that forms a seta.

Trichogen cell: The epidermal cell from which a seta develops.

Trichoid sensillum: A seta modified for reception of stimuli which is hair-like cuticular projection; a seta, hair or macrotrichium.

Trichome: A hair like outgrowth of a plant that may serve various functions, including defense against insect attack.

Trichoptera, order: (*tricho,* hair; *ptera,* wing) The order of insects comprising the caddisflies (Figs. 236, 237 & 238).

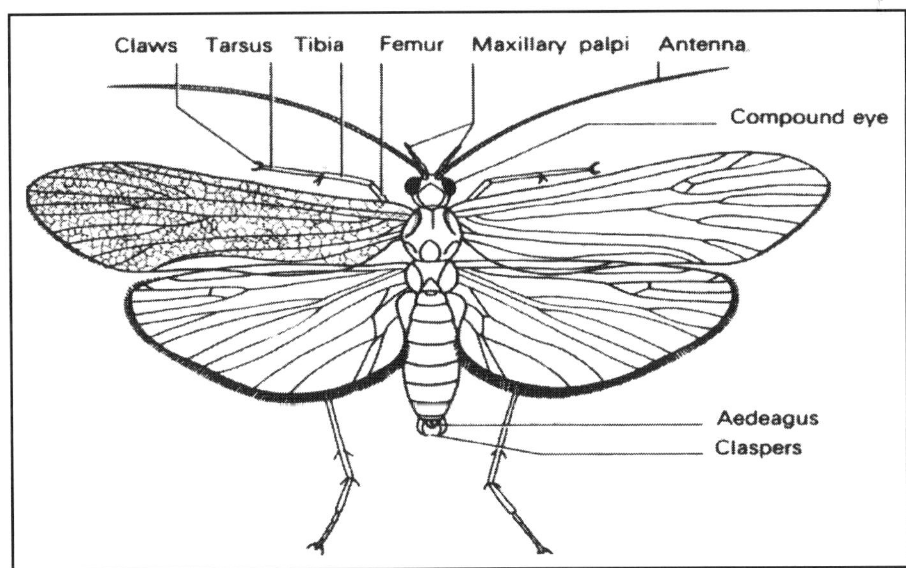

Fig. 236. Trichoptera, a caddisfly.

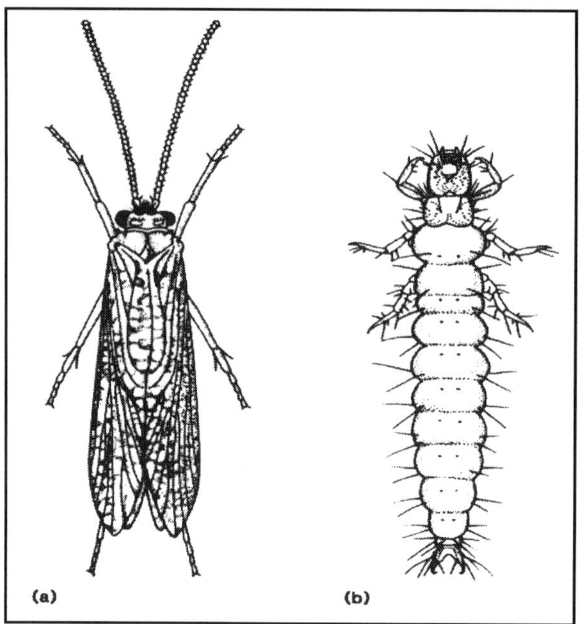

Fig. 237. Caddisfly, *Rhyacophila fenestra* (Rhyacophilidae). (a) Adult in resting posture. (b) Larva.

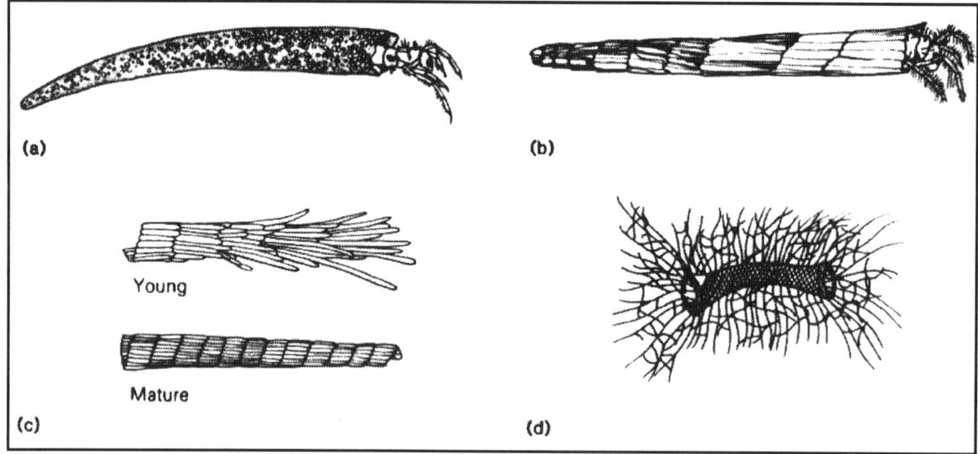

Fig. 238. Examples of caddisfly larval cases (a) Leptocella (b) Triaenodes (c) Agrypina (d) Plycentropus.

Trigonal brace: A strong forked vein in the basal part of the wing, formed by the fork of Cu_1 in the front wing and of M_{3+4} in the hind wing (Neuroptera, Ascalaphidae and Myrmeleontidae).

Trigonal vein: The vein forming the anterior fork of the trigonal brace (Neuroptera, Ascalaphidae and Myrmeleontidae).

Trinominal nomenclature: An extension of the binominal system of nomenclature to permit the designation of subspecies by a three-word name (cf. Binominal nomenclature, Subspecies, subspecific name, Nominotypical subspecies).

Tripectinate: Having three rows of comblike branches.

Triploblastic: Metazoans with three body layers (ecto, meso and endoderm) *e.g.*, all metazoans above the evolutionary level of coelentrata.

Tritocerebrum: The posterior pair of lobes of the brain that connect to the visceral nervous system; the most posterior part of the brain, which connects to the ventral nerve cord; or ventral lobe of the insect brain.

Triungulin: The active, tiny, first-instar larva of insects that undergo hyper metamorphosis e.g. Strepsiptera and certain parasitoid beetles such as Meloidae (Coleoptera).

Trivial movement: Displacement of insects within or close to the breeding habitat; nonmigratory movement; or movement of an animal within its normal habitat, not involving dispersal.

Trivial name: The second or third word in a binominal or trinominal name of an animal; the specific and subspecific components in the scientific designation of an animal (cf. Specific trivial name, Subspecific trivial name, Infrasubspecific trivial name).

Trivoltine: Three generations per year.

Trochanter: The second segment of the leg, between the coxa and the femur; or typically small segment of the insect leg that articulates with the coxa and usually forms an immovable attachment with the femur.

Trochantin: A small sclerite on the thoracic pleuron of many generalized pterygote insects immediately anterior to the base of the coxa; provides one of the coxal articulation points (Coleoptera).

Troglobite (also **troglobiont; troglodyte,** adj. troglodytic): Cave-dweller; *see also* **Cavernicolous.**

Tronchanter: The second segment of the leg, between the coxa and the femur.

Trophallaxis: Nutrient exchange between members of a social insect colony; may be from the mouth (stomodeal) or from the anus (proctodael); reciprocal feeding; the practice in some social insects of exchanging alimentary liquids.

Trophamnion: In parasitoids, the enveloping membrane surrounding the polyembryonically derived multiple individuals that arise from a single egg, derived from the host's haemolymph.

Trophic relationships: Feeding interactions among organisms. A description of who eats whom.

Trophic: Relating to food; of an egg of a social insect, degenerate and used in feeding other members of the colony.

Trophic level: Each organism requires nutrition for its survival and completing its life-cycle. If the food runs short, it makes its survival difficult in spite of the environmental factors remaining the same. An ecosystem presents complex of communities. The organisms which feed on the plants by the same number of steps are said to belong to the same trophic (food) level.

Trophocyte: The dominant metabolic and storage cell of the fat body.

Trophogenesis (adj. trophogenic): In social insects, the determination of caste type by differential feeding of the immature stages (in contrast to genetic determination of caste).

Trophogenic polymorphism: Polymorphism resulting from differences in the quantity or quality of food provided to the larvae.

Trophozoite: The first stage in the asexual cycle of a malaria parasite (*Plasmodium* spp.), derived from a merozoite (either from a liver cell or an erythrocyte).

Tropism: The orientation of an animal with respect to a stimulus, either positive (turning toward the stimulus) or negative (turning away from the stimulus).

True mouth: The opening at the base of the hypopharynx, within the preoral (cibarial) cavity formed by the mouthparts.

Truncate: Cut off square at the end.

Truss cell: The cell immediately behind the point of fusion of Sc and R_1 (Neuroptera).

Trypanosome: A protozoan parasite of the blood and lymph, usually transmitted from host to host by biting flies. The most notorious are probably the species of *Trypanosoma* that cause "sleeping sickness" in people in Africa.

Trypanosomiasis: A disease caused by *Trypanosoma* protozoans, transmitted to humans predominantly by reduviid bugs (Chagas' disease) or tsetse flies (sleeping sickness).

Tuber: An enlarged, fleshy underground stem, e.g., potato and nutsedges.

Tubercle: A small knoblike or rounded protuberance.

Tubular body: Array of microtubules in the distal process of a receptor cell; deformation of this structure initiates a nervous impulse.

Tylus: The clypeal region of the head (Hemiptera).

Tymbal: A sclerotized plate in the sound-producing organ of a cicada; a stretched elastic membrane capable of sound production when flexed.

Tympanal hood: One of a pair of tubercles or rounded prominences on the dorsal surface of the first abdominal segment (Lepidoptera).

Tympanic organ (s): The eardrum like structure of certain insects, consisting of a thin portion of integument and associated scolopidia that perceive sound; or auditory organs composed of a thin integumental area (tympanic membrane) and a group of chordotonal sensilla attached to the entad surfaces.

Tympanum (pl. tympana) or **tympanal organ:** Any organ sensitive to vibration, comprising a tympanic membrane (thin cuticle), an air sac and a chordotonal sensory organ attached to the tympanal membrane; an auditory membrane; or a vibrating membrane involved in hearing; the eardrum or tympanic membrane. (Gr. *Tympanon* = drum).

Tyndall blue: A blue color produced by colloidal-sized granules e.g. in the epidermals cells of some dragonflies and in neurosecretory cells.

Type designation: Determination of the type of a genus under Art. 30, Rule (g) of the International Rules of Zoological Nomenclature (cf. Type indication, Type selection).

Type indication: Determination of the type of a genus under Art.30, Rules (b), (c), and (d) of the International Rules of Zoological Nomenclature (cf. Type designation, Type selection).

Type locality: The locality at which a holotype, lectotype, or neotype was collected (cf. Topotype).

Type method: The method of preserving the identity of a taxonomic category by fixing an included zoological object as "type."

Type selection: Determination of the type of a genus under Art.30, Rule (g) of the International Rules of Zoological Nomenclature (cf. Type indication, Type designation).

Type species: The expression recommended by the International Comission on Zoological Nomenclature to refer to the concept of "a type species of a genus" [*see Bul. Zool. Nomencl.*, **4:** 300 (19500] (cf. Genotype).

Type: A zoological object which serves as the base for the name of a taxonomic category (*e.g.*, a specimen which is the name bearer for a species, a species which is the name bearer for a genus, etc.); forms designated when a species or group is described to be most representative or typical, to serve as the reference if there is any question about what the species or group includes. The type of a species or subspecies (the holotype) is a specimen, the type of a genus or subgenus is a species, and the type of a tribe, subfamily, family, or superfamily is a genus.

Typhlosole: A longitudinal fold projecting into the intestine in certain invertebrates to increase digestive surface. (Gr. *typhlos* = blind + *solan* = channel) *e.g.* in earthwotm.

Typhus fever: A human disease caused by the bacterium like microorganism *Rickettsia prowazekii* and transmitted by the body louse *Pediculus hunanus hunanus* (See epidemic typhus fever).

Typology: In taxonomy, the method of approach to classification which involves the postulate that all members of a taxonomic unit conform to a given morphological "type."

U

Ultra low volume spraying (ULV): The application of pesticides in extremely concentrated or undiluted form using 10 liters or less per hectare (1 gal per acre); highly concentrated pesticide applied without dilution at a rate of 0.5 gallons per acre or less.

Umbilical cord: The cord containing blood vessels supported by connective tissue that unites the embryo or fetus of a mammal with the mother during development in the uterus. (L. *umbilicus* = navel).

Ungulate: Having hoofs, as a deer or (L. *ungula* = hoof).

Unguligrade: Walking or adopted for walking on hoofs. (L. *ungula* = hoof + *gradior* = to step).

Ungue: Pretarsal claw.

Unguiculate: Having claws, as a cat. (L. *unguis* = claw).

Unguis (pl. ungues): Claw.

Unguitractor plate: The ventral sclerite of the pretarsus that articulates with the **claws**.

Uniform coverage: The even deposition of a pesticide applied over a target.

Uninominal nomenclature: The designation of a taxonomic category by a scientific name consisting of a single word; required for the categories above the species, but occasionally also advocated for the species.

Union: The point at which scion and stock are joined in grafting.

Unipolar neuron: Neuron with a single stalk from the cell body connecting with the axon and collateral.

Uniporous: Having a single opening; with a single pore.

Unisexual: Consisting of or involving only females.

Univoltine: Having one generation per year (cf., multivoltine), *see also* **Bivoltine; Moltivoltine.**

Urate cell: A cell of the fat body that stores uric acid (nitrogenous waste).

Urea: A minor component of insect nitrogenous excretion, $CO\ (NH_2)_2$.

Ureter: The duct carrying urine from kidney to the urinary bladder or to the cloaca. (Gr. *ouron* = urine).

Urethra: The duct by which urine is discharged from the bladder to the outside in mammals and connected with the *vasa deferentia* in the male (Gr. *ouron* = urine).

Uric acid: The primary nitrogenous waste excretion product in insects, $C_5H_4N_4O_3$.

Uricotelism: An excretory system based on uric acid excretion.

Urinogenital: Pertaining to the excretory and reproductive organs and functions. (L. urina = urine + genitalis = genital).

Urocyte (also **urate cell**): A cell that acts as a temporary store for urate excretion products.

Urogenital: See urinogenital.

Urogomphi (sing., **urogomphus**): Fixed or movable paired ceruslike processes on the last segment of a beetle larva (also called pseudocerci or corniculi).

Uropod: One of the terminal pair of abdominal appendages, usually lobelike (Crustacea).

Urticating: Stinging; nettling; setae or chemicals that cause skin irritation, either mechanical or chemical.

USDA: United States Department of Agriculture.

Uterus: The enlarged posterior portion of an oviduct in which eggs may be retained for development. (L. womb).

V

Vagina: A pouch-like or tubular genital chamber of the female genitalia; external opening of the female reproductive system; or the terminus of the female reproductive tract that opens to the outside; or the terminal portion of the female reproductive tract which receives the copulatory organ of a male in mating (L. sheath).

Valid name: The name of a taxonomic category which is available nomenclaturally and also is recognized as valid on Zoological grounds (cf. Available name).

Valvar lamina: The posterior margin (usually prolonged posteriorly) of the eighth abdominal sternite (female Odonata).

Valve: In animals, any structure that limits or closes an opening; the thin folds in veins, lymph vessels, or hearts or the circular muscles about a tubular exit; also, either external shell of bivalve mollusk, brachiopod, or some crustaceans.

Valvifer (s): A basal sclerite of a valve of the ovipositor, articulating with the tergum; or proximal structures of the ovipositor that bear the valvulae; in female insect genitalia, derivations of **gonocoxites** 8 and 9, supporting the valves of the ovipositor.

Valvulae: Distal structures of the ovipositor borne on the valvifers; the three pairs of processes forming the sheath and piercing structures of the ovipositor.

Vannal fold: See **anal fold.**

Vannal lobe: A lobe in the anal area of the wing, immediately distal to the jugal lobe when a jugal lobe is present (Hymenoptera).

Vannus: The anal area of the wing anterior to the jugal area; a fan-shaped lobe at the posterior margin of the hindwing of certain insects.

Vapor drift: The movement of pesticide vapors to areas where they may cause injury. See *drift.*

Vapor pressure: The property that determines how easily a chemical changes from the solid or liquid state to the gaseous state.

Variance: The Square of the standard deviation.

Variation: The occurrence of heritable or non-heritable difference among the individuals of a population or among populations.

Varietal control: The use of varieties (cultivars) of cultivated plants which are resistant to pests.

Variety: A term originally applied indiscriminately to various kinds of infraspecific forms, individuals as well as populations (i.e., subspecies); in modern usage, usually limited to discontinuous variants within a single interbreeding population (cf. Subspecies, Infrasubspecies form); members of a species that differ from other members of a species. Variety is sometimes used synonymously with cultivar. A loosely defined taxonomic category below the level of sub - species. Usually possesses some visibly different character.

Vas deferens (pl. vasa deferentia): The sperm duct that leads away from a testis; or a duct, into which the vas efferentia empty, that connects each testis to a seminal vesicle; the sperm duct from the vasa efferentia to the cloaca or ejaculatory duct The tube leading from the testis down which sperm passes (L. *deferens* = carrying out).

Vas efferens (pl. vasa efferentia): A short duct connecting a sperm tube in the testis to the vas deferens; or tiny ducts leading from each testicular follicle to a common lateral duct, the vas deferens (L. *ex* = out + *ferro* = varry).

Vascular wilts: Plant diseases that cause wilting by the plugging of and damage to the xylem (e.g., *Verticillium and Fusarium* wilts).

Vascular: Pertaining to vessels or ducts for conveying or circulating blood or lymph. (L. *dim* = of *vas* = *vessel*).

Vector: Any carrier that transfers a pathogen from one host to another host for example, *Anopheles* mosquitoes serve as vectors for malaria parasites, carrying the parasites from one human host to another; organisms able to transmit viruses or other pathogens either directly or indirectly. An insect (or other animal) which transmits a disease causing micro-organism from one plant or animal to another, e.g. anopheline mosquitoes are vectors of malaria. Direct virus vectors include insects, mites and nematodes.

Vectorial capacity: A mathematical expression of the probability of disease transmission by a particular vector.

Vegetative growth: The growth of stems, roots, and leaves rather than of flowers and fruits.

Vegetative: Asexual or somatic.

Vein (of a wing): A thickened sclerotized rod or rib supporting the wing membrane (collectively called the venation of a wing).

Vein: A vessel carrying blood from capillaries toward the heart. (L. *vena* = vein).

Venom: Inherently toxic insect (or other animal) secretion, harmful to humans and animals.

Venter: The ventral region or the lower or under surface of an organism.

Ventilate: To pass air or oxygenated water over a gas exchange surface.

Ventilation: Transport of oxygen to and carbon dioxide from the insect tissues.

Ventrad: Toward the ventral or underside of the body; downward.

Ventral diaphragm: A fibromuscular septum that is located ventral to the alimentary canal and dorsal to the ventral nerve cord; a ventral muscular sheath that assists in circulating the hemolymph around the nerve cord; a membrane lying horizontally above the nerve cord in the body cavity, separating the perineural sinus from the perivisceral sinus.

Ventral nerve cord: The chain of ventral ganglia.

Ventral tube: A tubular protrusion from beneath the first abdominal segment of springtails. In Collembola, a ventral sucker, also called collophore.

Ventral visceral system: Subsystem of the visceral nervous system associated with the ventral nerve cord.

Ventral: Pertaining to the underside of the body; or the lower surface of an organism; or towards the lower side or belly away from the back. Opposite to dorsal. (L. *venter* = belly).

Ventricle: The muscular chamber in a heart; also a cavity in the brain of a vertebrate. (L. *vertebrates* = jointed).

Ventricular ganglion: Small ganglionic masses found on either side of the posterior region of the foregut and connected via nerves with the hypocerebral ganglion.

Ventricular nerves: Nerves that connect the hypocerebral ganglion with the ventricular ganglion.

Ventriculus: The tubular somewhat enlarged part of the midgut, the main digestive section of the gut that serves as the insect's stomach.

Verbenone: An inhibitory pheromone produced by bark beetles of both sexes, inhibiting arrival of further individuals.

Vermiform larva: A legless (apodous), headless (acephalic), worm-like larva typical of some Diptera.

Vermiform: Worm-shaped, relatively long and slender. Refers to worm-shaped nematodes stages and insect larval types.

Vermin: Pests; usually rats, mice or insects.

Vernacular name: The colloquial designation of a taxonomic category (cf. scientific name).

Vernalization: The process by which exposure of seeds or plants to low temperatures stimulates germination or flower development.

Verruca: A wartlike prominence on some caterpillars, usually bearing tufts of setae.

Vertebrate: An animal with a bony spinal column. Examples include mammals, fish, birds, reptiles, and amphibians.

Vertebrate: Higher animal possessing a backbone, e.g. mammals, birds, fish.

Vertex: The top of the head, between the eyes and anterior to the occipital suture. The dorsal portion of the cranium bisected by the coronal suture.

Vertical classification: Classification based upon the historical development of groups of organisms as indicated by the fossil record; classification which, according to Simpson, unites ancestral and descendent groups and separates contemporaneous groups that are diverging from a common ancestry (cf. Classification, Horizontal classification).

Vertical resistance: In crop plants, the almost complete resistance to a specific species or strain of pest. Usually determined by a single gene. Some wheat varieties are totally resistant to Hessian fly (cf., horizontal resistance); or host-plant resistance to one or a narrow group of pest genotypes.

Vesicle: A small sac or bladder, or cyst sometimes extensible.

Vessel: A tubular structure that conveys fluid, espically blood or lymph. (L. *vascellum* = dim of vas, vessel).

Vestigial: Small, poorly developed, degenerate, nonfunctional structure that formerly was more fully developed or functional. Compare Rudimentary. (L. *vestigium* = footstep, a trace).

Viable: Capable of germination or growth; alive; or eggs capable of hatching out *e.g.* eggs of insects, ducks and fowls.

Vibration dance: A communication dance of honey bees.

Vibrissae, oral (see *oral vibrissae*).

Vicariance: The division of the range of a species by Earth history events (e.g. ocean or mountain formation).

Virion: The complete individual virus particle that includes the protein coat and the nucleic acid.

Viroid: An infectious ribonucleic acid that does not have a protein coat.

Virulent gene: A gene that allows a pest species to overcome resistance factors in a plant cultivar.

Virulent: Vigorously pathogenic; capable of producing disease (cf., avirulent).

Virus: Microscopic parasites composed of a protein capsule and a nucleic acid core that needs living cells to grow or replicate.

Viscera: The organs within the cranium, thorax, and abdomen, espically the letter. (L. internal organs).

Visceral (also **sympathetic**) **nervous system:** Collective term for the stomatogastric, ventral visceral, and caudal visceral systems; or a series of nerve fibers and ganglia closely associated with the gut and reproductive organs; the nerve system that innervates the gut, reproductive organs and tracheal system.

Visceral muscles: Muscles responsible for movements of internal organs; usually attached to other muscles.

Visceral skeleton: The surrounding framework of the jaws and gill arches and their derivatives in vertebrates.

Visceral: Pertaining to or affecting the internal organs of the body.

Vitamin: An organic substance necessary in trace amounts for normal metabolic functioning.

Vitellarium: The region of an ovariole in which oocytes undergo vitellogenesis.

Vitellarium: The structure within the ovariole in which **oocytes** develop and yolk is provided to them.

Vitellin: Yolk protein.

Vitelline membrane: The cell wall of the insect egg: a thin membrane lying beneath the chorion; the outer layer of an **oocytes,** surrounding the yolk.

Vitellogenesis: The process by which oocytes grow by yolk deposition; the process of deposition of nutrients in oocytes; yolk deposition; or yolk formation in the developing egg.

Vitellogenin (s): Protein produced in the fat body that is transferred to oocytes during vitellogenesis; or female-specific proteins synthesized by the fat body and taken up by maturing oocytes.

Vitellophage: Yolk cells considered responsible for the initial digestion of yolk.

Vitta (pl., vittae): A broad stripe.

Viviparae: Female aphids that reproduce viviparously.

Viviparity: The bearing of live young (i.e. post-egg hatching) by the female; the condition of giving birth to active young rather than laying eggs; *see also* **Adenotrophic viviparity; Haemocoelous viviparity; Ovoviviparity; Pseudoplacental viviparity.**

Viviparous insects: Female insects that gives "birth" to larvae.

Viviparous: (i) Bulbs, seeds, or plantlets that germinate before becoming detached from the parent plant. (ii) Giving birth to active young that undergo some growth and development in the mother e.g. common in aphids; or producing living young that have been nourished internally by the mother.

Vlave: Generally, any unidirectional opening flap or lid; in female genitalia, the blade-like structures comprising the ovipositor shaft (also called **gonapophysis**).

Volatility: The degree to which a substance, usually a liquid, changes to a gas at ordinary temperatures upon exposure to air.

Voltinism: The number of generations per year.

Voltinity: The number of generations that occur in one year.

Volume median diameter (VMD): A measure of spray droplet diameter at which half the spray volume has larger drops and half has smaller drops.

Volunteer crop: Growth of undesired plants self-seeded from the previous crop.

Volunteer: Crop plant growing accidentally from shed seed; not deliberately cultivated.

Vulva: The external opening of the copulatory pouch (bursa copulatrix) or vagina of the female genitalia.

Wettable powder (WP): An insecticide formulation consisting of finely ground solid particles of active ingredient plus inert carrier; forms a suspension when added to water.

Waggle dance: A communication dance of honey bees. A form of recruitment in the honey bee in which direction and distance to a food source are indicated.

Waiting period: See *time interval.*

Waiting period: The prescribed period between last application of a pesticide to crop and harvest.

Warble: The swelling on the host body caused by the larva of a warble fly.

Warm-blooded (Homoiothermous): Animals in which body temperature normally does not change with changes in atmospheric temperature *e.g.* birds and mammals.

Warning: Signal word used to designate a moderately toxic pesticide; oral LD_{50} = 50 to 500 mg/kg body formulated.

Wasmannian mimicry: A form of **mimicry** which allows an insect of another species to be accepted into a social insect colony.

Water soaked: A symptom of certain plant diseases or following herbicide application, where plant tissues appear wet due to loss of cell integrity.

Water soluble powder (SP): Pesticide dust dissolved in water to form a solution. See *soluble powder.*

Water-dispersable liquid (flowable, F): Wettable powder preparations suspended in an oil base when formulated.

Wax canals (or filaments): Fine tubules that transport lipids from the pore canals to the surface of the epicuticle.

Wax filament: Tiny canals, continuous with pore canals that penetrate the protein epicuticle and cuticulin layers and are thought to be involved in transportibg wax molecules to the epicuticle; also called wax canals.

Wax layer: An epicuticular lipid layer between the cuticulin layers that contribute to the permeability characteristics of the cuticle; may be absent.

Wax mirrors: Overlapping plates on the venter of the fourth to seventh abdominal segments of social bees that serve to direct the wax flakes that are produced beneath each mirror.

Wax: A complex lipid mixture giving water-proofing to the cuticle, providing covering, or building material.

Weather: Weather results from the combined action and influence of all the physical factors of the environment at any given time. It varies continually throughout days, months and years. It changes rapidly often violently over a period of several years.

Weed control: The actions taken to eradicate, inhibit, or limit weeds and to prevent them from interfering with crop profitability or agricultural operations.

Weed: A plant growing where it is not wanted; any plant 'in the wrong place', particularly used of plants away from their natural range, or invading human monocultural crops.

Weismans ring: See ring gland.

Wettable powder (WP, W): A pesticide formulation of active ingredient mixed with inert dust and a surfactant that mixes readily with water and forms a short-term suspension; or A dry powder pesticide formulation that is mixed with water for application by sprayer.

Wetting agent: A chemical that reduces surface tension and allows a pesticide to spread out more evenly coat a surface. See *surfactant.*

White ear head: The ears are dry and hollow, e.g., in wheat and rice.

Wick application: Selective herbicide application by moving a herbicide-soaked wick over the surface of weeds. Rope is often used for wick fabrication.

Wilt: (i) Loss of turgor or drooping of plant parts because of inadequate water supply or excessive transpiration. (ii) Generic descriptor for vascular plant diseases.

Wind pollination: Pollination in which pollen grains are carried by the wind rather than by insects.

Wind-borne: Pests that are disseminated by the wind.

Windrow: A row of hay, grain, alfalfa, beans, etc., cut and left in a field for drying before being baled or processed further.

Wing flexion: The folding of the wings posteriorly over the abdomen.

Wing pads: Externally apparent developing wings of hemimetabolous insect's e. g. nymphs.

Wing veins: Longitudinal and transverse supportive framework of the wings.

Wing: Forelimbs in birds are modified into wings and enable them to lead aerial life. In insects it is a part of exoskeleton and characterisitic feature of pterygota.

Winter annual: A plant that germinates in the fall, lives over the winter, and completes its growth, including seed production, the following spring.

Wireworm (Elateriform): The larva of Elateridae (Coleoptera); long; slender, well-sclerotized, thoracic legs but no prolegs, and few setae.

Wireworm: An elateriform larva; a larva that is slender, elongate, heavily sclerotized, with a few hairs on the body, with thoracic legs but without prolegs; the larva of a click beetle (family Elateridae).

Witch's broom: An abnormal growth of tufted, small, closely set branches, apparently caused by a fungus or virus, and found on various trees and shrubs.

Withholding period: The period of time following application of a pesticide, that livestock must be kept out of a pasture or fodder crop.

Wood borer: An insect that tunnels into live or dead wood.

Worker(s): Sterile individuals in the nests of social insects which forage for food and generally run the colony; in social insects, a member of the sterile caste that assists the reproductives.

Worm filarial: A roundworm (nematode endoparasite in gut or tissues of vertebrates.

Worm guinea: Nematode animal endoparasite in vertebrates; intermediate host is Cyclops.

Worm, glow: Non flying adult females of certain species and the larvae of elateried and lampyrid beetles are known by this name because they look like worms and glow *e.g. Lamprophorus* sp.

Worm, hook: A group of round worms (Nematode), have well-developed buccal-capsule armed with teeth. Anterior end is usually bent hence the name.

Worm, round: Worms belonging to phylum Nemathelminthes and class Nematoda, commenly known as nematodes; animal and plant parasites e.g. **Ascaris, Meloidogyne** sp.

Worm, segmented: Phylum Annelida, ring-like segments, earthworms, leeches are included in it.

Worm, ship: A wood boring mollusk of genus *Tcredo* causing considerable damage to wharfs.

Worm, tape: Body flat and long tape-like hence the name tape worm; internal parasite of vertebrates *e.g.* beef tape worm, pork tape worm.

Worm: A living creature with long slender body.

WP: See wettable powder.

X Y Z

Xerophyte: Plant adapted for survival in desert or physiologically dry soil.

Xylem: The water-conducting woody tissue in vascular plants.

Xylomycetophagy: Cultivation and feeding upon symbiotic fungi growing in woody tree tissues; for example, some Scolytidae.

Xylophage (adj. xylophagous)**:** An eater of wood.

Xylophagy: Feeding on xylem tissues.

Yellow fever: An infectious, mosquito-transmitted viral disease.

Yellow fever: An acute infectious disease caused by a virus transmitted by certain mosquitoes and marked by fever, jaundice, and albumin and globulin in the urine.

Yellows: Plant diseases, in which foliage is yellow and stunted, caused usually by a virus, MLO, or fungus.

Yield forming organ: That part of a plant which gives rise to the harvested yield.

Yolk: Nutrients (carbohydrates, proteins, and lipids) found in mature insect eggs; also called deutoplasm.

Y-vein: Two adjacent veins fusing distally, forming a **Y**-shaped figure (for example, the anal veins in the front wing).

Zero tolerance: No amount of pesticide residue permitted on commodity when it is offered for shipment.

Zoocecidia: Plant galls induced by animals such as insects, mites and nematodes, as opposed to those formed by the plant response to micro-organisms.

Zoogeography: Study of geographical distribution of animals in space.

Zoophagous: eating animals.

Zoophagous: The habit of feeding on living animals.

Zoophilic: Preferring other animals to humans, especially used of feeding preference of blood-feeding insects.

Zoospore: A motile, asexually produced spores; often released in water.

Zoraptera, order (*zor,* pure; *ptera,* wing)**:** An order of Pterygota that consists of rare, small and termite like insects. It is a small order of tropical insects related to the book-lice for which there is no common name zorapterans*.

Zygote: Diploid cell resulting from the fusion of two gametes; the fertilized egg; in malaria parasites (*Plasmidium* spp.) resulting from fusion of a microgamete and macrogamete.

References

Abrahamson, W. G. 1989. Plant-animal interaction. McGraw-Hill: New York.

Berridge, M. J. 1970. A structural analysis of intestinal absorption. In: Neville, A. C., ed. Insect ultrastructure. London: Royal Entomological Society, Symposium 5: p. 135-151.

Binnington, K. C. 1993. Ultrastructure of the attachment of the bacteria *Serratia entomophila* to foregut cuticle of *Costelytra zealandica* (Coleoptera: Scarabaedae) abd a review of nomenclature for insect epicuticular layers. *International Journal of Insect Morphology and Embryology,* 22 (2-4), 145-55.

Birch, M. C. and Haynes, K. F. 1982. Insect Pheromones, Studies in Biology no. 147. Edward Arnold, London.

Borror, D. J., Triplehorn, C. A. and Johnson, N. F. 1989. An Introduction to the Study of Insects, 6th edn, Saunders College Publishing, Philadelphia, etc.

Chapman, R. F. 1971. The insects – structure and function. 2nd ed. American Elsevier: New York.

Chu, H. F. 1949. How to know the Immature Insects, Wm. C. Brown Company Publishers, Dubuque, Iowa.

Claassen, P. W. 1931. Plecoptera nymphs of North America (north of Mexico). Thomas Say Foundation Publication No. 3. College Park, MD: Entomological Society of America.

Common, I. F. B. and Waterhouse, D. F. 1972. Butterflies of Australia, Angus and Robertson, Sydney.

Comstock, J. H. 1918. The wings of insects. Ithaca, NY: Cornell University Press.

Cook, B. J.; Holman, G. M. 1985. Peptides and Kinins. In: Kerkut, G. A.; Gilbert, L. I., eds. Comprehensive insect physiology, biochemistry, and pharmacology. Pergamon Press: New York: p. 531-593, vol. 11.

CSIRO 1970. The Insects of Australia, Ist edn, Melbourne University Press, Carlton.

Emersom, K. C.; Price, R. D. 1988. A new species of *Haemulomyzus* (Mallophaga: Haematomyzidae) of the bush pig, *Pomatochoerus porcus,* from Ethiopia, with comments on lice found on pigs. Proceedings of the Entomological Society of Washington 90: 338-342.

Engelmann, F. 1970. The physiology of insect reproductuction. Pergamon: New York.

Essig, E. O. 1942. Colledge entomology. MacMillan: New York.

Ferrar, P. 1987. A Guide to the Breeding Habits and Immature Stages of Diptera Cyclorrhapha, Part 2, Entomonograph volume 8, E. J. Brill/Scandinavian Science Press, Leiden, Copenhagen.

Frost, S. W. 1959. Insect life and Insect Natural History, 2nd edn, Dover Publication, Inc. New York.

Gilbert, L. I. 1964. Physiology of growth and development: endrocrine aspects. In: Rockstein, M., ed. The physiology of insecta. Academic Press: New York, Inc.: vol. 1.

Hadely, N. F. 1986. The arthropod cuticle, *Scientific American,* 255 (1): 98-106.

Hepburn, H. R. 1985. Structure of the integument, in *Comprehensive Insect Physiology, Biochemistry and Pharmacology, Volume 3, Integument, Respiration and Circulation,* (eds G. A. Kerkut and L. I. Gilbert), Pergamon Press, Oxford, New York, Toronto, Sydney, Paris, Frankfurt, pp. 1-58.

Herried, C. F., 1977. II. Biology. Macmillan: New York: p. 884.

James, M. T.; Harwood, R. F. 1969. Herms' medical entomology. Macmillan Publishing Co.: New York, Inc.

Johannsen, O. A.; Butt, F. H. 1941. Embryology of insects and myriapods. McGraw – Hill: New York.

Johannsen, O. A.; Butt, F. H. 1941. Embryology of insects and myriapods. McGraw-Hill: New York.

Katz, M., Despommier, D. D. and Gwadz, R. W. 1989. Parasitic Diseases, 2nd edn, Springer-Verlag, New York, Berlin, Heidelberg, London, Paris, Tokyo.

Kettle, D. S. 1984. Medical and Veterinary Entomology, Croom Helm, London and Sydney.

Link, E. 1909. Uber die Stirnaugen der hemimetabolen Insecten. Zoologische Jahrbucher, 27: 281-376, 4pls.

McAlpine, J. F. (ed.) 1987. Manual of Nearctic Diptera, Volumes 1 & 2, Research Branch, Agriculture Canada, Monograph Nos 27 & 28.

Michelsen, A. and Larsen, O. N. 1985. Hearing and sound, in Comprehensive Insect Physiology, Biochemistry, and Pharmacology, Volume 6, Nervous System: Sensory, (eds G. A. Kerkut and L. I. Gilbert), Pergamon Press, Oxford, New York, Toronto, Sydney, Paris, Frankfurt, pp. 495-556.

Novak, V. J. A. 1975. Insect Hormones, Chapman & Hall, London.

Packcard, A. S. 1898. A textbook of entomology. Macmillan: New York.

Palmer, M. A. 1914. Some notes on life history of ladybeetles. Annals of the Entomological Society of America, 7: 213-38.

Raabe, M. 1986. Insect reproduction: regulation of successive steps. *Advances in Insect Physiology,* 19, 29-154.

Riley, C. V. 1888. The habits of *Thalessa* and *Tremex.* U. S. Division of Entomology Ins. Life 1: 168-179.

Ross, H. H. 1962. How to collect and preserve insects. Illinois Natural History Survey, Circular 39.

Ross, H. H. 1965. A textbook of entomology. John Wiley & Sons, 3d. New York.

Schwabe, J. 1906. Beitrege zur Morphologie und Histologie der tympanalen Sinnesapparate der Orthopteren. Zoologica, Stuttgart, 50: 1-154.

Snodgrass, R. E. 1935. Principles of insect Morphology, McGraw-Hill Book Company, New York and London.

Snodgrass, R. E. 1935. Principles of insect morphology. McGraw – Hill, New York.

Snodgrass, R. E. 1956. Anatomy of the Honey Bee, Comstock Publishing Associates, Ithaca, New York.

Snodgrass, R. E. 1957. A revised interpretation of the external reproductive organs of male insects. Smithsonian Miscellaneous Collections, 135 (6): 1-60.

Snodgrass, R. E. 1959. The anatomical life of the mosquito. Smithsonian Institution, Miscellaneous Collections 138 (8).

Spradbery, J. P. 1973. Wasps: An Account of the Biology and Natural History of Solitary and Social Wasps, Sidgwick and Jackson, London.

Spencer, K. A. 1990. Host Specialization in the world Agromyzidae (Diptera), Kluwer Academic Publishers, Dord reeht, Boston, London.

Suzuki, N. 1985. Embryonic development of the scorpionfly, *Panorpodes paradoxa* (Mecoptera, Panorpodidae) with special reference to the larval eye development, in *Recent Advances in Insect Embryology in Japan,* (eds H. Ando and K. Miya), ISEBU Co. Ltd, Tsukubo, pp. 231-8.

Ursprung, H.; Nothiger, R. 1972. The biology of imaginal discs. Springer-Verlad: New York.

Vogel, R. 1921. Zur Kenntnis des Baues und der Funktion des Stachels und des Vorderdarmes der Kleiderlaus. Zoologische Jahrbuecher, Anatomie 42: p. 229-258.

Wadton, J. A. L. and Abbey, H. M. 1985. Seasonal cycles in *Nasutitermes exitiosus* (Hill) (Isoptera: Termitidae). *Sociobiology,* 10, 73-92.

West, L. S. 1951. The housefly: its natural history, medical importance, and control. Ithaca, NY: Comstock Publishing Company, Inc.

Whitham, T. G. 1978. Habitat selection by *Pemphigus* aphids in response to resource limitation and competition. Ecology 59: 1164 – 1176.

Wigglesworth, V. B. 1970. Insect hormones. Oliver & Boyd: Edinburgh, Scotland.

Wigglesworth, V. B. 1972. The Principles of Insect Physiology, 7 th edn, Chapman & Hall, London.

Zacharuk, R. Y. 1985. Antenna and sensilla, in Comprehensive Insect Physiology, Biochemistry, and Pharmacology, Volume 6, Nervous System: Sensory, (eds G. A. Kerkut and L. I. Gilbert), Pergamon Press, Oxford, New York, Toronto, Sydney, Paris, Frankfurt, pp. 1-69.

Zeitfracht Medien GmbH
Ferdinand-Jühlke-Straße 7
99095 Erfurt, Deutschland
produktsicherheit@kolibri360.de